Lecture Notes in Mathematics

Edited by A. Dold and B. Eckmann

516

Martin L. Silverstein

Boundary Theory for Symmetric Markov Processes

Springer-Verlag
Berlin · Heidelberg · New York 1976

Author
Martin L. Silverstein
Department of Mathematics
University of Southern California
University Park
Los Angeles, California 90007
USA

Library of Congress Cataloging in Publication Data

Silverstein, Martin L 1939-
Boundary theory for symmetric Markov processes.

(Lecture notes in mathematics ; 516)
Bibliography: p.
Includes index.
1. Markov processes. 2. Semigroups. 3. Symmetry groups. I. Title. II. Series: Lecture notes in mathematics (Berlin) ; 516.
QA3.L28 no. 516 [QA274.7] 510'.8s [519.2'33] 76-10683

AMS Subject Classifications (1970): 60J25, 60J45, 60J50

ISBN 3-540-07688-3 Springer-Verlag Berlin · Heidelberg · New York
ISBN 0-387-07688-3 Springer-Verlag New York · Heidelberg · Berlin

Printing and binding: Beltz Offsetdruck, Hemsbach/Bergstr.

Dedicated to the memory of my father

JOSEPH SILVERSTEIN

Introduction.

Let $P_t, t > 0$ be a submarkovian semigroup on a measurable space $\underline{\underline{X}}$. This means that each P_t maps bounded measurable functions into bounded measurable functions and that

$$(0.1) \qquad 0 \leq P_t f \leq 1 \quad \text{whenever} \quad 0 \leq f \leq 1$$

$$(0.2) \qquad P_t P_s = P_{t+s} \quad \text{for} \quad s, t > 0.$$

Also it is usually necessary to impose some regularity condition before one can hope to do any serious work. For example, that each P_t can be represented

$$(0.3) \qquad P_t f(x) = \int p_t(x, dy) f(y)$$

where the $p_t(x, d\cdot)$ are subprobabilities on $\underline{\underline{X}}$ and that at least for an appropriate class of functions f

$$(0.4) \qquad \mathrm{Lim}_{t \downarrow 0} P_t f(x) = f(x).$$

We agree that a second such semigroup $\tilde{P}_t$, $t > 0$ dominates the first if

$$(0.5) \qquad \tilde{P}_t f \geq P_t f \quad \text{whenever} \quad f \geq 0.$$

This volume is concerned with the general problem of analyzing and to some extent classifying submarkovian semigroups $\tilde{P}_t$, $t > 0$, which dominate a fixed submarkovian semigroup P_t, $t > 0$.

There are good technical reasons for restricting attention to the special case when both the P_t and $\tilde{P}_t$ are symmetric with respect to a given reference measure dx on $\underline{\underline{X}}$. This means that

$$(0.6) \qquad \int dx P_t f(x) g(x) = \int dx f(x) P_t g(x)$$

at least when f, g are bounded and integrable. The restriction will be in effect throughout the volume.

Once symmetry is imposed, it is convenient to modify the regularity condition (0.3) and (0.4). In place of (0.3) we assume that each P_t is continuous with respect to bounded almost everywhere convergence. This, together with the symmetry condition (0.6), guarantees that each P_t extends uniquely to a bounded symmetric contraction on the Hilbert space $L^2(\underline{\underline{X}}, dx)$. In place of (0.4) we assume that the extended operators P_t form a semigroup which is continuous relative to the strong operator topology on $L^2(\underline{\underline{X}}, dx)$. This means that

$$\text{(0.4')} \qquad \operatorname{Lim}_{t\downarrow 0} \int dx \{f(x) - P_t f(x)\}^2 = 0.$$

Actually, it suffices to assume this for bounded integrable f. The semigroup P_t, $t > 0$, has a generator A which is self-adjoint and nonpositive definite so that the nonnegative square root $(-A)^{\frac{1}{2}}$ is uniquely defined by the operator calculus. The pair $(\underline{\underline{F}}, E)$ where $\underline{\underline{F}}$ is the domain of $(-A)^{\frac{1}{2}}$ and where E is the symmetric bilinear form defined on $\underline{\underline{F}}$ by

$$\text{(0.7)} \qquad E(f, g) = \int dx (-A)^{\frac{1}{2}} f(x) (-A)^{\frac{1}{2}} g(x)$$

is called a Dirichlet space on $L^2(\underline{\underline{X}}, dx)$. Such pairs can be characterized by a contractivity property and a completeness property and the former guarantees in particular that the subspace $\underline{\underline{F}}_b$ of bounded functions in $\underline{\underline{F}}$ forms an algebra with the usual operations. If $L^2(\underline{\underline{X}}, dx)$ is a separable Hilbert space, then $\underline{\underline{F}}_b$ has a dense separable subalgebra and all of the relevant structure can be transferred to the maximal ideal space for the uniform closure of such a subalgebra. Once this is done $(\underline{\underline{F}}, E)$ becomes a regular Dirichlet space in the sense of A. Beurling and J. Deny [2]. That is, $\underline{\underline{F}} \cap C_{com}$ is dense in $\underline{\underline{F}}$ and also uniformly dense in C_{com}, the space of continuous functions with compact support. Therefore, it is natural to assume from the beginning that $\underline{\underline{X}}$ is a separable, locally compact Hausdorff space, that dx is Radon, and that $(\underline{\underline{F}}, E)$ is regular. Then one can develop a potential theory and in particular a notion of capacity and therefore of polar sets such that the P_t, $t > 0$, are the transition operators for a Markov process which satisfies the usual regularity conditions modulo a polar set. The precise meaning of the last phrase is that there exists a polar set N such that (0.3) is valid for all $t > 0$ and for all x outside N and that the Markov process can be defined for any starting point outside N and then it never hits N. The potential theory itself was developed first by H. Cartan [4, 5] for the special case of Brownian motion and then in a general context by Beurling and Deny [2]. M. Fukushima first applied this to the general theory of symmetric Markov processes and in particular in [18] he gave the first construction of the process in this context. A systematic treatment can be found in Sections 1 through 4 of the author's previous monograph [47] which will be referred to below as SMP.

In Chapter I we give a direct sample space construction for a large class of Markov processes whose transition operators $\tilde{P}_t$ are symmetric and dominate the given P_t. The basic idea is to start with a symmetric "process on the boundary" satisfying a particular contractivity condition, then insert "excursions into the interior" whose distribution depends on the Markov process for P_t, and finally to collapse the time scale so that ultimately no time is actually spent outside $\underline{\underline{X}}$. The latter step is needed to make $\tilde{P}_t$ symmetric relative to the original reference measure dx and not relative to dx plus the reference measure for the "process on the boundary". The first two sections contain a preliminary decomposition which motivates and guides the construction itself which is given in Section 3. We start with a symmetric Markov process and decompose it into a time changed process which "lives on" a closed set M and a collection of excursions into the complementary open set D. The conditional distribution for these excursions depends only on the absorbed process for D, on the part of the Levy measure which "connects" D and M, and on the manner in which M acts as a topological boundary for D. The basic idea of this decomposition and of using it to construct new processes is not new. It is the main theme of a paper by M. Motoo [41] which seems to have priority. We became aware of the decomposition through the work of M. Fukushima [16], who assumes symmetry and then uses Dirichlet spaces to put the results in an elegant form. Symmetry plays no role in Motoo's work. The decomposition is also used by E. B. Dynkin [10], K. Ito [31], P. A. Meyer [38] and quite recently by S. Watanabe [53] and it is closely related to the last exit decomposition of K. L. Chung [8, 9], of R. Getoor and M. Sharpe [23] and of A. Pittenger and C. T. Shih [44]. When we were actually working on Chapter I it seemed to us that the general decomposition, although clearly understood by several authors, had not been completely established except for special cases. We have quite recently become aware of some work of B. Maissoneuve and P. A. Meyer [37] which establishes the basic decomposition in a general context. Their techniques seem to be quite different from ours. It is clear that our assumption of symmetry is only a convenience in the context of Chapter I and indeed it is not assumed by the other authors mentioned (except for Fukushima).

Symmetry does play a crucial role in Chapter II and at present we can say very little about how the basic results carry over to a general nonsymmetric context. Perhaps the work of H. Kunita [34] contains the deepest results in this direction. In Section 5 we study the Dirichlet space $(\underline{\underline{F}}^{\sim}, E^{\sim})$ associated with the process constructed in Section 3, and we establish results which supplement the ones in Section 15, 20 and 21 in SMP.

The original process can die either by "wandering to infinity" in a finite time or by "jumping to the death point ∂" directly from the interior with an intensity governed by a measure $\kappa(dx)$ referred to in SMP as the "killing measure".

The simplest possibility is that new behavior is prescribed only after wandering to infinity. In this case the function space $\underline{\underline{F}}^{\sim}$ has an orthogonal decomposition into $\underline{\underline{F}}$ and the subspace of harmonic functions in $\underline{\underline{F}}^{\sim}$, that is, the functions in $\underline{\underline{F}}^{\sim}$ which are fixed by the hitting operator for the complement of any compact set. The harmonic part of $\underline{\underline{F}}^{\sim}$ together with the restriction of the Dirichlet norm $E^{\sim}$ to this harmonic part corresponds in a natural way to the original "Dirichlet space on the boundary". In fact the latter is contained in a specific Hilbert space which "lives on" $\underline{\underline{T}}$, the so-called "terminal boundary", and this depends only on the original process. This enables us to classify the possibilities for $(\underline{\underline{F}}^{\sim}, E^{\sim})$ in terms of Dirichlet spaces in this Hilbert space. There is a maximal one which corresponds to the so-called reflected space $(\underline{\underline{F}}^{ref}, E)$ defined directly in terms of the original Dirichlet norm E. The others are characterized by a contractivity property relative to the maximal one. The classification can also be given in terms of Dirichlet spaces on the Martin boundary whenever the latter is well defined. We refer to [46] where details are given for the special case of stable Markov chains. The basic results for the Dirichlet spaces $(\underline{\underline{F}}^{\sim}, E^{\sim})$ are established in SMP, Sections 14 and 15, and we do not add anything in this volume. However, we do give more precise information about the generator. We formulate the boundary conditions in terms of a "terminal normal derivative" which is defined in terms of difference quotients on the sample space. Also we sharpen our characterization in terms of local generators. The basic results are summarized in Section 9.

The construction of examples where there is new behavior also after a jump to the death point depends on an auxiliary set $\underline{\underline{M}}$ and jumping measures μ which give the conditional intensity for jumping to $\underline{\underline{M}}$ or to the terminal boundary $\underline{\underline{T}}$ instead of to the death point. Once $\underline{\underline{M}}$ and μ have been prescribed, the possibilities for $(\underline{\underline{F}}^{\sim}, E^{\sim})$ can be classified by Dirichlet spaces in a Hilbert space which "lives on" $\underline{\underline{M}} \cup T$, pretty much as for the previous case. It is useful to look also at the so-called "expanded process". This is the intermediate process obtained before collapsing the time spent on the boundary. A point on the boundary is said to be "singular" if the expanded process starting at this point spends an initial time segment on the boundary. Otherwise it is regular. It turns out that the singular points are precisely those which "disappear" for the final process. A detailed analysis of this phenomenon from the sample space point of view is given in Section 4. Also in Theorem 4.3 we give a direct analytic characterization of the singular boundary.

We say that $(\underline{\underline{F}}^{\sim}, E^{\sim})$ is an "extension" of $(\underline{\underline{F}}, E)$ if the original process can be recovered as the absorbed process for an appropriate subset. A direct characterization of extensions is given in SMP, Theorem 20.1. It turns out that $(\underline{\underline{F}}^{\sim}, E^{\sim})$ is an extension if and only if the singular boundary is empty. Section 20 in SMP contains a detailed analysis of $(\underline{\underline{F}}^{\sim}, E^{\sim})$ for the special case of extensions. Theorems 5.1 and 5.2 in this volume contain an independent treatment which covers everything in SMP except for showing that we obtain all of the possibilities. The condition (i) in SMP, Theorem 20.2, is formulated incorrectly, and this leads to a misconception in SMP about the possibility of classifying extensions in terms of modified reflected spaces. Again we obtain sharper results here for the generator, and the basic results at this level are summarized in Section 10.

Most of Section 5 is devoted to a detailed study of $(\underline{\underline{F}}^{\sim}, E^{\sim})$ when the singular boundary is nonempty. In SMP we only observed that this could happen, and in Section 24 we "hacked our way through" an elementary example.

The distinction between the regular and singular boundaries is significant both at the sample space level and the Dirichlet space level. However, we show in Section 11 that boundary conditions can be formulated in such a way that the distinction disappears at the level of the generator. This is why the distinction between the regular and singular boundary plays no role in the previous literature which focuses on boundary conditions for the generator.

The last three chapters contain a detailed treatment of examples which goes far beyond the material in SMP, Chapter 4.

The examples in Chapter III are closely related to Brownian motion. In Section 12 the state space is a finite interval. Our purpose is to illustrate as much as possible in this simple context, and so we study several cases in great detail. In Section 13 the state space is a half line. The main new feature here is that we have little "a priori" control over the regular boundary. This contradicts the misguided statement on page 24.1 in SMP that there is no real loss of generality in assuming that Δ has at most two points. Also we redo the example which is "hacked through" in SMP. In Section 14 the state space is a full Euclidean space, and there is no killing. Our main purpose is to demonstrate the connection between transience and the existence of a "normal derivative at infinity". In Section 15 the state space is the upper half plane. Our attitude is that this example has intrinsic interest, and we go into considerable detail. Section 16 contains a proof of a theorem of Beurling and Deny [2] which states that any Dirichlet norm on Euclidean space which is defined at least on C^∞_{com}, the space of smooth functions with compact support, must have the form

(0.8) $$E(f,f) = \Sigma_{i,j}\int \nu_{ij}(dx)\partial_i f(x)\partial_j f(x) + \int \kappa(dx) f^2(x) + \int\int J(dx,dy)\{f(x)-f(y)\}^2 .$$

In Section 17 the state space is the plane with the real axis deleted, and we indicate how our techniques might be applied to establish converses to this theorem.

Chapter IV is concerned with symmetric stable processes. The one-dimensional case is treated in Sections 18 through 20, and the multi-dimensional case in Section 21. The basic tool for obtaining precise information is a factorization of the local generator into the composition of a simple differential operator with an integral operator. Our guide for this is the work of J. Elliott [11, 12] and S. Watanabe [52]. We get the most information for the Cauchy process in Section 18 because the integral operator is the restricted Hilbert transform, and this can be related by a change of variables to the Hilbert transform for the circle.

In Chapter V the state space is again the upper half plane. But now it is viewed as a symmetric space with elements of the group $SL_2(R)$ acting in the usual way as linear fractional transformations. R. Getoor [22] and R. Gangolli in a more general context [20] used the Plancherel formula for bi-invariant functions to identify all possible Markovian semigroups which are invariant under this action. It turns out that every such semigroup is automatically symmetric with respect to the unique invariant measure, and this enables us in Section 23 to re-derive this result using Dirichlet space techniques. Of course the associated Dirichlet spaces are also invariant. In addition, there exist invariant Dirichlet spaces which do not correspond to invariant semigroups, and we construct some of them in Section 25. The key to our construction is the observation that a function is harmonic for one of the invariant semigroups if and only if it is harmonic in the classical sense. This follows directly from a special case of H. Furstenberg's results [19, Chapter III] which is reproduced here in Section 24.

Throughout the volume we work with a fixed symmetric Markov process with associated Dirichlet space which is assumed to be irreducible and (except in Sections 1 and 2) transient. The notations and results in the first eight sections in SMP will be taken for granted. (The remainder of SMP will be used only with specific references.) In addition, we impose two conditions which will be in effect throughout the volume.

0.1. Restriction. The killing measure $\kappa(dx)$ is absolutely continuous relative to the reference measure dx and therefore it can be represented $\kappa(dx) = \kappa(x)dx$.

0.2. Restriction. There exists a potential density $N(x,y)$. This means that the potential operator N can be represented

$$N\mu(x) = \int N(x,y)\mu(dy). \tag{0.9}$$

These restrictions have no intrinsic significance and are imposed only to simplify the notation and terminology. In SMP we distinguished the potential operator N acting on measures from the Green's operator G acting on functions. In this volume we suppress the symbol G and instead use the abuse of notation

$$Nf(x) = \int N(x,y)f(y)dy. \tag{0.10}$$

However, we will continue to use the symbol G_u for resolvent operator.

The most technical arguments in this volume are relevant only for the special case when the singular boundary $\Delta_s \neq \phi$. At least at a first reading, the reader may prefer to assume throughout that $\Delta_s = \phi$ and skip over the relevant portions in the text.

This monograph was begun at the University of Southern California and completed while the author was in residence as a visiting member at The Institute for Advanced Study.

I take this opportunity to thank the National Science Foundation for their generous support under grants GP 24239 and MPS75-01872, both for this monograph and the previous one [47].

Table of Contents

Chapter I. Sample Space Construction

Section 3 plays a central role. In it we construct the process whose associated Dirichlet space and infinitesimal generator are the object of study in the remainder of the volume. We start with an irreducible transient regular Dirichlet space $(\underline{\underline{F}}, E)$ on $L^2(\underline{\underline{X}}, dx)$ together with an auxiliary set $\underline{\underline{M}}$ and certain jumping measures $\mu(x, \cdot)$ which together determine the excursion space $(\underline{\underline{N}}, N)$. We then choose a Dirichlet space $(\underline{\underline{H}}, Q)$ satisfying certain contractivity properties relative to $(\underline{\underline{N}}, N)$, and we use $(\underline{\underline{H}}, \bar{Q})$ to construct a regularized space $\underline{\underline{X}} \cup \Delta \cup \{\delta\}$ which will be the actual state space. The Dirichlet space $(\underline{\underline{H}}, Q)$ itself determines a process which lives on $\Delta \cup \{\delta\}$. The approximate Markov process of SMP, Chapter 5 is modified slightly and then used to determine the conditional distribution for excursions into $\underline{\underline{X}}$ which are then "inserted" into the process on Δ. The basic idea for this insertion is clear from the work of M. Motoo [41], but the details are somewhat tedious. Therefore, we have included two sections which can be regarded as an introduction.

Sections 1 and 2 focus on the decomposition of any irreducible symmetric Markov process into a process which "lives on" a closed set M and excursions into the complementary open set D. Once this decomposition is understood, the pattern in Section 3 becomes rather transparent. A complete description of the decomposition is given in Section 1, and this is all that is required to understand Section 3. The proofs are given in Section 2, and they rely on certain techniques from SMP.

In the course of the construction in Section 3 we introduce certain auxiliary operators and establish various identities which will be important in later sections, and especially Sections 4 and 5.

In Section 4 we focus on the distinction between singular and regular points of the boundary. The main results are an analytical characterization of the singular part Δ_s and a proof that the constructed process never hits Δ_s. We will be interested in this distinction from a different point of view in Chapter II.

1. Description of the Decomposition.

In this and the next section, D is open with $M = \underline{\underline{X}} \setminus D$ nonpolar.

As reference measure for the time changed process on M we use the 1-balayage of dx onto M, that is,

$$\nu^M = \underline{\underline{\Pi}}_1^M dx \tag{1.1}$$

in the notation of SMP, Section 7. The time changed process itself is defined by

$$Y_\tau = X_{b(\tau)}$$

where, in the notation of SMP,

$$a(t) = a(\nu^M; t)$$

and $b(\tau)$ is the inverse process

$$b(\tau) = \inf\{t : a(t) > \tau\}$$

with the usual understanding that $b(\tau) = +\infty$ when the right side is empty. By convention $X_\infty = \partial$ and therefore $Y_\tau = \partial$ for $\tau \geq a(\infty)$. It is easy to check directly that the Y_τ form a Markov process with right continuous paths. In SMP, Section 8, we showed that this process is symmetric with respect to the reference measure ν^M and we identified the corresponding Dirichlet space $(\underline{\underline{H}}, Q)$ on $L^2(M, \nu^M)$ which is regular. It follows that $\{Y_\tau\}$ is identical with the process obtained from $(\underline{\underline{H}}, Q)$ by the procedure in SMP, Section 4, and therefore it too has all of the "usual" regularity properties "modulo a polar set".

The excursions will be points in an excursion space $\Omega^{(D)}$ which is defined by modifying slightly the inverse limit sample space of SMP, Section 5, with D here playing the role of $\underline{\underline{X}}$ there. For technical reasons we wish to regard jumps within $M \cup \{\partial\}$ also as excursions into D. Therefore we include the dead sequence δ which will account for this possibility. Also we adjoin two additional coordinates $X^{(D)}_{\zeta^*-0}$, $X^{(D)}_{\zeta}$ taking values in $M \cup \{\partial\}$. Notations for functions which depend only on the coordinates X_t, $\zeta^* < t < \zeta$, are exactly as in SMP, Section 5, except that we define

$$\zeta^*(\delta) = \zeta(\delta) = 0 \tag{1.2}$$

so that the set $[\zeta^* = \zeta = 0]$ exactly corresponds to the dead trajectory. Also we let $X^{(D)}_{\zeta^*}$, $X^{(D)}_{\zeta-0}$ be the appropriate limits

$$X^{(D)}_{\zeta^*} = \operatorname{Lim}_{t \downarrow \zeta^*} X_t \; ; \; X^{(D)}_{\zeta-0} = \operatorname{Lim}_{t \uparrow} X_t$$

in $\underline{\underline{X}} \cup \{\partial\}$ rather than in a one point compactification of D. Notice that $X^{(D)}_{\zeta^*}$, $X^{(D)}_{\zeta-0}$ depend only on the coordinates X_t, $\zeta^* < t < \zeta$, and not on the adjoined coordinates $X^{(D)}_{\zeta^*-0}$, $X^{(D)}_{\zeta}$.

Let T_1 be the set of jumping times within $M \cup \{\partial\}$ and T_2 the set of entrance times for true excursions into D. That is,

$$T_1 = \{t : X_t, X_{t-0} \in M \cup \{\partial\} \text{ and } X_{t-0} \neq X_t.\}$$

and T_2 is the set of left-hand endpoints for the intervals which are the connected components of the random open set

$$\{t > \sigma(M) : X_t \in D \text{ and } X_{t-0} \in D\}.$$

The set of entrance times is their union $T = T_1 \cup T_2$. We will use for indices the excursion times

$$\tau(t) = a(t) \text{ , defined for } t \in T.$$

Notice that for every excursion time

$$b(\tau) = \inf\{t > b(\tau-0) : X_t \in M\}.$$

The excursion w_τ is the point in $\Omega^{(D)}$ determined by

(1.3)
$$X^{(D)}_{\zeta^*-0}(w_\tau) = X_{b(\tau-0)}$$
$$X^{(D)}_{\zeta}(w_\tau) = X_{b(\tau)}$$
$$X_{\zeta^*+s}(w_\tau) = X_{b(\tau-0)+s} \quad \text{for } 0 < s < b(\tau)-b(\tau-0).$$

Clearly

(1.4)
$$\zeta(w_\tau)-\zeta^*(w_\tau) = b(\tau)-b(\tau-0)$$

when τ corresponds to an honest excursion, that is, $\tau = \tau(t)$ with $t \in T_2$. When τ corresponds to a jump within $M \cup \{\partial\}$, we define $\zeta^*(w_\tau) = \zeta(w_\tau) = 0$.

The measure $\mathbb{P}^{(D)}$ on $\Omega^{(D)}$ which gives the conditional distribution for excursions is determined by the following prescription.

1.1.1. On the set $[\zeta^* < \zeta]$ the variables X_t, $\zeta^* < t < \zeta$, are distributed exactly as in SMP, Theorem 5.2 with the absorbed Dirichlet space $(\underline{F}^D, E^D)$ (which is always transient) here playing the role of $(\underline{F}, E)$ there.

1.1.2. Conditioned on the set $[\zeta^* < \zeta]$ and on the variables X_t, $\zeta^* < t < \zeta$, the adjoined variables $X^{(D)}_{\zeta}$, $X^{(D)}_{\zeta^*-0}$ are independent and depend only on the corresponding limit variables $X^{(D)}_{\zeta-0}$, $X^{(D)}_{\zeta^*}$. The conditional distribution of $X^{(D)}_{\zeta}$ given $X^{(D)}_{\zeta-0}$ is the same as that of $X^{(D)}_{\zeta^*-0}$ given $X^{(D)}_{\zeta^*}$ and is determined by

$$X^{(D)}_{\zeta} = X^{(D)}_{\zeta-0} \quad \text{on } [X^{(D)}_{\zeta-0} \in M \cup \{\partial\}]$$

$$\text{Ⓔ}^{(D)}[X^{(D)}_{\zeta-0} \in D; X^{(D)}_{\zeta} = \partial;\ \varphi(X^{(D)}_{\zeta-0})] = \int_D dx\,\kappa(x)\varphi(x)$$

$$\text{Ⓔ}^{(D)}[X^{(D)}_{\zeta-0} \in D; X^{(D)}_{\zeta} \in M;\ \varphi(X^{(D)}_{\zeta-0})\psi(X^{(D)}_{\zeta})] = \iint_{D\times M} J(dx, dy)\varphi(x)\psi(y).$$

1.1.3. On the set $[\zeta^* = \zeta = 0]$ the joint distribution of $X^{(D)}_{\zeta-0}$, $X^{(D)}_{\zeta}$ is determined by

$$\text{Ⓔ}^{(D)}[\zeta^* = \zeta = 0;\ X^{(D)}_{\zeta^*-0} = \partial;\ X^{(D)}_{\zeta} \in M;\ \varphi(X^{(D)}_{\zeta})]$$

$$= \text{Ⓔ}^{(D)}[\zeta^* = \zeta = 0; X^{(D)}_{\zeta^*-0} \in M; X^{(D)}_{\zeta} = \partial;\ \varphi(X^{(D)}_{\zeta^*-0})]$$

$$= \int_M dx\,\kappa(x)\varphi(x)$$

$$\text{Ⓔ}^{(D)}[\zeta^* = \zeta = 0; X^{(D)}_{\zeta^*-0} \in M; X^{(D)}_{\zeta} \in M;\ \varphi(X^{(D)}_{\zeta^*-0})\psi(X^{(D)}_{\zeta})]$$

$$= \iint_{M\times M} J(dy, dz)\varphi(y)\psi(z).$$

///

The time reversal operator ρ on $\Omega^{(D)}$ is defined on variables X_t, $\zeta^* < t < \zeta$, as in SMP, page 5.7, and in addition, it interchanges $X^{(D)}_{\zeta}$ and $X^{(D)}_{\zeta^*-0}$. With the help of SMP, Theorem 5.3, it is easy to check time reversal invariance

(1.5) $$\text{Ⓔ}^{(D)}\xi \circ \rho = \text{Ⓔ}^{(D)}\xi.$$

Also the formula (5.8) in SMP corresponds here to

(1.6) $$\text{Ⓔ}^{D}\int_{\zeta^*}^{\zeta} dt\,\varphi(X_t) = \int_D dx\,\varphi(x).$$

What appears directly in our decomposition is not $\text{Ⓟ}^{(D)}$ itself, but instead certain conditioned measures determined by it. The off-diagonal bi-conditional measures $\text{Ⓟ}^{(D)}_{y,z}$, $y \neq z$, in M and $\text{Ⓟ}^{(D)}_{y,\partial}$, $y \in M$, are defined by conditioning jointly on the initial position $X^{(D)}_{\zeta^*-0}$ and terminal position $X^{(D)}_{\zeta}$. We will see in

Section 2 that the Lévy measure $J^M(dy, dz)$ and killing measure $\kappa^M(dy)$ for the time changed process can be represented

(1.7) $$J^M(dy, dz) = 1_M(y)J(dy, dz)1_M(z) + \nu^M(dy)\nu^M(dz)I(y \neq z)u^M_{0,\infty}(y, z)$$

(1.8) $$\kappa^M(dy) = 1_M(y)\kappa(y)dy + \pi^M_* p^{(D)}(y)\nu^M(dy)$$

where the functions $u^M_{0,\infty}(y, z)$, $\pi^M_* p^D(y)$ account for "true excursions" into D. Therefore, the off-diagonal bi-conditional measures are defined by

(1.9) $$Ⓔ^{(D)}[X^{(D)}_{\zeta^*-0}, X^{(D)}_\zeta \in M; X^{(D)}_{\zeta^*-0} \neq X^{(D)}_\zeta; \varphi(X^{(D)}_{\zeta^*-0})\psi(X^{(D)}_{\zeta-0})]$$
$$= \iint_{M\times M} J^M(dy, dz)\varphi(y)\psi(z)Ⓔ^{(D)}_{y,z}\xi$$

(1.10) $$Ⓔ^{(D)}[X^{(D)}_{\zeta^*-0} \in M, X^{(D)}_\zeta = \partial; \varphi(X^{(D)}_{\zeta^*-0})\xi] = \int_M \kappa^M(dy)\varphi(y)Ⓔ^{(D)}_{y,\partial}\xi$$

for $\varphi, \psi \geq 0$ on M and $\xi \geq 0$ on $\Omega^{(D)}$ and they are normalized

(1.11) $$Ⓟ^{(D)}_{y,z}(\Omega^{(D)}) = 1 \text{ for } y \neq z ; Ⓟ^{(D)}_{y,\partial}(\Omega^{(D)}) = 1.$$

Existence modulo the obvious exceptional sets is guaranteed for example by the measure theoretic results in Parthasarathy [42, Chapter 3]. The diagonal bi-conditioned measures $P^{(D)}_{y,y}$ cannot be normalized in general. We will see in Section 2 that they need only be considered for the atoms of ν^M, and so we can define them by the formula

(1.12) $$Ⓔ^{(D)}[X^{(D)}_{\zeta^*-0}, X^{(D)}_\zeta \in M; X^{(D)}_{\zeta^*-0} = X^{(D)}_\zeta; \varphi(X^{(D)}_{\zeta-0})\xi]$$
$$= \Sigma_y \nu^M(\{y\})\varphi(y)Ⓔ^{(D)}_{y,y}\xi$$

with the sum taken over the atoms of ν^M.

When stating our results, it is convenient to consider separately the *off-diagonal entrance times*

$$T_{od} = \{t \in T : X_{t-0} \neq X_{b(\tau(t))}\}$$

and the diagonal ones

$$T_d = \{t \in T : X_{t-0} \neq X_{b(\tau(t))}\}.$$

The off-diagonal excursions are taken care of in

Theorem 1.1. i) *The set of off-diagonal excursion times* $\{\tau(t) : t \in T_{od}\}$ *is measurable with respect to the time changed process* $\{Y_\tau, \tau \geq 0\}$.

ii) *Conditioned on the time changed process* $\{Y_\tau, \tau \geq 0\}$ *and on the diagonal excursion times* $\{\tau(t) : t \in T_d\}$, *the off-diagonal excursions* $\{w_{\tau(t)} : t \in T_{od}\}$ *are mutually independent and depend only on the corresponding initial and terminal positions* $Y_{\tau(t)-0}, Y_{\tau(t)}$. *The conditional distribution of* $w_{\tau(t)}$ *is the normalized bi-conditional excursion probability* $\textcircled{P}^{(D)}_{Y_{\tau(t)-0}, Y_{\tau(t)}}$. *This is true relative to the* $\textcircled{P}_x$ *for quasi-every* $x \in \underline{\underline{X}}$. ///

Before discussing the diagonal excursions we must introduce some additional machinery. For z an atom of ν^M consider

$$T_z = \{t \in T_d : X_{t-0} = X_{b(\tau(t))} = z\}$$

and the "local time"

$$a_z(t) = \int_0^t a(ds) I(X_s = z)$$

together with the inverse local time

$$b_z(\tau) = \inf\{t : a_z(t) > \tau\}.$$

The point is that the random set T_z will have a simple structure relative to the local time $\{a_z(t)\}$, but not relative to original time or to the $a(\cdot)$ clock. Clearly,

$$T_d = U_z T_z$$

with z running over the atoms of z. The difficulty with treating T_z (or rather $\{a_z(t) : t \in T_z\}$) directly is that it need not be a discrete set. To get around this, we introduce an increasing sequence of compact subsets A_n of D whose interiors expand to fill D. The point is that for each n

$$\{t \in T_z : \sigma(w_{\tau(t)}, A) < +\infty\}$$

has no finite limit point and therefore can be ordered $e(z,n,1), e(z,n,2), \ldots$. Let

$$a(z,n,i) = a_z(e(z,n,i))$$

be the entrance times according to the local time clock and then let

$$\tau(z,n,i) = \tau(e(z,n,i))$$

$$w_{z,n,i} = w_{\tau(z,n,i)}$$

$$r(z,n,i) = b(\tau(z,n,i)).$$

Thus $\tau(z,n,i)$ is the level of the $a(\cdot)$ clock at time $e(z,n,i)$ and $w_{z,n,i}$, $r(z,n,i)$ are the corresponding excursion and terminal times.

1.2. Terminology. Fix $\gamma > 0$ and $0 \leq \ell \leq +\infty$ and let $\{R_i\}_{i=1}^{\infty}$ be a sequence of mutually independent variables, each exponentially distributed at the rate γ, that is, $P[R_i > t] = e^{-\gamma t}$ for $t > 0$. The sequence $\{X_n\}_{n=1}^{\infty}$ defined by

$$\begin{aligned} X_n &= R_1 + \ldots + R_n \quad \text{when } R_1 + \ldots + R_n < \ell \\ &= +\infty \quad \text{when } R_1 + \ldots + R_n \geq \ell \end{aligned}$$

will be called a sequence of holding times with rate γ truncated at ℓ. ///

Theorem 1.2. i) Conditioned on the time changed process $\{Y_\tau\}$ the families of diagonal excursion times (according to corresponding local clocks) $\{a(z,n,i)\}_{n,i=1}^{\infty}$ are mutually independent as z varies. For fixed z and n the times $\{a(z,n,i)\}_{i=1}^{\infty}$ form a sequence of holding times with rate

$$P_{z,z}^{(D)}(\sigma(A_n) < +\infty) = P^{(D)}(X_{\zeta^*-0}^{(D)} = X_{\zeta}^{(D)} = z\,;\sigma(A_n) < +\infty)/\nu^M(\{z\})$$

truncated at $a_z(\zeta)$.

ii) Conditioned on the time changed process $\{Y_\tau\}$ and all the excursion times $\{\tau(z,n,i)\}_{z,n,i}$ the excursions $\{w_{\tau(t)} : t \in T\}$ are mutually independent and depend only on the initial and return positions $Y_{\tau-0}, Y_\tau$.

iii) The conditional distribution of the diagonal excursion $w_{z,n,i}$ is $P_{z,z}^{(D)}(\;\,|\sigma(A_n) < +\infty)$ on the set where $\tau(z,n,i) < +\infty$. ///

The excursion time $\tau(z,n,i)$ is the level of the $a(\cdot)$ clock at the instant that the a_z clock reaches $a(z,n,i)$. Therefore, the sequence $\{\tau(z,n,i)\}$ is a function of the sequence $\{a(z,n,i)\}$ together with the time changed process $\{Y_\tau\}$ and so its conditional distribution is specified by Theorem 1.2, i).

2. Proof of the Decomposition.

We begin with some notations. Together with the usual hitting operators

$$H_u^M \varphi(x) = \textcircled{E}_x e^{-u\sigma(M)} \varphi(X_{\sigma(M)})$$

(see pages 4.16 through 4.22 in SMP) we consider the relative hitting operators

$$H_u^{M;A} \varphi(x) = \textcircled{E}_x [\sigma(M) < \sigma(A) ; e^{-u\sigma(M)} \varphi(X_{\sigma(M)})]$$

$$H_u^{A;M} \varphi(x) = \textcircled{E}_x [\sigma(A) < \sigma(M) ; e^{-u\sigma(A)} \varphi(X_{\sigma(A)})]$$

with A a subset (usually compact) of D. Here and below it is understood that for $u = 0$ we replace the exponential factor by the indicator of the set where the corresponding stopping time is finite. Often we suppress the subscript 0. By SMP, Theorem 7.3(iv), the measures $H_u^M(x,dy)$, $H_u^{M;A}(x,dy)$ are for quasi-every x respectively in D, $D-A$ absolutely continuous with respect to ν^M and therefore the densities $h_u^M(x,y)$, $h_u^{M;A}(x,y)$ are well defined modulo the usual ν-null set by

$$H_u^M(x,dy) = h_u^M(x,y)\nu^M(dy)$$

$$H_u^{M;A}(x,dy) = h_u^{M;A}(x,y)\nu^M(dy).$$

For $0 \leq u < v$ the Feller density and linking Feller kernel are defined by

$$u_{u,v}^M(y,z) = (v-u)\int_D dx\, h_u^M(x,y) h_v^M(x,z)$$

$$U_{u,v}^{M\#A}(y,dz) = (v-u)\int_{D-A} dx\, h_u^{M;A}(x,y) H_v^{A;M}(x,dz).$$

By (20.13') in SMP the Feller density is symmetric in u, v and the roles of u, v can be reversed in the integral for the linking Feller kernel. Also for

$0 \leq u < v < w$

$$u^M_{u,w}(y,z) = u^M_{u,v}(y,z) + u^M_{v,w}(y,z)$$

$$U^{M\#A}_{u,w}(y,dz) = U^{M\#A}_{u,v}(y,dz) + U^{M\#A}_{v,w}(y,dz),$$

and therefore the definitions

$$u^M_{u,\infty}(y,z) = \operatorname{Lim}_{v\uparrow\infty} u^M_{u,v}(y,z)$$

$$U^{M\#A}_{u,\infty}(y,dz) = \operatorname{Lim}_{v\uparrow\infty} U^{M\#A}_{u,v}(y,dz)$$

make sense. We will use also the corresponding Feller operators

$$U^M_{u,v}\varphi(y) = \int_M u^M_{u,v}(y,z)\varphi(z)\nu^M(dz)$$

$$U^{M\#A}_{u,v}\varphi(y) = \int_A U^{M\#A}_{u,v}(y,dz)\varphi(z)$$

and the adjoint operators

$$\pi^M_u f(y) = \int_D dx\, f(x) h^M_u(x,y)$$

mapping functions on D to functions on M. Clearly

$$U^M_{u,v} = (v-u)\pi^M_u H^M_v = (v-u)\pi^M_v H^M_u.$$

Occasionally we will write

$$\pi^M \mu(y) = \int_D \mu(dx) h^M_u(x,y)$$

for μ a measure on D.

Next we introduce some machinery for connecting the standard sample space Ω and the excursion space $\Omega^{(D)}$. For $\omega \in \Omega$ the truncated path $s^D\omega$ is defined by

$$X_t(s^D\omega) = X_t(\omega) \quad \text{for } t < \sigma(M) \wedge \zeta$$

$$= X_{\sigma(M)\wedge\zeta} \quad \text{for } t \geq \sigma(M) \wedge \zeta.$$

For σ a random time on $\Omega^{(D)}$ such that $\zeta^* \leq \sigma \leq \zeta$ and for $\omega \in \Omega^{(D)}$ the shifted trajectory $\theta_\sigma\omega$ in Ω is defined by

$$X_t(\theta_\sigma\omega) = X_{\sigma+t}(\omega) \quad \text{for } \sigma+t < \zeta$$

$$= X_\zeta^{(D)} \quad \text{for } \sigma+t \geq \zeta.$$

The reversed shifted trajectory $\theta_\sigma^*\omega$ is defined by

$$X_t(\theta_\sigma^*\omega) = X_{\sigma-t-0}(\omega) \quad \text{for } \sigma-t > \zeta^*$$

$$= X_{\zeta^*-0}^{(D)}(\omega) \quad \text{for } \sigma-t \leq \zeta^*.$$

Also the shifted hitting and last exit times are defined by

(2.1) $$\sigma_t(A) = \inf\{s > t : X_s \in A\}$$

$$\sigma_t^*(A) = \sup\{s < t : X_s \in A\}$$

with the usual understanding that these times are respectively $+\infty$ and $-\infty$ when not otherwise defined.

The technique of SMP (20.37) can be adapted to establish

(2.2) $$\text{Ⓔ}^{(D)}[\zeta^* < \zeta; X_{\zeta^*-0}^{(D)}, X_\zeta^{(D)} \in M; \varphi(X_{\zeta^*-0}^{(D)})\psi(x_\zeta^{(D)})]$$

$$= \int_M \nu^M(dy) \int_M \nu^M(dz) u_{0,\infty}^M(y,z)\varphi(y)\psi(z)$$

$$(2.3)\qquad \text{Ⓔ}^{(D)}[\zeta^* < \zeta ; X^{(D)}_{\zeta^*-0} \in M ; \sigma(A) < +\infty ; \varphi(X^{(D)}_{\zeta^*-0})\xi \cdot \theta_{\sigma(A)}]$$

$$= \int_M \nu^M(dy)\varphi(y)U^{M\#A}_{0,\infty} E \cdot \xi \cdot s^D$$

$$(2.4)\qquad \text{Ⓔ}^{(D)}[\zeta^* < \zeta ; X^{(D)}_{\zeta^*-0} \in M ; X^{(D)}_{\zeta} = \partial ; \varphi(X^{(D)}_{\zeta^*-0})]$$

$$= \int_M \nu^M(dy)\varphi(y)\pi^M_* p^D(y)$$

where

$$(2.5)\qquad p^D(x) = \text{Ⓔ}_x[\sigma(M) = +\infty ; X_{\zeta-0} = \partial] ; \pi^M_* p^D = \operatorname{Lim}_{v\uparrow\infty} v\pi^M_v p^{(D)} .$$

The function p^D here corresponds to $p^D + r^D + s^D$ in (10.7) but to $\tilde{p}$ in (20.25) in SMP. The calculation (20.27) in SMP (after being corrected) shows that $v\pi^M_v p^D$ increases with v and therefore $\pi^M_* p^D$ is well defined. We give a proof here for (2.3) only. The arguments for (2.2) and (2.4) are analogous and details can easily be filled in with the help of SMP, (20.37) and (20.38). For $v > 0$

$$\int_M \nu^M(dy)\varphi(y)U^{M\#A}_{0,v}\text{Ⓔ}\,\xi \circ s^D$$

$$= v\int_{D\setminus A} dx H^{M;A}_v \varphi(x) H^{A;M}\text{Ⓔ}\,\xi \circ s^D$$

$$= \text{Ⓔ}^{(D)}\int_{\zeta^*}^{\zeta} dt\, I(X_t \in D\setminus A) v H^{M;A}_v \varphi(X_t) H^{A;M}\text{Ⓔ}\,\xi \circ s^D$$

$$= \text{Ⓔ}^{(D)}\int_{\zeta^*}^{\zeta} dt\, I(X_t \in D\setminus A) I(\sigma_t(A) < \sigma_t(M))\xi \circ \theta_{\sigma_t(A)} v H^{M;A}_v \varphi(X_t)$$

and by time reversal invariance the last expression

$$= \text{Ⓔ}^{(D)}\int_{\zeta^*}^{\zeta} dt\, I(X_t \in D\setminus A) I(\sigma^*_t(A) > \sigma^*_t(M))\xi \circ \theta^*_{\sigma^*_t(A)} v H^{M;A}_v \varphi(X_t).$$

By 1.1.2 and the interpretation of $J(dx, dy)$ on $D\times M$ as an intensity for "jumping" from D to M (see (10.11) and (10.18) in SMP), this

$$= \text{Ⓔ}^{(D)} \int_{\zeta^*}^{\zeta} dt\, I(X_t \in D\setminus A) I(\sigma_t^*(A) > \sigma_t^*(M)) \xi \circ \theta_{\sigma_t^*(A)} \, v e^{-v(\zeta - t)}$$

$$\cdot I(\sigma_t(A) = +\infty ; X_\zeta^{(D)} \in M) \varphi(X_\zeta^{(D)})$$

which after another application of time reversal invariance

$$= \text{Ⓔ}^{(D)} I(X_{\zeta^*-0}^{(D)} \in M) \varphi(X_{\zeta^*-0}^{(D)}) \int_{\zeta^*}^{\sigma(A)} dt\, v e^{-v(t-\zeta^*)} I(\sigma(A) < +\infty) \xi \circ \theta_{\sigma(A)}$$

and (2.3) follows after passage to the limit $v \uparrow \infty$.

The relations (1.7) and (1.8) follow from Theorems 10.1 and 10.5 together with the calculations (20.37) and (20.38) in SMP.

In this section we will need also the conditioned measures $\text{Ⓟ}_y^{(D)}$, $y \in M$, defined by conditioning only on the initial position $X_{\zeta^*-0}^{(D)}$. In general, they cannot be normalized. Since the distribution of $X_{\zeta^*-0}^{(D)}$ on the set $[\zeta^* \neq \zeta = 0]$ need not be absolutely continuous with respect to ν^M, it is technically convenient to restrict everything to the set $[\zeta^* < \zeta]$. The defining formula is

$$(2.6) \qquad \text{Ⓔ}^{(D)}[X_{\zeta^*-0}^{(D)} \in M ; \zeta^* < \zeta ; \varphi(X_{\zeta^*-0}^{(D)}) \xi] = \int_M \nu^M(dy) \varphi(y) \text{Ⓔ}_y^{(D)} \xi .$$

The following operators are defined for $\varphi \geq 0$ on M and for $y \in M$.

$$(2.7) \qquad R^A \varphi(y) = N^{\underline{\underline{X\setminus A}}} \varphi \cdot \nu^M(y)$$

$$R_{(1)}^A \varphi(y) = N_1^{\underline{\underline{X\setminus A}}} \varphi \cdot \nu^M(y)$$

$$R_{(1)} \varphi(y) = N_1 \varphi \cdot \nu^M(y).$$

Here $N^{\underline{\underline{X\setminus A}}}$ is the potential operator for the absorbed Dirichlet space $(\underline{\underline{F}}^{\underline{\underline{X\setminus A}}}, E)$

in the notation of SMP, Section 7. Similarly N_1 is the potential operator for $(\underline{\underline{F}}, E_1)$ and $N_1^{\underline{\underline{X}}\setminus A}$ is the potential operator for $(\underline{\underline{F}}^{\underline{\underline{X}}\setminus A}, E_1)$. Thus for example

$$R_{(1)}^{(A)}\varphi(y) = Ⓔ_y\int_0^{\sigma(A)} a(dt)e^{-t}\varphi(X_t)$$

$$= Ⓔ_y\int_0^{b(\sigma(A))} d\tau e^{-b(\tau)}\varphi(Y_\tau).$$

Our main preliminary result is

Lemma 1.1. *For $\varphi \geq 0$ on M, for $y \in M$ and for $\xi > 0$ on the subset $[\zeta^* < \zeta]$ of $\Omega^{(D)}$*

$$Ⓔ_y \Sigma_{t\in T} e^{-t}\varphi(X_{t-0})\xi(w_{\tau(t)}) = R_{(1)}\varphi Ⓔ_{\cdot}^{(D)}\xi(y). \qquad ///$$

Proof. We begin by fixing a compact subset A of D, defining

$$e(A) = \inf\{t \in T : \sigma(A, w_{\tau(t)}) < +\infty\}$$

and then establishing

$$Ⓔ_y e^{-e(A)}\varphi(X_{e(A)-0})\xi \circ \theta_{\sigma(A)}(w_{\tau(e(A))}) = R_{(1)}^{A}\varphi Ⓔ_{\cdot}^{(D)}[\sigma(A) < +\infty ; \xi \circ \theta_{\sigma(A)}] \tag{2.8}$$

By the strong Markov property it suffices to consider the special case $\xi = \psi(X_0)$ in (2.8) and then by the usual "stripping away" argument for excessive functions and by (2.3) we need only establish

$$\int \nu^M(dy)f(y)Ⓔ_y[e^{-e(A)}\varphi(X_{e(A)-0})\psi(X_{\sigma(A)})] = \int \nu^M(dy)R_{(1)}^{A}f(y)\varphi(y)U_{0,\infty}^{M\#A}\psi(y) \tag{2.9}$$

with $f \geq 0$ on M. To prove (2.9) we apply the technique of (20.38) in SMP. Let

Ⓔ be the expectation functional for the approximate Markov process of SMP, Section 5, corresponding to $(\underline{\underline{F}}, E)$ and for present purposes only let $(e(i), r(i))$ be the excursions as defined in SMP, Section 13, except that A plays the role of M and $\underline{\underline{X}}$-A the role of D. Also let

$$\sigma(i) = \inf\{t > e(i) : t < r(i) \text{ and } X_t \in M\}$$

with the understanding that $\sigma(i) = r(i)$ when not otherwise defined. Then by SMP, Theorem 6.3, the left side of (2.9) equals

(2.10) $$Ⓔ\int_{\zeta^*}^{\zeta} a(dt) f(X_t) e^{-\{e_t(A)-t\}} \varphi(X_{e_t(A)-0}) \psi(X_{\sigma_t(A)})$$

where $\sigma_t(A)$ is defined in (2.1) and $e_t(A)$ is defined by analogy. Regarding the interior integral as a sum over excursions from A and applying time reversal invariance, we can rewrite (2.10) as

(2.11) $$Ⓔ\Sigma_i \psi(X_{e(i)-0}) \varphi(X_{\sigma(i)}) \int_{\sigma(i)}^{r(i)} a(dt) f(X_t) e^{-\{t-\sigma(i)\}}$$

$$= \operatorname{Lim}_{v\uparrow\infty} Ⓔ\Sigma_i \psi(X_{e(i)-0}) \int_{e(i)}^{\sigma(i)} ds\, v e^{-v\{\sigma(i)-s\}} \varphi(X_{\sigma(i)}) \int_{\sigma(i)}^{r(i)} a(dt) f(X_t) e^{-\{t-\sigma(i)\}}$$

$$= \operatorname{Lim}_{v\uparrow\infty} Ⓔ\Sigma_i \psi(X_{e(i)-0}) \int_{e(i)}^{\sigma(i)} ds\, v H_v^{M;A} \varphi R_{(1)}^{A} f(X_s)$$

$$= \operatorname{Lim}_{v\uparrow\infty} \int_{D\setminus A} dx\, v H_v^{M;A} \varphi R_{(1)}^{A} f(x) H^{A;M} \psi(x)$$

which equals the right side of (2.9). This proves (2.9) and therefore (2.8). Now (2.8) immediately implies

(2.12) $$Ⓔ_y \Sigma_{t\in T} e^{-t} I(\sigma(A, w_{\tau(t)}) < +\infty) \varphi(X_{t-0}) \xi \circ \theta_{\sigma(A)}(w_{\tau(t)})$$

$$= R_{(1)} \varphi Ⓔ_{\cdot}^{(D)} [\sigma(A) < +\infty ; \xi \circ \theta_{\sigma(A)}]$$

since the two sides of (2.12) are the minimal solutions g of the integral equation

2.8

$$g(y) = g_o(y) + \textcircled{E}_y e^{-r(A)} g(X_{r(A)})$$

with g_o the corresponding side of (2.8). Of course

$$r(A) = \inf\{t > \sigma(A) : X_t \in M\}.$$

Finally, the lemma follows upon passage to the limit in A. ///

Remark. Lemma 1.1 is closely related to the basic result (1.8) in [10]. However, the setting and also the techniques for the proof are quite different.

A special case of (2.3) is

$$\textcircled{E}^{(D)}[\zeta^* < \zeta ; X^{(D)}_{\zeta^*-0} \in M ; X^{(D)}_{\zeta} \in M ; \sigma(A) < +\infty ; \varphi(X^{(D)}_{\zeta^*-0})\psi(X^{(D)}_{\zeta})]$$

$$= \int_M \nu^M(dy)\varphi(y) U^{M\#A}_{0,\infty} H^M \psi(y)$$

$$= \operatorname{Lim}_{v\uparrow\infty} \int_{D\setminus A} dx\, v H^{M;A}_v \varphi(x) H^{A;M} H^M \psi(x)$$

for A a compact subset of D, and this guarantees that the joint distribution of $X^{(D)}_{\zeta^*-0}, X^{(D)}_{\zeta}$ relative to $\textcircled{P}^{(D)}$ is, when restricted to M, absolutely continuous with respect to $\nu^M \times \nu^M$. This together with Lemma 2.1 permits us to restrict attention to atoms of ν^M when studying the diagonal excursions.

We turn now to the proof of Theorem 1.1. Part (i) is obvious. To prove (ii) we begin by fixing B a closed subset of M, setting

$$\tau = a(\sigma(B) \wedge \zeta)$$

and establishing

(2.13) $$Ⓔ_y[b(\tau-0) < +\infty ; \varphi(Y_{\tau-0}) ; \sigma(A, w_\tau) < +\infty ; \xi \circ \theta_{\sigma(A)}(w_\tau) ; \psi(Y_\tau)]$$

$$= R^B \varphi Ⓔ^{(D)}_{\cdot}[\sigma(A) < +\infty ; \xi \circ \theta_{\sigma(A)} ; \psi(X^{(D)}_\zeta)](y)$$

where $y \in M\text{-}B$, where A is a compact subset of D, where $\varphi \geq 0$ on $M\text{-}B$ and $\varphi = 0$ on B, where $\xi \geq 0$ on Ω, and where $\psi \geq 0$ on $B \cup \{\partial\}$ and $\psi = 0$ on $M\text{-}B$. The operator R^B is defined by (2.7) with A replaced by B. Again we apply the technique of SMP, (20.38), but now $(e(i), r(i))$ represent time intervals for excursions w_i from M. After smoothing as for (2.8) we can represent the left side of (2.13) as

(2.14) $$Ⓔ\int_{\zeta^*}^{\zeta} a(dt) f(X_t) \Sigma_i I(t < e(i) < t+\sigma_t(B) ; \sigma(A, w_i) < +\infty) \varphi(X_{e(i)-0}) g(X_{\sigma(A, w_i)})$$

where

$$g(x) = Ⓔ_x[\xi\psi(X^{(D)}_\zeta)].$$

Notice that the validity of replacing (2.13) by (2.14) depends on the fact that $\psi = 0$ on $M \backslash B$. Arguing as in (2.11) we see that (2.14)

$$= \text{Lim}_{v \uparrow \infty} \int_{D \backslash A} dx H^{A;M} g(x) v H_v^{M;A} \varphi R^B f(x)$$

$$= \int \nu(dy) f(y) R^B \varphi U_{0,\infty}^{A\# M} g(y)$$

and (2.13) follows with the help of (2.3). Passage to the limit in A in (2.13) establishes

(2.15) $$Ⓔ_y[b(\tau-0) < +\infty ; \varphi(Y_{\tau-0}) \xi(w_\tau) \psi(Y_\tau)] = R^B \varphi Ⓔ^{(D)}_{\cdot}(\xi\psi(X^{(D)}_\zeta))$$

for $\xi \geq 0$ on $\Omega^{(D)}$ and vanishing on the set $[\zeta = \zeta^*]$ (corresponding to "pseudo-excursions"). The results in SMP, Section 10, imply

(2.16) $$\text{Ⓔ}_y[b(\tau-0) < +\infty ; \varphi(Y_{\tau-0})\psi(Y_\tau)]$$
$$= N^{\underline{\underline{X}}\setminus B} \varphi \int J^M(d\cdot, dy)\psi(y) + N^{\underline{\underline{X}}\setminus B} \varphi \cdot \kappa^M \psi(\partial)$$

which supplements (2.15). Then with the help of (2.15) and (2.16) it is easy to show that conditioned on the past prior to $b(\tau-0)$ (that is, on the trajectory variables $X_t I(t < b(\tau-0))$ and on the future succeeding $b(\tau)$ (that is, on shifted variables $\xi \circ \theta_{b(\tau)}$) the excursion w_τ depends only on $X_{b(\tau)-0} = Y_{\tau-0}$ and $X_{b(\tau)} = Y_\tau$, and in the manner prescribed by the theorem. The complete Theorem 1.1 follows after a routine "piecing together" argument.

In proving Theorem 1.2 we proceed somewhat indirectly. For z an atom of ν^M let $(\Omega^z, \text{Ⓟ}^z)$ be exactly analogous to $(\Omega^{(D)}, \text{Ⓟ}^{(D)})$ in Section 1 with $\{z\}$ playing the role of M, except that we <u>restrict Ⓟ^{z} to the set where</u> $X_{\zeta^*-0} = z$. We will omit the clumsy superscript $\underline{\underline{X}}-\{z\}$ for terminal and initial variables, and simply agree that we are always considering limits in $\underline{\underline{X}} \cup \{\partial\}$ rather than in the one point compactification of $\underline{\underline{X}}\setminus\{z\}$. Let T_z^0 be the set of entrance times for excursions from $\{z\}$. That is, T_z^0 is the set of left hand end-points for the intervals which are the connected components of the random open set

$$\{t : \sigma(\{z\}) < t < \zeta, X_t \neq z \text{ and } X_{t-0} \neq z\}.$$

Notice that T_z^0 contains together with diagonal excursions from z a possible terminal off-diagonal excursion terminating at ∂. For $t \in T_z^0$ let w_t be defined as in (1.3). Let $e^0(z,n,i)$, $a^0(z,n,i)$, $\tau^0(z,n,i)$, $w^0_{z,n,i}$ and $r^0(z,n,i)$ be defined in the same way as $e(z,n,i)$, $a(z,n,i)$, $\tau(z,n,i)$, $w_{z,n,i}$ and $r(z,n,i)$ except that T_z^0 plays the role of T_z. The $e^0(z,n,i)$ do not necessarily exhaust T_z^0 (since there may be excursions from z which never enter D), but their union does contain

$$T_z = \{t \in T_z^0 ; \sigma(M-\{z\}, \tau(t)) = +\infty\}$$

which is the subset of direct interest. The proof of (2.8) can easily be adapted to establish

(2.17)
$$\textcircled{E}_z[a^0(z,n,1) < +\infty ; \xi \circ \theta_{\sigma(A_n)}(w^0_{z,n,1})]$$

$$= N^{\underline{\underline{X}}\setminus A_n}(z,z)\textcircled{E}^z[\sigma(A_n) < +\infty ; \xi \circ \theta_{\sigma(A_n)}].$$

It is clear from the strong Markov property that the excursion times $\{a^0(z,n,i)\}_{i=1}^{\infty}$ form a sequence of holding times with some rate γ truncated at $a_z(\zeta)$. To determine γ note that

$$\textcircled{P}_z[a^0(z,n,1) < +\infty] = \textcircled{E}_z(1-e^{-\gamma a_z(\zeta)})$$

$$= \gamma\textcircled{E}_z\int_0^{a_z(\zeta)} d\tau e^{-\gamma\tau}$$

$$= \gamma\textcircled{E}_z\int_0^{\zeta} a(dt)I(X_t = z)I(t < e^0(z,n,1))$$

$$= \gamma\textcircled{E}_z\{a_z(\zeta) \wedge a^0(z,n,1)\}.$$

Putting $\xi = 1$ in (2.17) we get

$$\textcircled{P}_z[a^0(z,n,1) < +\infty]$$

$$= N^{\underline{\underline{X}}\setminus A_n}(z,z)\textcircled{P}^z(\sigma(A_n) < +\infty)$$

$$= \nu^M(\{z\})^{-1}\textcircled{E}_z\{a_z(\zeta) \wedge a^0(z,n,1)\}\textcircled{P}^z(\sigma(A_n) < +\infty)$$

and therefore

$$\gamma = \textcircled{P}^z(\sigma(A_n) < +\infty)/\nu^M(\{z\}).$$

As in the proof of Theorem 1.1, the relation (2.17) can be used to show that conditioned on the past up to time $e^0(z,n,i)$, on the future after time $r^0(z,n,i)$, and on the event $[a^0(z,n,i) < +\infty]$, the trajectory $w^0_{z,n,i}$ depends only on the terminal position $X_{r^0(z,n,i)}$ and is distributed according to the conditional distribution

$$\textcircled{P}^z[\,\cdot\,|\sigma(A_n) < +\infty ; X_\zeta = X_{r^0(z,n,i)}].$$

Consider the excursion time of interest $a(z,m,1)$ and the corresponding excursion $w_{z,m,1}$. On the event $[a(z,m,1) < +\infty]$ it is true that for sufficiently large $n > m$ the excursion time $\tau(z,m,1)$ is also the first i such that $\sigma_{A_m}(w^0_{z,n,i}) < +\infty$, that $X_t(w^0_{z,n,i}) \in D$ for $\sigma(A_n) \le t < \zeta$, and that $X_\zeta(w^0_{z,n,i}) = z$. It follows easily that conditioned on the set $[a(z,m,1) < +\infty]$ and on the appropriate past and future the conditional distribution of $w_{z,m,1}$ is the same as the conditional probability

$$\textcircled{P}^z[\quad |\sigma(A_m) < +\infty ; \sigma(M\setminus\{z\}) = +\infty ; X_\zeta = z].$$

Of course 1 can be replaced by a general index i and it is clear that the $\{a(z,m,k)\}_{i=1}^{\infty}$ form a sequence of holding times truncated at $a_z(\zeta)$ with rate

$$(2.18) \qquad \gamma = \textcircled{P}^z[\sigma(M\setminus\{z\}) = +\infty ; \sigma(A_m) < +\infty ; X_\zeta = z]/\nu^M(\{z\}).$$

There is no difficulty replacing z by a general starting point $x \in \underline{\underline{X}}$ and so Theorem 1.2 will follow with the help of Theorem 1.1 if we can establish the identity

$$(2.19) \qquad \begin{aligned} &\textcircled{E}^z[\sigma(M\setminus\{z\}) = +\infty ; X_\zeta = z ; \eta] \\ &= \{\textcircled{E}^{(D)}_{z,z}\eta\}\nu^M(\{z\}) \\ &= \textcircled{E}^{(D)}[X^{(D)}_{\zeta^*-0} = X^{(D)}_\zeta = z ; \eta] \end{aligned}$$

for $\eta \geq 0$ a function on $\Omega^{(D)}$ which is identified in the obvious way with a corresponding function on Ω^z. But this is easily proved with the help of Theorem 1.1. Integrating both sides of Theorem 1.1 with respect to ν^M we get

$$(2.20) \qquad ⓔ^{(1)} \Sigma_{t\in T} e^{-(t-\sigma(M))} \xi(w_{\tau(t)}) = ⓔ^{(D)}[X^{(D)}_{\zeta^*-0} \in M; \xi],$$

where $ⓔ^{(1)}$ is the approximate Markov process of SMP, Section 5, with $(\underline{\underline{F}}, E_1)$ playing the role of $(\underline{\underline{F}}, E)$. But if T' is the set of entrance times for excursions from z then clearly

$$\Sigma_{t\in T} e^{-(t-\sigma(M))} \eta I(X^{(D)}_{\zeta^*-0} = X^{(D)}_{\zeta} = z)(w_{\tau(t)})$$

$$= \Sigma_{t\in T'} e^{-(t-\sigma(\{z\}))} \eta I(X_{\zeta} = z) I(\sigma(M\setminus\{z\}) = +\infty)(w_{\tau(t)})$$

and so (2.19) follows from (2.20). ///

3. Sample Space Construction.

We assume given a transient regular Dirichlet space $(\underline{\underline{F}}, E)$ on $L^2(\underline{\underline{X}}, dx)$ and construct a large class of symmetric Markov processes whose transition semigroups dominate in the sense of (0.5) the semigroup associated with $(\underline{\underline{F}}, E)$. We do not get all such processes (see Case 7 in Section 12), but we do get all those for which the associated Dirichlet space is an extension as defined in SMP, paragraph 20.2. In particular, we get all those having the same local generator in the sense of SMP, Section 15. (But see Section 6 below for a better formulation.)

We begin by introducing a boundary which will be used to prescribe the relevant Dirichlet space and then discarded in favor of a boundary which is suitable for the sample space construction.

The terminal set is the subset

$$\underline{\underline{T}} = [\zeta < +\infty ; X_{\zeta-0} = \partial]$$

of the standard sample space Ω. Two subsets of $\underline{\underline{T}}$ are called equivalent if their symmetric difference is $\textcircled{P}_x$ null for quasi-every $x \in \underline{\underline{X}}$. The terminal algebra is the sigma algebra $\textcircled{T}$ of subsets A of $\underline{\underline{T}}$ such that $\theta_t A = A$ on $[t < \zeta]$, divided by this equivalence relation. A terminal variable is any $\textcircled{T}$ measurable random variable Φ. The terminal algebra $\textcircled{T}$ and terminal variables Φ are viewed also as being defined on the sample space Ω_∞ by the obvious prescription.

Remark 1. For special cases such as [46] the terminal algebra $\textcircled{T}$ can be identified with a sigma-algebra on the so-called "Martin boundary", two sets being equivalent if their difference is null for harmonic measure. ///

Remark 2. It follows from the proof of Theorem 7.3(iv) in SMP that two subsets of $\underline{\underline{T}}$ are equivalent if and only if their symmetric difference is null for the reference measure ν defined below by (3.1).

The auxiliary space is a given separable locally compact Hausdorff space $\underline{\underline{M}}$, disjoint from $\underline{\underline{X}}$. The jumping measures are subprobabilities $\mu(x, d\cdot)$ defined on $\underline{\underline{T}} \cup \underline{\underline{M}}$, indexed by $x \in \underline{\underline{X}}$, and satisfying the usual regularity conditions. The residue $r(x) = 1-\mu(x, \underline{\underline{T}} \cup \underline{\underline{M}})$. Temporarily we work with pairs (Φ, φ) with Φ a terminal variable and with φ Borel on $\underline{\underline{M}}$. The reference measure ν is determined on $\underline{\underline{T}} \cup \underline{\underline{M}}$ by

(3.1)
$$\int_{\underline{\underline{T}} \cup \underline{\underline{M}}} \nu(dy)(\Phi, \varphi)(y)$$
$$= \int_{\underline{\underline{X}}} dx \, Ⓔ_x[\underline{\underline{T}}; e^{-\zeta}\Phi]$$
$$+ \int_{\underline{\underline{X}}} dx G_1\{\kappa \int_{\underline{\underline{T}} \cup \underline{\underline{M}}} \mu(\cdot, dz)(\Phi, \varphi)(z)\}(x).$$

This will be our definition of ν for the general theory. When treating examples we will often define ν in a more natural manner, but the alternative will always dominate and be dominated by a constant multiple of the measure (3.1) so that none of the arguments below will be effected by the substitution. The excursion form N is defined on pairs (Φ, φ) by

(3.2)
$$N((\Phi, \varphi), (\Phi, \varphi))$$
$$= \tfrac{1}{2} Ⓔ[X_{\zeta^*-0} = \partial; X_\zeta = \partial; \{\Phi - \Phi \cdot \rho\}^2]$$
$$+ \tfrac{1}{2} \int\int (\mu\kappa N\kappa\mu)(dy, dz)\{(\Phi, \varphi)(y) - (\Phi, \varphi)(z)\}^2$$
$$+ \int dx \kappa(x) \int \mu(x, dy) \int Ⓟ_x(dw)\{(\Phi, \varphi)(y) - \Phi(w)\}^2$$
$$+ \int dx \kappa(x) r(x) Ⓔ_x \Phi^2$$
$$+ \int dx \kappa(x) r(x) \int N\kappa\mu(x, dy)\{(\Phi, \varphi)(y)\}^2.$$

The excursion space is the set $\underline{\underline{N}}$ of (Φ, φ) belonging to $L^2(\underline{\underline{T}} \cup \underline{\underline{M}}, \nu)$ for which (3.2) is finite. Of course Ⓔ is the expectation functional for the appropriate Markov process of SMP, Section 5, and ρ is the time reversal operator. On the right side of (3.2) we have introduced some notations which are intended to be self-explanatory. For example, $\mu\kappa N\kappa\mu$ is the measure on the product space $(\underline{\underline{T}} \cup \underline{\underline{M}}) \times (\underline{\underline{T}} \cup \underline{\underline{M}})$ determined by

$$\iint (\mu\kappa N\kappa\mu)(dy, dz)(\Phi, \varphi)(y)(\Psi, \psi)(z)$$
$$= \int dx\kappa(x) \int \mu(x, dy)(\Phi, \varphi)(y) N\{\kappa \int \mu(\cdot\, , dz)(\Psi, \psi)(z)\}(x).$$

The terms in (3.2) correspond to various kinds of excursions from the boundary. The first term corresponds to excursions which enter and exit at the active or passive boundary--there is no contribution from excursions which both enter and exit at the passive boundary. The second corresponds to excursions which enter by jumping from the boundary to the interior and exit by jumping back to the boundary. The third, to excursions which enter by jumping from the boundary and exit at the active or passive boundary, or vice versa. The fourth, to excursions which enter by jumping from the dead point and exit at the active boundary, or vice versa. The fifth, to excursions which enter by jumping from the dead point and exit by jumping to the boundary, or vice versa. There is no contribution from excursions from the dead point to the dead point or to the passive boundary.

Before continuing we establish a general formula which will be used several times below. But first we introduce some terminology and prove a simple preliminary result.

3.1. Notation. For each positive integer n the truncation $\tau_n(f)$ is defined by

$$\tau_n f(x) = f(x) \qquad \text{if } |f(x)| \le n$$

$$= n \operatorname{sgn} f(x) \quad \text{if } |f(x)| > n. \qquad ///$$

For f belonging to the local Dirichlet space $\underline{\underline{F}}_{loc}$ the local Dirichlet measure $<A_c f>(dx)$ is defined by SMP, (14.1) and (11.29). Also by SMP, (11.30), this measure is contractive. That is, if f' is a normalized contraction of $f \in \underline{\underline{F}}^{loc}$, then

(3.3) $$<A_c f'>(dx) \le <A_c f>(dx).$$

Consider a function f which is defined and finite quasi-everywhere and such that the truncations $\tau_n(f)$ are all in $\underline{\underline{F}}^{loc}$. By (3.3) the local Dirichlet measures increase with n and therefore

(3.4) $$<A_c f>(dx) = \operatorname{Lim}_{n\uparrow\infty} <A_c \tau_n f>(dx)$$

is well defined as a Borel measure, although in general it need not be Radon or even sigma-finite.

3.2. Definition. A function f on $\underline{\underline{X}}$ is locally controlled if

3.2.1. F is defined and finite quasi-everywhere and the truncations $\tau_n f$ all belong to the local Dirichlet space $\underline{\underline{F}}_{loc}$.

3.2.2. The measure $<A_c f>(dx)$ defined by (3.4) is a Radon measure.

3.2.3. The integral

$$J<f, g> = \iint J(dx, dy)\{f(x)-f(y)\}\{g(x)-g(y)\}$$

converges absolutely for any $g \in \underline{\underline{F}} \cap C_{com}(\underline{\underline{X}})$. ///

We put

$$(3.5) \qquad D(f,f) = \int \langle A_c f \rangle (dx)$$

whenever this makes sense and in particular when f is locally controlled.

It is natural to work with the requirement "f is locally controlled" rather than "$f \in \underline{\underline{F}}^{loc}$" since the latter imposes the artificial restriction that locally $f \in L^2(\kappa)$.

Clearly $\underline{\underline{F}}^{ref}$ is the set of locally controlled f for which

$$(3.6) \qquad D(f,f) + \tfrac{1}{2}J\langle f,f\rangle + \int dx\kappa(x)f^2(x) < +\infty.$$

Also the __enveloping space__ $\underline{\underline{F}}^{env}$ (see SMP, Section 18) is the set of locally controlled f for which

$$(3.7) \qquad D(f,f) + \tfrac{1}{2}J\langle f,f\rangle < +\infty.$$

The simple preliminary result is

__Theorem 3.1.__ __For each $f \in \underline{\underline{F}}^{env}$, the enveloping space, there is a unique terminal variable $\gamma_T f$ such that__

$$(3.8) \qquad \mathrm{Lim}_{t\uparrow\zeta} f(X_t) = \gamma_T f \text{ on } [\zeta < +\infty ; X_{\zeta-0} = \partial]. \qquad ///$$

__Proof.__ The theorem is a consequence of the results in SMP, Section 18. The point is this. The resurrected space $\underline{\underline{F}}^{res}$ is the subset of functions in $\underline{\underline{F}}^{env}$ which are the limits quasi-everywhere of a sequence in $\underline{\underline{F}}$ for which the norms (3.7) are uniformly bounded. By SMP, Theorems 18.1 and 18.2, the function space $\underline{\underline{F}}^{res}$ is the extended Dirichlet space corresponding to the minimal process which can be recovered from the given $(\underline{\underline{F}}, E)$ process by suppressing the killing. The terminal set for this resurrected process contains in a natural way the terminal set for $(\underline{\underline{F}}, E)$ and so it follows from SMP, Proposition 4.5, and the Tchebychev type estimate SMP, (3.9), together with SMP, (5.9), that actually $\gamma_T f = 0$ for $f \in \underline{\underline{F}}^{res}$. Again by SMP, Theorem 18.1, the function space $\underline{\underline{F}}^{env}$ is the reflected space associated with the resurrected process and so by SMP, Theorem 14.4, the terminal variable $\gamma_T f$ exists for general $f \in \underline{\underline{F}}^{env}$. ///

3.3. Convention. From now on $\gamma_T f$ is defined by (3.8) whenever this makes sense. ///

Now we are ready for the general formula.

Theorem 3.2. *Let f have a representation*

$$f = f_0 + N\kappa\mu(\Phi, \varphi) + Ⓔ\,\Phi \tag{3.9}$$

with (Φ, φ) *defined on* $\underline{\underline{T}} \cup \underline{\underline{M}}$ *and with* $f_0 \in \underline{\underline{F}}_{(e)}$. *Then* $\Phi = \gamma_T f$ *and*

$$\begin{aligned} &D(f,f) + \tfrac{1}{2}J\langle f, f\rangle + \int dx\kappa(x)r(x)f^2(x) \\ &\quad + \int dx\kappa(x)\int\mu(x,dy)\{f(x) - (\gamma_T f, \varphi)(y)\}^2 \\ &\quad = E(f_0, f_0) + N((\gamma_T f, \varphi), (\gamma_T f, \varphi)). \end{aligned} \tag{3.10}$$

Conversely, if $f \in \underline{\underline{F}}^{env}$ *and if there exists* φ *defined on* $\underline{\underline{M}}$ *such that the left side of* (3.10) *converges, then f has a representation* (3.9) *with* $\Phi = \gamma_T f$ *and* (3.10) *is valid.* ///

In (3.9) we are using the symbol $\mu(\Phi, \varphi)$ to denote the function

$$\mu(\Phi, \varphi)(x) = \int\mu(x, dy)\{\Phi, \varphi\}(y).$$

We will often use such notations below without special comment.

The direct part of Theorem 3.2 was stated and proved in SMP as

Lemma 20.7(ii). In fact the converse is an easy consequence of Lemma 20.7(i). In showing this it suffices to consider the special case when both f and φ are bounded since we can apply the truncation τ_n simultaneously to f and φ and then pass to the limit. If D is open with compact closure and if $M = \underline{\underline{X}}\setminus D$, then clearly f has a representation

$$f = f^D + N^D \kappa\mu(\Phi, \varphi) + H^M f \tag{3.11}$$

with f^D in the absorbed space $\underline{\underline{F}}^D$ of SMP, Section 7. As $D \uparrow \underline{\underline{X}}$ clearly $H^M f \to \textcircled{E}_{\cdot} \gamma_T f$ and $N^D \kappa\mu(\Phi, \varphi) \to N\kappa\mu(\Phi, \varphi)$ quasi-everywhere and therefore $f_0 = \operatorname{Lim} f^D$ exists quasi-everywhere and (3.9) is valid, of course with $\Phi = \gamma_T f$. Finally, it follows from SMP, Lemma 20.7(i) that the norms $E(f^D, f^D)$ are bounded independent of D and so $f_0 \in \underline{\underline{F}}_{(e)}$ by paragraph 1.6.1' in SMP. This completely proves Theorem 3.2.

The special case $f_0 = 0$ gives the formula

$$\begin{aligned} N((\Phi, \varphi), (\Phi, \varphi)) &= D(h,h) + \tfrac{1}{2}J\langle h,h\rangle + \int dx\,\kappa(x)r(x)h^2(x) \\ &\quad + \int dx\,\kappa(x)\int \mu(x, dy)\{h(x) - (\Phi, \varphi)(y)\}^2 \end{aligned} \tag{3.12}$$

where h is defined by

$$h(x) = N\kappa\mu(\Phi, \varphi)(x) + \textcircled{E}_x \Phi. \tag{3.13}$$

Pairs (Φ, φ) are only specified modulo ν null sets. A second pair (Φ', φ') is said to be a normalized contraction of (Φ, φ) if there exist versions such that

$$|\Phi'(y)| \leq |\Phi(y)| \quad \text{for } y \in \underline{\underline{T}}$$

$$|\varphi'(y)| \leq |\varphi(y)| \quad \text{for } y \in \underline{\underline{M}}$$

$$|(\Phi', \varphi')(y) - (\Phi', \varphi')(z)| \leq |(\Phi, \varphi)(y) - (\Phi, \varphi)(z)| \quad \text{for } y, z \in \underline{\underline{T}} \cup \underline{\underline{M}}.$$

The behavior of the process while on the boundary is prescribed by a pair $(\underline{\underline{H}}, Q)$ where $\underline{\underline{H}}$ is a linear subset of $\underline{\underline{N}}$ which is closed under normalized contractions and where Q is a bilinear form on $\underline{\underline{H}}$ such that

3.4.1. $Q - N$ is contractive on $\underline{\underline{H}}$.

3.4.2. $(\underline{\underline{H}}, Q)$ is a Dirichlet space on $L^2(\underline{\underline{T}} \cup \underline{\underline{M}}, \nu^0)$.

The meaning of 3.4.1 is that if (Φ', φ') is a normalized contraction of (Φ, φ) then

$$(3.14) \qquad Q((\Phi', \varphi'), (\Phi', \varphi')) - N((\Phi', \varphi'), (\Phi', \varphi')) \leq Q((\Phi, \varphi), (\Phi, \varphi)) - N((\Phi, \varphi), (\Phi, \varphi)).$$

The idea behind 3.4.1 is that Q must contain the part of the Dirichlet form for excursions from the boundary and in addition may contain more which will correspond to behavior on the boundary. The significance of 3.4.1 was first established by M. Fukushima [16] for the special case of Brownian motion on a bounded domain in Euclidean space. The idea behind 3.4.2 is that for the process to be constructed the pair $(\underline{\underline{H}}, Q)$ must correspond to the time changed process determined by ν in the sense of SMP, Section 8, and therefore must be an "honest Dirichlet space". The crucial property to be checked is "closability" as defined in [17]. If a sequence of pairs converges strongly to 0 in the Hilbert space $L^2(\underline{\underline{T}} \cup \underline{\underline{M}}, \nu)$ and is Cauchy relative to the Q norm, then also the Q norms converge to 0. If this property is satisfied, then we can always satisfy 3.4.2 by taking an appropriate completion--

but not otherwise. This property is always satisfied (modulo mild regularity conditions) when Q has no diffusion form in the sense of SMP, Section 11. In general, efforts to verify this property seem to lead to deep analytical problems. We will discuss this for a special case in Section 15 below.

The choice of the pair $(\underline{\underline{H}}, Q)$ completely determines the resolvent and therefore the transition semigroup and Dirichlet space for the process, and indeed it would be easy to construct this resolvent directly as in the proof of the converse to Theorem 20.2 in SMP. However, in order to give a direct sample space construction we must first use $\underline{\underline{H}}$ to regularize $\underline{\underline{X}} \cup \underline{\underline{T}} \cup \underline{\underline{M}}$ in the sense of SMP, Section 19. We begin by extending pairs $h = (\Phi, \varphi)$ to $\underline{\underline{X}}$ by the prescription

$$h(x) = ⓔ_x[\underline{\underline{T}}; e^{-\zeta}\Phi] + G_1\{\kappa \int \mu(\cdot, dy)(\Phi, \varphi)(y)\}(x). \tag{3.15}$$

The restriction of h to $\underline{\underline{X}}$ determines Φ uniquely by the martingale convergence theorem, but not necessarily φ. The subcollection $\underline{\underline{H}}_{bl}$ of bounded and ν-integrable functions in $\underline{\underline{H}}$ form an algebra with operations defined pointwise on $\underline{\underline{T}} \cup \underline{\underline{M}}$. We assume

3.5. Regularity Condition. There is a countable dense subset $\underline{\underline{H}}_0$ of $\underline{\underline{H}}_{bl}$ forming a subalgebra over the rationals such that every member of $\underline{\underline{H}}_0$ is continuous on $\underline{\underline{X}}$. ///

There is no real loss of generality in imposing 3.5 since it can always be guaranteed if we are willing to replace the given state space $\underline{\underline{X}}$ by a quasi-homeomorphic image as in SMP, Section 19.

Let Δ be the space of real valued continuous nontrivial homomorphisms on $\underline{\underline{H}}_0$ and let δ be the trivial one. $\underline{\underline{H}}_0$ is viewed as a function space on $\underline{\underline{X}} \cup \Delta \cup \{\delta\}$ in the obvious way and then $\underline{\underline{X}} \cup \Delta \cup \{\delta\}$ is given the coarsest topology which contains the given topology on $\underline{\underline{X}}$ together with the topology generated by the function space $\underline{\underline{H}}_0$. In general $\underline{\underline{X}} \cup \Delta \cup \{\delta\}$ is not compact or even locally compact (see Section 13 below), but it is metrizable. The singleton δ will be referred to as the dead point; it must be distinguished from the dead point ∂ for $\underline{\underline{X}}$.

From now on we use Δ as a boundary rather than $\underline{\underline{T}} \cup \underline{\underline{M}}$ and we work with functions φ on Δ rather than pairs (Φ, φ) on $\underline{\underline{T}} \cup \underline{\underline{M}}$. The jumping measures $\mu(x, d\cdot)$ and the measure ν are transferred to Δ via their action on $\underline{\underline{H}}_0$. For example ν on Δ is determined by

(3.16) $$\int_\Delta \nu(dy)h(y) = \int_{\underline{\underline{X}}} dx\, h(x) \quad \text{for } h \in \underline{\underline{H}}_0$$

with the understanding that h is extended to $\underline{\underline{X}}$ by (3.15). The pair $(\underline{\underline{H}}, Q)$ is now regarded as a Dirichlet space on $L^2(\Delta, \nu)$. In general the function space $\underline{\underline{N}}$ cannot be identified with a subset of $L^2(\Delta, \nu)$. (See Case 2 in Section 12.)

We introduce some machinery now which is analogous to (and in a sense even identical with) machinery introduced in Sections 1 and 2. The sets $\underline{\underline{X}}$, Δ here play the role of D, M there and the superscript "ex" here plays the role of (D) there.

By the supermartingale convergence theorem $\mathrm{Lim}_{t\uparrow\zeta} h(X_t)$ exists for $h \in \underline{\underline{H}}_0$. Also it is clear that $\mathrm{Lim}_{t\uparrow\zeta} G_1 \kappa\mu\varphi(X_t) = 0$ on the set $[X_{\zeta-0} = \partial]$ and therefore on this set the limit determines a homomorphism on $\underline{\underline{H}}_0$. Thus

$$X^{ex}_{\zeta-0} = \mathrm{Lim}_{t\uparrow\zeta} X_t$$

is well defined as a point in $\underline{\underline{X}} \cup \Delta \cup \{\delta\}$ modulo the usual exceptional sets on Ω and Ω_∞. Similarly $X^{ex}_{\zeta^*}$ is well defined on Ω_∞. Clearly

$$[X_{\zeta-0} = \partial] = [X^{ex}_{\zeta-0} = \delta] \cup [X^{ex}_{\zeta-0} \in \Delta]$$

and since $\underline{\underline{T}}$ is contained in the set $[\zeta < +\infty]$ also

$$X^{ex}_{\zeta-0} = \delta \text{ when } \zeta = +\infty.$$

Similar identities are valid for $X^{ex}_{\zeta^*}$.

With the same convention for $u = 0$ as in Section 1 the hitting operators H_u, $u \geq 0$, are defined by

$$H_u \varphi(x) = Ⓔ_x e^{-u\zeta} I(X^{ex}_{\zeta-0} \in \Delta) \varphi(X^{ex}_{\zeta-0})$$
$$+ Ⓔ_x e^{-u\zeta} I(X_{\zeta-0} \in \underline{\underline{X}}) \mu \varphi(X_{\zeta-0})$$

for $x \in X$ and $\varphi \geq 0$ on Δ. For $\varphi \in \underline{\underline{H}}_{(e)}$, the extended Dirichlet space, the formulae (3.2) and (3.12) can be replaced by

(3.2')
$$N(\varphi,\varphi)$$
$$= \tfrac{1}{2} Ⓔ[X_{\zeta^*} = \partial ; X_{\zeta-0} = \partial ; \{\varphi(X^{ex}_{\zeta-0}) - \varphi(X^{ex}_{\zeta^*})\}^2]$$
$$+ \tfrac{1}{2}\int\int_{\Delta\times\Delta} (\mu\kappa N\kappa\mu)(dy, dz)\{\varphi(y) - \varphi(z)\}^2$$
$$+ \int dx\kappa(x)\int_\Delta \mu(x, dy)\int Ⓟ_x(dw)\{\varphi(y) - \varphi(X^{ex}_{\zeta-0}(w))\}^2$$
$$+ \int dx\kappa(x) r(x)\int Ⓟ_x(dw)\{\varphi(X^{ex}_{\zeta-0}(w))\}^2$$
$$+ \int dx\kappa(x) r(x) N\kappa\mu\varphi^2(x)$$

(3.12')
$$N(\varphi, \varphi)$$
$$= D(H\varphi, H\varphi) + \tfrac{1}{2} J\langle H\varphi, H\varphi\rangle + \int dx\kappa(x) r(x)\{H\varphi(x)\}^2$$
$$+ \int dx\kappa(x)\int\mu(x, dy)\{H\varphi(x) - \varphi(y)\}^2.$$

Again by SMP, Theorem 7.3(iv), (or rather by the proof) we can write

$$H_u(x, dy) = h_u(x, y)\nu(dy).$$

For $0 \leq u < v$ the Feller density is defined by

$$u_{u,v}(y, z) = (v-u)\int_{\underline{\underline{X}}} dx h_u(x, y)h_v(x, z).$$

As in SMP, (20.13'), the Feller density is symmetric in y, z and for $0 \leq u < v < w$

$$u_{u,w}(y, z) = u_{u,v}(y, z) + u_{v,w}(y, z)$$

so that in particular the definition

$$u_{u,\infty}(y, z) = \operatorname{Lim}_{v \uparrow \infty} u_{u,v}(y, z)$$

makes sense. We also introduce the operators

$$U_{u,v}\varphi(y) = \int u_{u,v}(y, z)\varphi(z)\nu(dz)$$

and the two bilinear forms

$$U_{u,v}(\varphi, \varphi) = \int \nu(dy)\int \nu(dz)u_{u,v}(y, z)\varphi(y)\varphi(z)$$

$$U_{u,v}\langle\varphi, \varphi\rangle = \int \nu(dy)\int \nu(dz)u_{u,v}(y, z)\{\varphi(y) - \varphi(z)\}^2$$

for φ defined on Δ. The adjoint operators π_u, $u \geq 0$, map functions on $\underline{\underline{X}}$ into functions on Δ via

$$\pi_u f(y) = \int_{\underline{\underline{X}}} dx f(x)h_u(x, y).$$

As in Section 2 we define also

$$\pi_u \lambda(y) = \int \lambda(dy)h_u(x, y)$$

for λ a measure on $\underline{\underline{X}}$. The defining relation (3.16) implies that for $\varphi \geq 0$

$$(3.17) \qquad \int dx H_1 \varphi(x) = \int \nu(dy)\varphi(y)$$

and it follows by the Cauchy-Schwarz inequality that H_1 is a contraction from $L^2(\Delta,\nu)$ to $L^2(\underline{\underline{X}}, dx)$ and therefore also π_1 is from $L^2(\underline{\underline{X}}, dx)$ to $L^2(\Delta,\nu)$. The boundedness of H_u and π_u for $u > 0$ follows from the resolvent like identities

$$H_u = H_v + (v-u)G_u H_v$$

$$\pi_u = \pi_v + (v-u)\pi_v G_u$$

which can be established by elementary calculations. It is worth observing that for $v < +\infty$

$$(3.18) \qquad U_{u,v}\varphi = (v-u)\pi_u H_v \varphi = (v-u)\pi_v H_u \varphi .$$

Finally, let

$$p(x) = Ⓔ_x[X^{ex}_{\zeta-0} = \delta].$$

As for p^D in Section 1 the corrected calculation (20.7) in SMP shows that $v\pi_v p$ increases with v and therefore

$$\pi_* p(y) = \mathrm{Lim}_{v\uparrow\infty} v\pi_v p(y)$$

is well defined. As for (2.2) and (2.4) in Section 2 the technique of SMP, (20.37) and (20.38) can be used to establish the identity

$$(3.19) \qquad N(\varphi,\varphi) = \tfrac{1}{2}U_{0,\infty}\langle\varphi,\varphi\rangle + \int \nu(dy)\{\pi r\kappa(y) + \pi_* p(y)\}\varphi^2(y).$$

The first term corresponds to excursions from Δ to Δ; the second to excursions which enter by jumping from the death point δ to the interior of X and exit at Δ

or vice versa; the third to excursions which enter at δ, wander into $\underline{\underline{X}}$ and exit at Δ or vice versa. The first and third term in (3.19) account for the first three terms in (3.2). The second accounts for the last two terms.

We introduce now the Markov process which corresponds to the regular Dirichlet space $(\underline{\underline{H}}, Q)$ on $L^2(\Delta, \nu)$. The trajectory variables will be denoted by Y_τ and a superscript Δ will be used for the corresponding sample space probabilities, Lèvy kernel and killing measure. The resolvent operators will be denoted R_α, $\alpha > 0$. Since Q-N is contractive on $\underline{\underline{H}}$, it follows from (3.19) and from SMP, Section 12, that

$$J^\Delta(dy, dz) \geq \nu(dy)\nu(dz)I(y \neq z)u_{0,\infty}(y, z)$$

$$\kappa^\Delta(dy) \geq \nu(dy)\{\pi r \cdot \kappa(y) + \pi_* p(y)\}.$$

Let $J^{\Delta,0}(dy, dz)$ and $\kappa^{\Delta,0}(dy)$ be defined by

$$J^\Delta(dy, dz) = J^{\Delta,0}(dy, dz) + \nu(dy)\nu(dz)I(y \neq z)u_{0,\infty}(y, z) \tag{3.20}$$

$$\kappa^\Delta(dy) = \kappa^{\Delta,0}(dy) + \nu(dy)\{\pi r \ \kappa(y) + \pi_* p(y)\}.$$

Our construction of the <u>excursion space</u> $(\Omega^{ex}, Ⓟ^{ex})$ here exactly parallels the construction of $(\Omega^{(D)}, Ⓟ^{(D)})$ in Section 1. Two additional variables $X^{ex}_{\zeta^*-0}$, X^{ex}_ζ are adjoined to the inverse limit sample space Ω^0_∞ of SMP, Section 5, and the measure $Ⓟ^{ex}$ is determined as follows.

3.6.1. On the set $[\zeta^* < \zeta]$ the variables X_t, $\zeta^* < t < \zeta$, are distributed according to the approximate Markov process in SMP, Section 5.

3.6.2. Conditioned on the set $[\zeta^* < \zeta]$ and on the variables X_t, $\zeta^* < t < \zeta$, the adjoined variables $X^{ex}_{\zeta^*-0}$, X^{ex}_ζ are independent and depend only on the corresponding limit variables $X^{ex}_{\zeta^*}$, $X^{ex}_{\zeta-0}$ respectively. The conditional distribution of X^{ex}_{ζ} given $X^{ex}_{\zeta-0}$ is the same as that of $X^{ex}_{\zeta^*-0}$ given $X^{ex}_{\zeta^*}$ and is determined by

$$X^{ex}_{\zeta} = X^{ex}_{\zeta-0} \text{ on } [X^{ex}_{\zeta-0} \in \Delta \cup \{\delta\}]$$

$$\text{Ⓔ}^{ex}[X^{ex}_{\zeta-0} \in \underline{\underline{X}}; X^{ex}_{\zeta} = \delta; \varphi(X^{ex}_{\zeta-0})] = \int_{\underline{\underline{X}}} dx\kappa(x) r(x)\varphi(x)$$

$$\text{Ⓔ}^{ex}[X^{ex}_{\zeta-0} \in \underline{\underline{X}}; X^{ex}_{\zeta} \in \Delta; \varphi(X^{ex}_{\zeta-0})\psi(X^{ex}_{\zeta})] = \int_{\underline{\underline{X}}} dx\kappa(x)\int_{\Delta}\mu(x, dy)\varphi(x)\psi(y).$$

<u>3.6.3</u>. The joint distribution of X^{ex}_{ζ}, $X^{ex}_{\zeta^*-0}$ conditioned on the set $[\zeta^* = \zeta = 0]$ is determined by

$$\text{Ⓔ}^{ex}[\zeta^* = \zeta = 0; X^{ex}_{\zeta^*-0} = \delta; X^{ex}_{\zeta} \in \Delta; \varphi(X^{ex}_{\zeta})]$$

$$= \text{Ⓔ}^{ex}[\zeta^* = \zeta = 0; X^{ex}_{\zeta^*-0} \in \Delta; X^{ex}_{\zeta} = \delta; \varphi(X^{ex}_{\zeta^*-0})]$$

$$= \int_{\Delta}\kappa^{\Delta,0}(dy)\varphi(y)$$

$$\text{Ⓔ}^{ex}[\zeta^* = \zeta = 0; X^{ex}_{\zeta^*-0} \in \Delta; X^{ex}_{\zeta} \in \Delta; \varphi(X^{ex}_{\zeta^*-0})\psi(X^{ex}_{\zeta})]$$

$$= \iint_{\Delta\times\Delta} J^{\Delta,0}(dy, dz)\varphi(y)\psi(z). \qquad ///$$

Before continuing we note that (3.2') can be replaced by the more compact formula

(3.2'')
$$N(\varphi, \varphi) = \tfrac{1}{2}\text{Ⓔ}^{ex}\{\varphi(X^{ex}_{\zeta}) - \varphi(X^{ex}_{\zeta^*-0})\}^2.$$

The biconditional and conditional measures $\text{Ⓟ}^{ex}_{y,z}$, Ⓟ^{ex}_{y} are defined exactly as Sections 1, 2. Again, and for the same reason as in Section 2, the $Ⓟ_y^{ex}$ are defined only on the set $[\zeta^* < \zeta]$.

Now we are ready to put all this together using Sections 1 and 2 as a guide. Let

$$S_{od} = \{\tau > 0 : Y_\tau \neq Y_{\tau-0}\}$$

be the set of off-diagonal excursion times for the process $\{Y_\tau\}$ and adjoin the off-diagonal excursion $\{w_\tau : \tau \in S_{od}\}$ so that conditioned on the process $\{Y_\tau\}$ they are mutually independent and such that w_τ has $Ⓟ^{ex}_{Y_{\tau-0}, Y_\tau}$ as its conditional distribution. To handle the diagonal excursions we introduce for z an atom of ν

(3.21) $$a_z \circ b(\tau) = \int_0^\tau ds I(Y_s = z)$$

$$a \circ b_z(\sigma) = \inf\{\tau > 0 : a_z \circ b(\tau) > \sigma\}.$$

We are using clumsy notations here in order to be consistent with Section 2. The idea is that $a_z \circ b(\tau)$ is the level of local time at z when the total time spent on the boundary is τ. The undefined $b_z(\sigma)$ is the time for the process to be constructed when the level of local time at z passes σ and then $a \circ b_z(\sigma)$ is the total amount of time spent on the boundary at this instant. Now fix an increasing sequence A_n, $n \geq 1$, of compact subsets of $\underline{\underline{X}}$ whose interiors increase to $\underline{\underline{X}}$. Adjoin the diagonal excursion times $\{a(z, n, i)\}_{n, i=1}^\infty$ and corresponding excursions $\{w_{z,n,i}\}_{n,i=1}^\infty$ so that conditioned on the process $\{Y_\tau\}$ they are independent of the off-diagonal excursions adjoined above and also the collections are mutually independent as z runs over the atoms of ν. For a fixed atom z the conditional distribution of the $a(z, n, i)$ and $w_{z,n,i}$ are determined as follows. Let $\{\sigma(1, i)\}_{i=1}^\infty$ form a sequence of holding times with rate $Ⓟ^{ex}_{z,z}(\sigma(A\ ) < +\infty)$ and truncated at $a_z \circ b(\zeta^\Delta)$. For $n \geq 2$ let $\{\sigma(n, i)\}_{i=1}^\infty$ form a sequence of holding times with rate

$\textcircled{P}^{ex}_{zz}(\sigma(A_n) < +\infty) - \textcircled{P}^{ex}_{zz}(\sigma(A_{n-1}) < +\infty)\}$, also truncated at $a_z \circ b(\zeta^{\Delta})$, which is independent of $\sigma(m, j)$ for $m < n$. On the set $[\sigma(n, i) < +\infty]$ let the diagonal excursion $w^0_{n,i}$ be independent of everything else and have the conditional probability $\textcircled{P}^{ex}_{z,z}[\,\cdot\,|\sigma(A_n) < +\infty; \sigma(A_{n-1}) = +\infty]$ as distribution. (For $n = 1$ omit the condition $\sigma(A_{n-1}) = +\infty$.) Finally, for each n let $\{a(z, n, i)\}_{i=1}^{\infty}$ be the times $\{\sigma(m, j)\}_{m=1,j=1}^{n \quad \infty}$ arranged in increasing order and let $w_{z,n,i}$ be the corresponding $w^0_{m,j}$. It is easy to check that $\{a(z, n, i)\}_{i=1}^{\infty}$ is for each n a sequence of holding times with rate $\textcircled{P}^{ex}_{z,z}(\sigma(A_n) < +\infty)$ truncated at $a_z \circ b(\zeta^{\Delta})$ and that on the set $[a(z, n, i) < +\infty]$ the distribution of $w_{z,n,i}$ is precisely $\textcircled{P}^{ex}_{z,z}[\,\cdot\,|\sigma(A_n) < +\infty]$. Let

$$\tau(z, n, i) = a \circ b_z(a(z, n, i)) \tag{3.22}$$

and then

$$S_d = \cup_{z,n,i}\{\tau(z, n, i)\}\,;\, S = S_d \cup S_{od},$$

and define w_τ for $\tau \in S_d$ in the obvious way. Before going on to construct the process itself, we take care of some preliminaries.

Lemma 3.3. For $\xi \geq 0$ on $\Omega^{ex} \cap [\zeta^* < \zeta]$, for $\alpha > 0$, and for $y \in \Delta$

$$\textcircled{E}^{\Delta}_{y} \Sigma_{\tau \in S} e^{-\alpha\tau} \xi(w_\tau) = R_\alpha \textcircled{E}^{ex}_{\cdot}(\xi)(y). \tag{3.23}$$ ///

Proof. It suffices to consider separately the cases when ξ vanishes off the sets

$$[X^{ex}_{\zeta^*-0}, X^{ex}_{\zeta} \in \Delta; X^{ex}_{\zeta} \neq X^{ex}_{\zeta^*-0}]$$

$$[X^{ex}_{\zeta^*-0} \in \Delta; X^{ex}_{\zeta} = \delta]$$

$$[X^{ex}_{\zeta^*-0}, X^{ex}_{\zeta} \in \Delta; X^{ex}_{\zeta^*-0} = X^{ex}_{\zeta} = z; \sigma(A_n) < +\infty].$$

For the first case the left side of (3.23)

$$= \textcircled{E}^{\Delta}_{y} \Sigma_{\tau} I(Y_{\tau-0} \neq Y_{\tau}; Y_{\tau} \in \Delta) e^{-\alpha\tau} \textcircled{E}^{ex}_{Y_{\tau-0}, Y_{\tau}}(\xi)$$

$$= N^{\Delta}_{\alpha}\{\int J^{\Delta}(d\cdot, dz) \textcircled{E}^{ex}_{\cdot, z} \xi\}(y)$$

which equals the right side. (Here and below N^{Δ}_{α} is the α-potential operator for the Dirichlet space $(\underline{\underline{H}}, Q)$.) For the second case the left side

$$= \textcircled{E}^{\Delta}_{y} I(Y_{\zeta-0} \in \Delta) e^{-\alpha\zeta} \textcircled{E}^{ex}_{Y_{\zeta-0}, \delta}(\xi)$$

$$= N^{\Delta}_{\alpha} \kappa^{\Delta} \cdot \textcircled{E}^{ex}_{\cdot, \delta}(\xi)$$

which again equals the right side. For the third case the left side equals

$$\textcircled{E}^{\Delta}_{y} \Sigma_i e^{-\alpha\tau(z,n,i)} \textcircled{E}^{ex}_{z,z}(\xi \mid \sigma(A_n) < +\infty). \tag{3.24}$$

But clearly

$$\textcircled{E}^{\Delta}_{y} \Sigma_i e^{-\alpha\tau(z,n,i)} = \textcircled{E}^{\Delta}_{y} \int_0^{\zeta} d\tau e^{-\alpha\tau} I(Y_{\tau} = z) \textcircled{E}^{ex}_{z,z}(\sigma(A_n) < +\infty)$$

and so the left side of (3.23) for the third case

$$= \textcircled{E}^{\Delta}_{y} \int_0^{\zeta} d\tau e^{-\alpha\tau} I(Y_{\tau} = z) \textcircled{E}^{ex}_{z,z} \xi$$

$$= \textcircled{E}^{\Delta}_{y} \int_0^{\zeta} d\tau e^{-\alpha\tau} I(Y_{\tau} = z) \textcircled{E}^{ex}_{z} \xi,$$

which equals the right side. ///

<u>Lemma 3.4.</u> <u>For</u> $u \geq 0$ <u>and for</u> λ <u>a Radon measure on</u> $\underline{\underline{X}}$ <u>charging no polar set</u>

3.19

$$\pi_u \lambda(y) = \textcircled{E}_y^{ex} \int_{\zeta^*}^{\zeta} a(\lambda; dt) e^{-u(t-\zeta^*)} \quad \text{a.e.} \quad \nu(dy).$$ ///

Proof. For $\varphi \geq 0$ on Δ

$$\int \nu(dy)\varphi(y)\textcircled{E}_y^{ex} \int_{\zeta^*}^{\zeta} a(\lambda; dt) e^{-u(t-\zeta^*)}$$

$$= \textcircled{E}^{ex}[X^{ex}_{\zeta^*-0} \in \Delta; \varphi(X^{ex}_{\zeta^*-0}) \int_{\zeta^*}^{\zeta} a(\lambda; dt) e^{-u(t-\zeta^*)}]$$

$$= \textcircled{E}^{ex}[\int_{\zeta^*}^{\zeta} a(\lambda, dt) e^{-u(\zeta-t)} I(X_\zeta \in \Delta)\varphi(X_\zeta)]$$

$$= \textcircled{E}^{ex}[\int_{\zeta^*}^{\zeta} a(\lambda, dt) H_u \varphi(X_t)]$$

$$= \int \lambda(dx) H_u \varphi(x)$$

and the lemma follows. ///

For the special case when λ is absolutely continuous, Lemma 3.4 reduces to

(3.25) $$\pi_u f(y) = \textcircled{E}_y^{ex} \int_{\zeta^*}^{\zeta} dt e^{-u(t-\zeta^*)} f(X_t).$$

For the special case $f = u1$ we get

(3.26) $$u\pi_u 1(y) = E_y^{ex}\{1 - e^{-u(\zeta-\zeta^*)}\}$$

which together with Lemma 3.3 gives

Corollary 3.5. For $\alpha, u > 0$ and for quasi-every y

$$\textcircled{E}_y^{\Delta} \Sigma_{\tau \in S} e^{-\alpha\tau}\{1 - e^{-u[\zeta(w_\tau) - \zeta^*(w_\tau)]}\} = uR_\alpha \pi_u 1(y).$$ ///

Since $\alpha R_\alpha \pi_1 1 \leq 1$ we conclude from Corollary 3.5 that for every $\tau > 0$

and for quasi-every y

$$\Sigma_{\sigma\in S} I(\sigma \leq \tau)\{1-e^{-[\zeta(w_\sigma)-\zeta^*(w_\sigma)]}\} < +\infty$$

[a.e. P_y^Δ] which guarantees

$$\Sigma_{\sigma\in S} I(\sigma \leq \tau\,;\,\zeta(w_\sigma)-\zeta^*(w_\sigma) < +\infty)\{\zeta(w_\sigma)-\zeta^*(w_\sigma)\} < +\infty.$$

The event $\zeta(w_\sigma)-\zeta^*(w_\sigma) = +\infty$ cannot occur for $\sigma < \zeta^\Delta$ and so it follows that the process

$$b(\tau) = \Sigma_{\sigma\in S} I(\sigma \leq \tau)\{\zeta(w_\sigma)-\zeta^*(w_\sigma)\}$$

is nondecreasing and right continuous and $b(0) = 0$. Also $b(\tau) < +\infty$ for $\tau < \zeta^\Delta$ although possibly $b(\zeta^\Delta) = +\infty$. Of course $b(\tau)$ will be the time for the process to be constructed when the time for $Y_.$ passes τ.

For $\varphi \geq 0$ on Δ and for $y \in \Delta$ let

(3.27) $$R_{(u)}\varphi(y) = Ⓔ_y^\Delta \int_0^{\zeta^\Delta} d\tau\, e^{-ub(\tau)}\varphi(Y_\tau)$$

$$R_{(u)\alpha}\varphi(y) = Ⓔ_y^\Delta \int_0^{\zeta^\Delta} d\tau\, e^{-ub(\tau)-\alpha\tau}\varphi(Y_\tau).$$

These auxiliary operators correspond to ones introduced in SMP, Section 20, but our point of view here is different. The proof of Lemma 3.3 can easily be refined to establish

<u>Lemma 3.6</u>. <u>For</u> $\xi \geq 0$ <u>on</u> $\Omega^{ex}\cap[\zeta^* < \zeta]$, <u>for</u> $u \geq 0$ <u>for</u> $\alpha \geq 0$ <u>and for quasi-every</u> $y \in \Delta$

(3.28) $$R_{(u)\alpha}\{Ⓔ_.^{ex}\xi\}(y) = Ⓔ_y^\Delta \Sigma_{\tau\in S} e^{-\alpha\tau-ub(\tau-0)}\xi(w_\tau).$$ ///

Lemma 3.7. For $0 \leq u < v < +\infty$ and for $\varphi \geq 0$ on Δ and for quasi-every $y \in \Delta$

(3.29) $$U_{u,v}\varphi(y) = E_y^{ex}[X_\zeta^{ex} \in \Delta; \{e^{-u(\zeta-\zeta^*)} - e^{-v(\zeta-\zeta^*)}\}\varphi(X_\zeta^{ex})].$$

For $v < +\infty$ also

(3.30) $$U_{u,v}\varphi(y) = E_y^{ex}[X_\zeta^{ex} \in \Delta; (v-u)\int_{\zeta^*}^{\zeta} dt e^{-v(\zeta-t)-u(t-\zeta^*)}\varphi(X_\zeta^{ex})]$$

$$= E_y^{ex}[X_\zeta^{ex} \in \Delta; (v-u)\int_{\zeta^*}^{\zeta} dt e^{-u(\zeta-t)-v(t-\zeta^*)}\varphi(X_\zeta^{ex})]. \qquad ///$$

Proof. The identity (3.29) follows directly from Lemma 3.4 together with (3.18) and then (3.30) follows from (3.29). ///

Lemma 3.8. For $0 \leq u < v < +\infty$, for $\alpha > 0$ and for $\varphi \geq 0$ on Δ

(3.31) $$R_{(u)}\varphi = R_{(v)}\varphi + R_{(u)}U_{uv}R_{(v)}\varphi$$
$$= R_{(v)}\varphi + R_{(v)}U_{uv}R_{(u)}\varphi$$

(3.32) $$R_{(u)\alpha}\varphi = R_{(v)\alpha}\varphi + R_{(u)\alpha}U_{uv}R_{(v)\alpha}\varphi$$
$$= R_{(v)\alpha}\varphi + R_{(v)\alpha}U_{uv}R_{(u)\alpha}\varphi. \qquad ///$$

Proof. It suffices to consider the second equation with φ bounded so that everything converges. Then

$$R_{(u)\alpha}\varphi(y) - R_{(v)\alpha}\varphi(y)$$

$$= E_y^{\Delta}\int_0^{\zeta^{\Delta}} d\tau e^{-\alpha\tau}\varphi(Y_\tau)\{e^{-ub(\tau)} - e^{-vb(\tau)}\}$$

3.22

$$= \text{Ⓔ}_y^\Delta \int_0^{\zeta^\Delta} d\tau e^{-\alpha\tau} \varphi(Y_\tau) \Sigma_{\sigma \leq \tau} (v-u) \int_{b(\sigma-0)}^{b(\sigma)} ds e^{-v[b(\tau)-s]-us}$$

$$= \text{Ⓔ}_y^\Delta \Sigma_\sigma e^{-\alpha\sigma - ub(\sigma-0)} \int_{b(\sigma-0)}^{b(\sigma)} ds (v-u) e^{-v[b(\sigma)-s]-u[s-b(\sigma-0)]}$$

$$\cdot \int_\sigma^{\zeta^\Delta} d\tau e^{-\alpha(\tau-\sigma)-v[b(\tau)-b(\sigma)]} \varphi(Y_\tau)$$

$$= \text{Ⓔ}_y^\Delta \Sigma_\sigma e^{-\alpha\sigma - ub(\sigma-0)} \int_{b(\sigma-0)}^{b(\sigma)} ds (v-u) e^{-v[b(\sigma)-s]-u[s-b(\sigma-0)]} R_{(v)\alpha} \varphi(Y_\sigma)$$

$$= R_{(u)\alpha} U_{u,v} R_{(v)\alpha} \varphi(y) \; ,$$

and similarly with u, v interchanged. ///

Corollary 3.9. For $0 < u < +\infty$ the operators $\{R_{(u)\alpha}, \alpha > 0\}$ form a symmetric resolvent on $L^2(\Delta, \nu)$ with corresponding Dirichlet space $(\underline{\underline{H}}, Q_{(u)})$ where

$$Q_{(u)}(\varphi, \varphi) = Q(\varphi, \varphi) + U_{0,u}(\varphi, \varphi).$$

The operator $R_{(u)}$ is bounded from $L^2(\Delta, \nu)$ into $\underline{\underline{H}}$ and

$$Q_{(u)}(R_{(u)}\varphi, \psi) = \int \nu(dy) \varphi(y) \psi(y)$$

for $\varphi \in L^2(\Delta, \nu)$ and for $\psi \in \underline{\underline{H}}$. Also

$$uR_{(u)} \uparrow_u 1.$$ ///

Proof. The estimate for the quadratic form

$$U_{0,1}(\varphi, \varphi) \leq U_{0,1}(1, \varphi^2)$$
$$\leq \int dx H_1 \varphi^2(x)$$
$$\leq \int \nu(dy) \varphi^2(y)$$

shows that the operator $U_{0,1}$ is a contraction on $L^2(\Delta,\nu)$ and it follows directly that actually

$$U_{0,u} \text{ for } 0 < u \leq 1 \; ; \; (1/u)U_{0,u} \text{ for } u > 1$$

are contractions on $L^2(\Delta,\nu)$. The identity

$$Q_{(u)}(\varphi,\varphi) = \{Q(\varphi,\varphi)-N(\varphi,\varphi)\} + U_{0,u}(1,\varphi^2) + \{N(\varphi,\varphi)-\tfrac{1}{2}U_{0,u}<\varphi,\varphi>\} \tag{3.33}$$

together with 3.4.1 and (3.19) shows that $(\underline{\underline{H}},Q_{(u)})$ is a Dirichlet space on $L^2(\Delta,\nu)$ and with the help of

$$\begin{aligned} R_\alpha\varphi &= R_{(u)\alpha}\varphi + R_\alpha U_{0,u}R_{(u)\alpha}\varphi \\ &= R_{(u)\alpha}\varphi + R_{(u)\alpha}U_{0,u}R_\alpha\varphi \end{aligned} \tag{3.34}$$

which is a special case of (3.32) it is easy to identify $\{R_{(u)\alpha}, \alpha > 0\}$ as the associated resolvent. (See SMP, calculation (12.5).) To complete the proof we use the full strength of 3.4.1 and refine (3.33) to

$$\begin{aligned} Q_{(u)}(\varphi,\varphi) &= \{Q(\varphi,\varphi)-N(\varphi,\varphi)\} \\ &+ \{U_{0,u}(1,\varphi^2) + \int\nu(dy)[u\,\pi_u p(y)+u\pi_u Nr\;\kappa(y)]\varphi^2(y)\} \\ &+ \{N(\varphi,\varphi)-\tfrac{1}{2}U_{0,u}<\varphi,\varphi>-\int\nu(dy)[u\pi_u p(y)+u\pi_u Nr\;\kappa(y)]\varphi^2(y)\}. \end{aligned} \tag{3.35}$$

Since the terms inside the middle { } can be collected to give

$$u\int dxH_u\varphi^2(x) = \int\nu(dy)\varphi^2(y)u\pi_u 1(y),$$

this implies on the one hand that $Q_{(u)}$ dominates the standard inner product on $L^2(\Delta,\nu)$ and so by the spectral theorem $R_{(u)}$ is bounded from $L^2(\Delta,\nu)$ into $\underline{\underline{H}}$

and implies on the other hand that

$$Q_{(u)}(\varphi, \varphi) - \int \nu(dy) u \pi_u 1(y) \varphi^2(y)$$

is contractive and so $u R_{(u)} \pi_u 1 \le 1$ by SMP, Theorem 12.1.   ///

Now we are ready to define the process $\{X_t^{\sim}\}$ with initial point in Δ. The clock $b(\cdot)$ keeps time for the process on Δ with the excursions being used to fill in the gaps. To be precise

(3.36)
$$X_{b(\tau-0)}^{\sim} = X_{\zeta^*}^{ex}(w_\tau)$$

$$X_t^{\sim} = X_{\zeta^* + t - b(\tau-0)}(w_\tau) \quad \text{for } t \epsilon\ (b(\tau-0), b(\tau)).$$

which defines $X_t^{\sim}$ for t belonging to the "gap set"

(3.37)
$$E = \cup_{\tau \epsilon S} [b(\tau-0), b(\tau)).$$

We still must take care of t belonging to the <u>complementary set</u>

(3.38)
$$F = [0, b(\zeta^\Delta)) - E.$$

For this we introduce the inverse process

(3.39)
$$a(t) = \inf\{\tau > 0 : b(\tau) > t\}$$

with the usual understanding that $a(t) = +\infty$ for $t \ge b(\zeta^\Delta)$ and we supplement (3.36) with

(3.36')
$$X_t^{\sim} = Y_{a(t)} \quad \text{for}\ t\ \epsilon\ F$$

$$X_t^{\sim} = \delta \quad \text{for } t \ge b(\zeta^\Delta).$$

An interesting feature of this construction is that part of the original process $\{Y_\tau\}$ can be lost. If a time t belongs to the complementary set F then there exists a sequence of distinct excursion times τ_n in S such that $b(\tau_n) \downarrow t$. Thus if $\tau = a(t)$ for some $t \in F$ then $b(\tau+\varepsilon) > b(\tau)$ for all $\varepsilon > 0$ and $\tau = a[b(\tau)]$. If τ belongs to an interval where $b(\cdot)$ is flat, then certainly Y_τ does not correspond to any $\tilde{X}_t$. We will study this phenomenon more closely in Section 4.

From now on we write

$$\textcircled{P}_y^{\sim} = \textcircled{P}_y^{\Delta} \quad \text{for } y \in \Delta$$

to indicate that we are viewing these now as sample space probabilities for the $\tilde{X}_t$ process. Next we construct the $\textcircled{P}_x^{\sim}$, $x \in \underline{\underline{X}}$, using the $\textcircled{P}_x$. First we adjoin the variable X_ζ^{ex} taking values in $\Delta \cup \{\delta\}$ according to the following rule which is exactly analogous to 3.6.2.

3.7.1. Conditioned on the variables $\{X_t\}_{t<\zeta}$ the adjoined variable X_ζ^{ex} depends only on $X_{\zeta-0}^{ex} = \mathrm{Lim}_{t\uparrow\zeta} X_t$ (interpreted as a limit in $\underline{\underline{X}} \cup \Delta \cup \{\delta\}$) and its distribution is given by

$$X_\zeta^{ex} = X_{\zeta-0}^{ex} \quad \text{on } [X_{\zeta-0}^{ex} \in \Delta \cup \{\delta\}].$$

$$\textcircled{P}_x^{\sim}[X_\zeta^{ex} = \delta \mid X_{\zeta-0}^{ex} ; X_{\zeta-0}^{ex} \in \underline{\underline{X}}] = r(X_{\zeta-0}^{ex})$$

$$\textcircled{E}_x^{\sim}[\varphi(X_\zeta^{ex}) ; X_\zeta^{ex} \in \Delta \mid X_{\zeta-0}^{ex} ; X_{\zeta-0}^{ex} \in \underline{\underline{X}}] = \mu\varphi(X_{\zeta-0}^{ex}). \qquad ///$$

The variables $\tilde{X}_t$, $t \geq 0$, are then constructed according to the following prescription.

3.7.2. $\tilde{X}_t = X_t$ for $t < \zeta$.

3.7.3. Conditioned on $X^{\sim}_t$, $t < \zeta$, and on X^{ex}_ζ the variables $\{X^{\sim}_{\zeta+t}\}_{t\geq 0}$ depend only on X^{ex}_ζ and are distributed as follows:

(i) If $X^{ex}_\zeta = \delta$, then $X^{\sim}_{\zeta+t} = \delta$ for all $t \geq 0$.

(ii) If $X^{ex}_\zeta \in \Delta$, then $\{X^{\sim}_{\zeta+t}\}_{t\geq 0}$ has the same law as $\{X^{\sim}_t\}_{t\geq 0}$ relative to $Ⓟ^{ex}_{X^{\sim}_\zeta}$. ///

Thus we have completely defined the process $\{X^{\sim}_t\}_{t\geq 0}$ with sample space probabilities $P^{\sim}_x$ for all x in $\underline{\underline{X}} \cup \Delta$. We define corresponding transition operators and resolvent operators by

(3.40)
$$P^{\sim}_t f(x) = Ⓔ^{\sim}_x f(X^{\sim}_t)$$

$$G^{\sim}_u f(x) = Ⓔ^{\sim}_x \int_0^{\zeta^{\sim}} dt e^{-ut} f(X^{\sim}_t)$$

and we are ready to prove

Theorem 3.10. *The operators $G^{\sim}_u$, $u > 0$, form a symmetric resolvent on $L^2(\underline{\underline{X}}, dx)$. Moreover,*

(3.41)
$$G^{\sim}_u = G_u + H_u R_{(u)} \pi_u. \qquad ///$$

Proof. The first sentence follows exactly as in SMP, pages 20.13, 20.14, once we establish (3.41). In establishing (3.41) we need only consider initial points $y \in \Delta$ and then it follows from

$$G^{\sim}_u f(y) = E^{\sim}_y \Sigma_\tau e^{-ub(\tau-0)} \int_{b(\tau-0)}^{b(\tau)} dt f(X_t) e^{-u\{t-b(\tau-0)\}}$$

which by Lemma 3.4 and Lemma 3.6

$$= R_{(u)} \pi_u f. \qquad ///$$

3.8. Definition. $(\underline{\underline{F}}^{\sim}, E^{\sim})$ is the Dirichlet space on $L^2(\underline{\underline{X}}, dx)$ which corresponds to the resolvent $\{G_u^{\sim}, u > 0\}$.

In Section 4 we will analyze the Dirichlet space $(\underline{\underline{F}}^{\sim}, E^{\sim})$ in considerable detail.

It follows from Theorem 3.10 by standard arguments involving Laplace inversion that the $X_t^{\sim}$ form a Markov process. Also it is easy to check directly that the trajectories are right continuous. We know from SMP, Section 4, that $\{X_t^{\sim}\}$ has a version taking values possibly in a regularization of $\underline{\underline{X}}$ which is a Hunt process modulo a polar set in the sense of SMP. If $\underline{\underline{X}} \cup \Delta$ is locally compact and if $(\underline{\underline{F}}^{\sim}, E^{\sim})$ is regular on $\underline{\underline{X}} \cup \Delta$, then again by SMP, Section 4, the process $\{X_t^{\sim}\}$ itself, since it is right continuous, must be this version. In fact it is true in general that $\{X_t^{\sim}\}$ itself is the "decent version". A simple proof is given in an appendix at the end of Chapter II.

4. The Singular Boundary Δ_s.

We continue to use the notation of Section 3. Terms such as "nearly open" and "quasi-every" on Δ are meant relative to the Dirichlet space $(\underline{\underline{H}}, Q)$.

4.1. Definition. A point $y \in \Delta$ is regular if

$$\tilde{P}_y[a(0) = 0] = 1.$$

The subset of regular points is denoted by Δ_r. The complement $\Delta_s = \Delta - \Delta_r$ is the set of singular points. ///

Evidently a point $y \in \Delta$ is regular if and only if the process $b(\cdot)$ immediately starts to increase for the process $\tilde{X}_t$ starting at y. This is equivalent to the condition that the expanded process, introduced in Section 5, does not spend an initial time interval on Δ. Thus y is regular in the sense of Definition 4.1 if and only if relative to the expanded process it is regular for $\underline{\underline{X}}$ in the sense of SMP, Definition 7.1. We will see that the process $\tilde{X}_t$ actually "lives on" $\underline{\underline{X}} \cup \Delta_r \cup \{\delta\}$. In particular $\tilde{X}_\zeta \neq X^{ex}_\zeta$ when the latter is in Δ_s.

Lemma 4.1. The singular set Δ_s is nearly open and $\tilde{P}_y[a(0) = 0] = 0$ for quasi-every $y \in \Delta_s$. ///

Proof. It suffices to apply the strong Markov property and 0-1 law (SMP, Corollary 4.11) after observing that the conditional distribution of $a(0)$ given the process $\{Y_\tau\}$ is determined by

$$(4.1)\quad P^\Delta_y[a(0) > \sigma \mid \{Y_\tau\}] = \exp\{-\Sigma_z a_z \cdot b(\sigma) P^{ex}_{z,z}(\Omega^{ex}) - \Sigma_{\tau \in S_{od}} I(\tau \leq \sigma)\eta(Y_{\tau-0}, Y_\tau)\}$$

where z runs over the atoms of ν and where η is the function on $\Delta \times \Delta$ minus the diagonal and on $\Delta \times \{\delta\}$ determined by

(4.2) $$\nu(dy)\nu(dz)I(y \neq z)u_{0,\infty}(y,z) = J^{\Delta}(dy,dz)\eta(y,z)$$

$$\nu(dy)\{\pi r\ \kappa(y)+\pi_{*}p(y)\} = \kappa^{\Delta}(dy)\eta(y,\delta).$$ ///

Next we introduce some more auxiliary operators and establish some identities. For $f \geq 0$ on $\underline{\underline{X}}$ and for $y \in \Delta$ let

(4.3) $$Mf(y) = \textcircled{E}_y^{\sim}[\tilde{X}_0 \in \underline{\underline{X}}; f(\tilde{X}_0)].$$

This operator can be nontrivial when $\Delta_s \neq \phi$. For $\varphi \geq 0$ on Δ and for $y \in \Delta$ let

(4.4) $$R_{(\infty)}\varphi(y) = \operatorname{Lim}_{u\uparrow\infty} R_{(u)}\varphi(y)$$

$$= \textcircled{E}_y^{\sim}\int_0^{a(0)} d\tau\, \varphi(Y_\tau)$$

$$D\varphi(y) = \textcircled{E}_y^{\sim}\varphi(Y_{a(0)})$$

$$D_u\varphi(y) = \textcircled{E}_y^{\sim} e^{-ub[a(0)]}\varphi(Y_{a(0)}).$$

If y is regular, then $a(0) = 0$ and the first expression is 0 while the other two reduce to $\varphi(y)$. If y is singular, then $a(0)$ is either the first excursion time itself or else a limit point of excursion times (of course relative to the clock on the boundary). The operator $R_{(\infty)}$ measures the time spent "on the boundary" before $a(0)$, and D gives the position on the boundary at $a(0)$. Also D_u gives information about the length of a possible first excursion. In general D is not the hitting operator for the regular boundary Δ_r. This will be important in Section 5.

In the sum which appears on the right in (3.28) the variable $b(\tau-0) = 0$ only for a possible first excursion which occurs if and only if $b[a(0)] > 0$. Thus a passage to the limit $u\uparrow\infty$ with $\alpha = 0$ gives

(4.5) $$R_{(\infty)}\textcircled{E}^{ex}\xi\}(y) = \textcircled{E}_y^{\sim}(b[a(0)] > 0; \xi(w_{a(0)})).$$

<u>Lemma 4.2.</u> <u>For φ bounded and quasi-continuous, for $u \geq 0$ and for quasi-every y</u>

(4.6) $$D_u\varphi(y) = \text{Lim}_{v\uparrow\infty} R_{(v)}U_{uv}\varphi(y).$$

<u>Also for $u > 0$ the operators D_u and $R_{(v)}U_{uv}$ are contractive with respect to the norm $Q_{(u)}$ and therefore (4.6) becomes convergence relative to $Q_{(u)}$ for φ in $\underline{\underline{H}}$ if we replace the $R_{(v)}U_{uv}$ by Cesàro sums for a sequence of $v\uparrow\infty$.</u> ///

<u>Remark.</u> For the definition of quasi-continuity see SMP, paragraph 3.7. The point is to guarantee that $\varphi(Y_\cdot)$ be continuous whenever $Y_\cdot$ itself is. It suffices for example that $\varphi \in \underline{\underline{H}}$ or $\varphi \in C(\Delta)$. ///

<u>Proof.</u> As in the proof of Lemma 3.8

$$R_{(v)}U_{uv}\varphi(y) = \textcircled{E}_y^{\sim}\Sigma_\tau e^{-vb(\tau-0)}(v-u)\int_{b(\tau-0)}^{b(\tau)}dte^{-v[t-b(\tau-0)]-u[b(\tau)-t]}\varphi(Y_t)$$

$$= \textcircled{E}_y^{\sim}\Sigma_\tau e^{-ub(\tau)}\varphi(Y_\tau)\int_{b(\tau-0)}^{b(\tau)}dt(v-u)e^{-(v-u)t}.$$

$$= \textcircled{E}_y^{\sim}\int_0^\infty dt(v-u)e^{-(v-u)t}\Sigma_\tau I(b(\tau-0) < t < b(\tau))e^{-ub(\tau)}\varphi(Y_\tau).$$

The interior sum contributes only one term which converges as $t\downarrow 0$ to $e^{-ub[a(0)]}\varphi(Y_{a(0)})$ and (4.6) follow directly. To prove the rest of the theorem we need only show that each $R_{(v)}U_{uv}$ is contractive relative to $Q_{(u)}$. (See paragraph 1.6.1' in SMP.) For $v > u$ and $\varphi \in L^2(\nu)$

(4.7) $$Q_{(u)}(R_{(v)}U_{u,v}R_{(u)}\varphi, R_{(v)}U_{u,v}R_{(u)}\varphi)$$

$$= \int\nu(dy)U_{u,v}R_{(v)}\varphi(y)\{R_{(u)}\varphi(y) - R_{(v)}\varphi(y)\}.$$

But the identity

$$U_{u,v}(\psi,\psi) = (v-u)\int dx H_u\psi(x)\{1-(v-u)G_v\}H_u\psi(x)$$

shows that $U_{u,v}$ is positive definite and so the right side of (4.7) is

$$\leq \int \nu(dy) U_{u,v} R_{(v)}\varphi(y) R_{(u)}\varphi(y)$$

$$= \int \nu(dy)\{R_{(u)} - R_{(v)}\}\varphi(y)\varphi(y)$$

which since $R_{(v)}$ is positive definite is

$$\leq \int \nu(dy) R_{(u)}\varphi(y)\varphi(y)$$

$$= Q_{(u)}(R_{(u)}\varphi, R_{(u)}\varphi)$$

and the theorem is proved since such functions $R_{(u)}\varphi$ are dense in $\underline{\underline{H}}$. ///

It follows directly from Lemma 4.2 and from (3.31) that for $u > 0$

(4.8) $$R_{(u)} = R_{(\infty)} + D_u R_{(u)}.$$

In general $D_u \neq R_{(\infty)} U_{u,\infty}$ so that (3.31) is not valid for $v = +\infty$. Indeed

(4.9) $$R_{(\infty)} U_{u,\infty}\varphi(y) = \textcircled{E}_y^{\sim}[b[a(0)] > 0 ; e^{-ub[a(0)]}\varphi(Y_{a(0)})].$$

This will be important in Section 5 for the case when the singular boundary is non-empty. (See the proof of Lemma 5.6.)

Now we are ready to characterize the singular boundary.

<u>Theorem 4.3.</u> (i) <u>Let</u> $\varphi \in L^2(\Delta,\nu)$ <u>be such that</u> $\varphi > 0$ [a.e. ν]. <u>Then</u> $R_{(\infty)}\varphi \in L^2(U_{0,\infty} 1\cdot\nu)$ <u>and the singular boundary</u> $\Delta_s = [R_{(\infty)}\varphi > 0]$ <u>quasi-everywhere.</u>

(ii) *If* $\psi \in \underline{\underline{H}} \cap L^2(\Delta, U_{0,\infty} 1 \cdot \nu)$, *then* $\psi = 0$ *quasi-everywhere on the regular boundary* Δ_r. ///

Proof. For (i) let $\psi = R_{(\infty)}\varphi$. For $u > 1$

$$\int \nu(dy)\varphi(y)R_{(1)}\varphi(y) \geq \int \nu(dy)\varphi(y)R_{(u)}\varphi(y)$$

$$= Q_{(u)}(R_{(u)}\varphi, R_{(u)}\varphi)$$

$$\geq Q(R_{(u)}\varphi, R_{(u)}\varphi)$$

and it follows after passage to the limit $u \uparrow \infty$ that $R_{(\infty)}\varphi \in \underline{\underline{H}}$. Similarly

$$\int \nu(dy)\varphi(y)R_{(1)}\varphi(y) \geq U_{0,u}(R_{(u)}\varphi, R_{(u)}\varphi) + \tfrac{1}{2}U_{0,u} < R_{(u)}\varphi, R_{(u)}\varphi > \geq U_{0,u}(1, \psi^2)$$

and therefore $\psi \in L^2(U_{0,\infty} 1 \cdot \nu)$. Finally it follows directly from Definition 4.1 that $\Delta_s = [\psi > 0]$ quasi-everywhere. To prove (ii) we need a preliminary result. Let $\nu_\infty = \{U_{0,\infty} 1 + \pi_* p + \pi r \kappa\} \cdot \nu$, $\underline{\underline{H}}_\infty = \underline{\underline{H}} \cap L^2(\Delta, \nu_\infty)$ and for $\varphi \in \underline{\underline{H}}_\infty$ put

(4.10) $$Q_{(\infty)}(\varphi, \varphi) = Q(\varphi, \varphi) + U_{0,\infty}(\varphi, \varphi).$$

Clearly $(\underline{\underline{H}}_\infty, Q_{(\infty)})$ is a Dirichlet space on $L^2(\Delta, \nu_\infty)$ and indeed since $Q_{(\infty)}(\varphi, \varphi) - \int \nu_\infty(dy)\varphi^2(y)$ is contractive, it follows from SMP, Theorem 12.1, that the Green's operator S_∞ actually maps $L^2(\Delta, \nu_\infty)$ into $\underline{\underline{H}}_\infty$ and satisfies $S_\infty 1 \leq 1$. In particular S_∞ is determined by

(4.11) $$Q_{(\infty)}(S_\infty \varphi, \psi) = \int \nu_\infty(dy)\varphi(y)\psi(y)$$

for $\varphi \in L^2(\Delta, \nu_\infty)$ and $\psi \in \underline{\underline{H}}_\infty$. We want to show that

(4.12) $$S_\infty \varphi = 0 \quad \text{q.e. on } \Delta_r.$$

for φ in $L^2(\Delta, \nu_\infty)$. To show this we introduce also for $u > 0$ the Green's operator

S_u which corresponds to the Dirichlet space $(\underline{\underline{H}}_\infty, Q_{(u)})$. Clearly the collection of bounded functions in $\underline{\underline{H}}_\infty$ forms an ideal in $\underline{\underline{H}}$ and so by SMP, Theorem 20.1, the Dirichlet space $(\underline{\underline{H}}_\infty, Q_{(u)})$ corresponds modulo random time changed to an absorbed space for the Dirichlet space $(\underline{\underline{H}}, Q_{(u)})$. (See Remark 2 below.) It follows in particular that for $\varphi \geq 0$

$$S_u\varphi \leq R_{(u)}\{U_{0,\infty}1+\pi_* p+\pi r\ \kappa\}\varphi. \tag{4.13}$$

In proving (4.12) we can restrict attention to $\varphi \geq 0$ such that $\{U_{0,\infty}1+\pi_* p+\pi r\ \kappa\}\varphi$ is bounded. But then (4.12) follows from (4.13) and the definition of $R_{(\infty)}$ since clearly $S_\infty\varphi \leq S_u\varphi$ for all $u > 0$. Finally, conclusion (ii) follows upon observing that any ψ as specified in (ii) belongs to $\underline{\underline{H}}_\infty$ and therefore can be approximated relative to the $Q_{(1)}$ (and indeed the $Q_{(\infty)}$) norm by functions in the range of S_∞.

///

__Corollary 4.4.__ *For the special case when* $(\underline{\underline{H}}, Q) = (\underline{\underline{N}}, N)$, *the singular set* $\Delta_s = [U_{0,\infty}1 < +\infty]$ *modulo a* ν*-null set.*

__Proof.__ The point is that for this special case $\underline{\underline{H}} \cap L^2(U_{0,\infty}1\cdot\nu) = L^2(U_{0,\infty}1\cdot\nu)$.

///

__Remark 1.__ By Theorem 4.3(i) the singular set is always contained in $[U_{0,\infty}1 < +\infty]$ modulo a ν-null set. In general, this containment can be proper, as we show at the end of Section 13 below. Also SMP, Theorem 20.2(i), is incorrect. Such φ can exist if we are not in the special case of Corollary 4.4. The crucial error in the proof is at the bottom of page 20.9. The application of Theorem 7.3(i) depends on knowing that φ itself is the restriction to Δ of a function in $\underline{\underline{F}}^\sim$. From the falsity of Theorem 20.2(i) it follows that in the language of SMP, Section 20, a given extension need not be subordinate to a modified reflected space which is actually an extension. ///

Remark 2. Our proof of the basic result (4.12) depended heavily on SMP, Theorem 20.1. The latter is correct but the proof given in SMP misplaces the emphasis somewhat. The existence of a function space $\underset{\equiv 0}{\widetilde{B}}$ satisfying 20.1.1 and 20.1.2 depends crucially on the ideal property of $\underset{\equiv b}{F}$. To see what happens when the ideal property is not satisfied, consider the case when $\underset{\equiv}{\widetilde{F}}$ corresponds to reflecting Brownian motion and $\underset{\equiv}{F}$ to periodic Brownian motion on a closed and bounded interval as in SMP, Section 22, and in Section 12 below. ///

Theorem 4.5. For quasi-every x in $\underset{\equiv}{X} \cup \Delta \cup \{\delta\}$ the process $\{\widetilde{X}_t\}$ takes values only in the set $\underset{\equiv}{X} \cup \Delta_s \cup \{\delta\}$.

Proof. It is clear from the discussion following (3.36') that if $\tau = a(t)$ for some $t \in F$ then $b(\tau+\epsilon) > b(\tau)$ for all $\epsilon > 0$ and then $t = b(\tau)$. For such τ it can never happen that $Y_\tau \in \Delta_s$ and therefore $\widetilde{X}_t \in \Delta_s$. This is easily proved by applying to the first such τ a form of the strong Markov property for the process $\{Y.\}$ which takes into account the excursions. On the other hand it is clear that $\widetilde{X}_t \in \underset{\equiv}{X}$ for $b(\tau-0) < t < b(\tau)$ with $\tau \in S$. Therefore we need only check that $X^{ex}_{\zeta^*}(w_\tau) \in \Delta_s$ is impossible for $\tau \in S$ such that $b(\tau-0) < b(\tau)$. For this it suffices to show that

(4.14)
$$\textcircled{P}^{ex}[X^{ex}_{\zeta^*} \in \Delta_s] = \textcircled{P}^{ex}[X^{ex}_{\zeta-0} \in \Delta_s] = 0.$$

Now consider any $\varphi \geq 0$ in $L^2(\Delta, \nu)$ such that

$$E^{ex}[X^{ex}_{\zeta-0} \in \Delta; \varphi(X^{ex}_{\zeta-0})] > 0.$$

Clearly $H_1\varphi$ is nontrivial and it is not true that for quasi-every x in $\underset{\equiv}{X}$ we have $\text{Lim}_{t\uparrow\zeta} H_1\varphi(X_t) = 0$ on the set $[X_{\zeta-0} = \partial]$. By SMP, Proposition 4.5, the function

$H_1\varphi$ cannot belong to $\underline{\underline{F}}$ and since $H_1\varphi$ is in $L^2(\underline{\underline{X}}, dx)$ it must be that

$$(4.15) \qquad U_{1,\infty}(\varphi, \varphi) = \operatorname{Lim}_{v\uparrow\infty} v \int dx H_1\varphi(x)\{1 - vG_{v+1}\} H_1\varphi(x)$$

$$= +\infty.$$

The theorem follows since if (4.14) failed, then by Theorem 4.3(i) we could choose φ to violate (4.15). ///

Finally, we use Theorem 4.5 to prove one more important identity.

Lemma 4.6. For $f \geq 0$ on $\underline{\underline{X}}$, for $u > 0$, and for quasi-every $y \in \Delta$

$$MG_u f(y) = R_{(\infty)} \pi_u f(y). \qquad ///$$

Proof. By Lemma 3.6 and Lemma 3.4

$$R_{(v)} \pi_u g(y) = \tilde{Ⓔ}_y \Sigma_\tau e^{-vb(\tau-0)} \int_{b(\tau-0)}^{b(\tau)} dt g(\tilde{X}_t) e^{-u[t-b(\tau-0)]}$$

which converges to

$$\tilde{Ⓔ}_y \int_0^{b(a(0))} dt g(\tilde{X}_t) e^{-ut}$$

as $v \uparrow \infty$. The interior integral is nonzero if and only if $b[a(0)] > 0$ which is the case if and only if there exists a first excursion into $\underline{\underline{X}}$. The lemma will follow if we can show that when the first excursion $w_{a(0)}$ exists, then the initial point $X^{ex}_{\zeta^*}(w_{a(0)})$ belongs to $\underline{\underline{X}}$ and not to Δ. By the very definition of Δ_s we need only consider the case when y belongs to the singular boundary Δ_s. By Theorem 4.5 we need only rule out the possibility that the initial point $X^{ex}_{\zeta^*}(w_{a(0)})$ belongs to the regular boundary Δ_r. This can happen only if $a(0)$ is the first hitting time for Δ_r and the left hand limit $Y_{a(0)-0} \in \Delta_r$. But then we are dealing with a predictable stopping time in the sense of SMP, Definition 4.4, and therefore by SMP,

Theorem 4.12, the excursion $w_{a(0)}$ is diagonal and therefore $Y_{a(0)-0} = Y_{a(0)} = z$ an atom of ν. Finally, if $P^{ex}_{z,z}(\Omega^{ex}) < +\infty$ then the first diagonal excursion from z cannot happen when $a_z \cdot b(\tau) = 0$ and if $P^{ex}_{z,z}(\Omega^{ex}) = +\infty$ then there is no first diagonal excursion from z, so the lemma is proved. ///

Chapter II. Associated Dirichlet Spaces and Boundary Conditions

The results in this chapter supplement Chapter II in SMP. For the reader's convenience we have kept the treatment here more or less independent of SMP although we do refer to the latter for specific results. We formulate everything in terms of three cases in increasing order of generality.

For the first case the jumping measures $\mu(x, d\cdot)$ in Section 3 are null, and there is no need to introduce an auxiliary set $\underline{\underline{M}}$. This means that new behavior occurs only after the trajectory for the original process "wanders to infinity" in a finite time interval. At the Dirichlet space level the basic results are established in SMP, Sections 14 and 15 and then summarized in the "First Structure Theorem". We say nothing about this here, but in Section 9 we give more precise information about the generator $A^{\sim}$. In particular, we formulate boundary conditions in terms of a "normal derivative" which is defined directly as a limit of difference quotients on an appropriate sample space. The main result is in Section 7 where we establish a general analogue to "Green's formula". In SMP we treated boundary conditions only on an "ad hoc" basis.

For the second case the measures $\mu(x, d\cdot)$ in Section 3 can be non-null, but the singular boundary $\Delta_s = \phi$. This corresponds exactly to the case when $(\underline{\underline{F}}^{\sim}, E^{\sim})$ is an "extension" in the sense of SMP, Theorem 20.1. Except for a serious error (see the remark following Corollary 4.4), the basic results at the Dirichlet space level are established in SMP in Section 20 and collected in the "Second Structure Theorem". We give an independent treatment in the first part of Section 5 which covers everything in SMP except for showing that we obtain all the possibilities. Just as for the first case, we also include here some results about the generator $A^{\sim}$ which are not given in SMP. These are collected in Section 10.

Most of the hard work in this chapter concerns the third case, when the singular boundary $\Delta_s \neq \phi$. Results at the Dirichlet space level are established in the second part of Section 5 and at the generator level in Section 11. The significance of the singular boundary is apparent at both the sample space and Dirichlet space levels. However, the main result in Section 11 is that things can be formulated so that the significance disappears at the generator level. One direct consequence is that when $\Delta_s \neq \phi$ it is much easier to describe the generator $A^{\sim}$ than the Dirichlet space $(\underline{\underline{F}}^{\sim}, E^{\sim})$. (This is especially clear in Section 21.)

Our point of view in SMP, Section 21, was quite different. We showed that the most general symmetric semigroup which dominates the given one can be obtained by first "suppressing jumps to the death point" and/or replacing them by "jumps within the state space" and then taking an extension. Our construction in Section 3 only gives a special case (see Case 6 in Section 12), but it is easy to see how it can be modified to handle the general situation.

5. The Associated Dirichlet Space.

In this section we study the Dirichlet space $(\underline{\underline{F}}^{\sim}, E^{\sim})$ on $L^2(\underline{\underline{X}}, dx)$ which is associated with the process $\{X_t^{\sim}\}$ constructed in Section 3. We first specialize to the case when the singular set Δ_s is absent.

Theorem 5.1. *Assume that* $\underline{\underline{H}} \cap L^2(U_{0,\infty} 1 \cdot \nu) = \{0\}$ *and therefore* $\Delta = \Delta_r$. *Then* $(\underline{\underline{F}}^{\sim}, E^{\sim})$ *is an extension of* $(\underline{\underline{F}}, E)$ *in the sense of* SMP, *Section* 20, *and every extension of* $(\underline{\underline{F}}, E)$ *can be so obtained. In particular* $\underline{\underline{F}}$ *is a closed subspace of* $\underline{\underline{F}}^{\sim}$ *and* $E^{\sim}(f, f) = E(f, f)$ *for* $f \in \underline{\underline{F}}$. *Moreover,*

(i) *for* f *in the extended space* $\underline{\underline{F}}^{\sim}_{(e)}$ *there is a unique function* γf *in the extended space* $\underline{\underline{H}}_{(e)}$ *(relative to* Q*) on* Δ *such that the difference* $F - H\gamma f$ *is in the extended space* $\underline{\underline{F}}_{(e)}$ *and*

$$E^{\sim}(f, f) = E(f - H\gamma f, f - H\gamma f) + Q(\gamma f, \gamma f).$$

Also, if $\varphi \in \underline{\underline{H}}_{(e)}$, *then* $H\varphi \in \underline{\underline{F}}^{\sim}_{(e)}$ *and* $\gamma H\varphi = \varphi$ *and of course*

$$Q(\varphi, \varphi) = E^{\sim}(H\varphi, H\varphi).$$

(ii) *Let* $u > 0$. *For* $f \in \underline{\underline{F}}^{\sim}$ *the restriction* $\gamma f \in \underline{\underline{H}}$, *the difference* $f - H_u \gamma f \in \underline{\underline{F}}$ *and*

$$E^{\sim}_u(f, f) = E(f - H_u \gamma f, f - H_u \gamma f) + Q_{(u)}(\gamma f, \gamma f).$$

If $\varphi \in \underline{\underline{H}}$, *then* $H_u \varphi \in \underline{\underline{F}}$ *and* $\gamma H_u \varphi = \varphi$ *and*

$$Q_{(u)}(\varphi, \varphi) = E^{\sim}_u(H_u \varphi, H_u \varphi). \qquad ///$$

Proof. Fix $g \in L^2(\underline{\underline{X}}, dx)$ and let $f = G_1 g$. From the calculation

$$v\int dx\{G_1 g(x) - v\tilde{G}_{1+v} G_1 g(x)\} G_1 g(x)$$

$$= v\int dx\{G_1 g(x) - vG_{1+v} G_1 g(x)\} G_1 g(x) - v^2 \int dx H_{1+v} R_{(1+v)} \pi_{1+v} G_1 g(x) G_1 g(x)$$

$$= \int dx v G_{1+v} g(x) f(x) - \int dx H_1 R_{(1+v)} \pi_1 v G_{1+v} g(x) v G_{1+v} g(x)$$

it follows after passage to the limit $v \uparrow \infty$ that

(5.1) $$\tilde{E}_1(f,f) = \int dx g(x) f(x) - \int \nu(dy) R_{(\infty)} \pi_1 g(y) \pi_1 g(y).$$

By hypothesis $\underline{\underline{H}} \cap L^2(U_{0,\infty} 1\cdot\nu) = \{0\}$ and by Theorem 4.3(i) we have $R_{(\infty)} = 0$ and therefore

(5.2) $$E^{\sim}(f,f) = E(f,f).$$

From the fact that $\tilde{P}_t$ dominates P_t it follows automatically that $\underline{\underline{F}}_b$ is an ideal in $\tilde{\underline{\underline{F}}}_b$ and therefore $(\underline{\underline{F}}^{\sim}, E^{\sim})$ is an extension of $(\underline{\underline{F}}, E)$. (See SMP, (20.43).) The remainder of the theorem follows as in SMP, pages 20.16 and 20.17. (See also the proof of Theorem 5.8 below). ///

Clearly Theorem 3.2 is applicable to any $f \in \tilde{\underline{\underline{F}}}_{(e)}$ and yields

$$D(f,f) + \tfrac{1}{2} J\langle f,f\rangle + \int dx \kappa(x) r(x) f^2(x) + \int dx \kappa(x) \int \mu(x,dy)\{f(x) - \gamma f(y)\}^2$$
$$= E(f - H\gamma f, f - H\gamma f) + N(\gamma f, \gamma f).$$

This together with Theorem 5.1(i) gives

(5.3) $$E^{\sim}(f,f) = D(f,f) + \tfrac{1}{2} J\langle f,f\rangle + \int dx \kappa(x) r(x) f^2(x)$$
$$+ \int dx \kappa(x) \int \mu(x,dy)\{f(x) - \gamma f(y)\}^2 + \{Q - N\}(\gamma f, \gamma f).$$

We show now that in an appropriate sense (5.3) actually characterizes $\tilde{\underline{\underline{F}}}_{(e)}$. Consider a function $\varphi \in \underline{\underline{H}}_{(e)}$ such that $\varphi(X^{ex}_{\zeta-0}) = 0$ whenever $X^{ex}_{\zeta-0} \in \Delta$. Then $H\varphi = N\kappa\mu\varphi$ and therefore

$$U_{0,\infty}(1,\varphi^2) = \mathrm{Lim}_{v\uparrow\infty} v\int dx\{1-vG_v)H1(x)N\kappa\mu\varphi^2(x)$$

$$\leq \mathrm{Lim}_{v\uparrow\infty}\int dx vG_v\kappa\mu\varphi^2(x)$$

$$\leq \int dx\kappa(x)\mu\varphi^2(x).$$

It follows then from Theorem 4.3 (ii) that if $\varphi \in \underline{\underline{H}}_{(e)}$ exists such that

(5.4) $$\mathrm{Lim}_{t\uparrow\zeta} f(X_t) = \varphi(X^{ex}_{\zeta-0}) \quad \text{on } [X_{\zeta-0} = \partial]$$

modulo a Ⓟ$_x$ null set for q.e. x in $\underline{\underline{X}}$ and such that

(5.5) $$\int dx\kappa(x)\int\mu(x,dy)\{f(x)-\varphi(y)\}^2 < +\infty,$$

then φ is necessarily unique. Finally, consider f locally controlled for which

(5.6) $$D(f,f)+\tfrac{1}{2}J<f,f>+\int\kappa(dx)r(x)f^2(x) < +\infty$$

and such that there exists $\varphi \in \underline{\underline{H}}_{(e)}$ satisfying (5.4),(5.5). Let $g = f-H\varphi$. Since (5.5) is valid also with f replaced by $H\varphi$ we have $\int dx\kappa(x)g^2(x)\mu 1(x) < +\infty$ and since (5.6) is valid also with f replaced by $H\varphi$, then g belongs to the reflected space $\underline{\underline{F}}^{ref}$. Finally, (5.4) is valid also with f replaced by $H\varphi$, and therefore by SMP, Theorem 14.4, actually g belongs to $\underline{\underline{F}}_{(e)}$. Thus $f = g+H\varphi$ belongs to $\underline{\underline{\tilde{F}}}_{(e)}$ and $\gamma f = \varphi$. We have proved

Theorem 5.2. A locally controlled function f belongs to the extended Dirichlet space $\underline{\underline{\tilde{F}}}_{(e)}$ if and only if it satisfies (5.6) and there exists $\varphi \in \underline{\underline{H}}_{(e)}$ satisfying (5.4) and (5.5). In this case such φ is unique, $\varphi = \gamma f$, and (5.3) is valid. ///

From now on in this section we assume that $\underline{\underline{H}} \cap L^2(U_{0,\infty}1\cdot\nu) \neq \{0\}$ and therefore $\Delta_s \neq \phi$.

Lemma 5.3. The operator μM is symmetric with respect to κ. Equivalently $dx\kappa(x)\mu M(x, dy)$ is a symmetric measure on $\underline{\underline{X}} \times \underline{\underline{X}}$.

Proof. For $f, g \geq 0$ on $\underline{\underline{X}}$ and for $u > 0$ we have by Lemma 4.6

$$\int dx\kappa(x)uG_u f(x)\mu MuG_u g(x) = \int dxf(x)uG_u \kappa\mu R_{(\infty)}u\pi_u g(x). \tag{5.7}$$

By (4.14)

$$H_u \varphi = G_u \kappa\mu\varphi \tag{5.8}$$

when $\varphi = 0$ on Δ_r and so the right side of (5.7)

$$= \int dxf(x)uH_u R_{(\infty)}u\pi_u g(x)$$

which is symmetric in f and g. The lemma will follow if we can show that for $f, g \in C_{com}(\underline{\underline{X}})$ the left side of (5.7) converges as $u \uparrow \infty$ to

$$\int dx\kappa(x)f(x)\mu Mg(x).$$

Since $\mu MuGug(x)$ is uniformly bounded and converges pointwise quasi-everywhere to $\mu Mg(x)$, it suffices by the concept of uniform integrability defined in [29] to show that

$$\operatorname{Lim}_{u\uparrow\infty} u\int dx\kappa(x)G_u f(x) = \int dx\kappa(x)f(x).$$

But this follows from the identity

$$u\int dx\kappa(x)G_u f(x) = Ⓔ I(X_{\zeta^*} \in \underline{\underline{X}})\int_{\zeta^*}^{\zeta} dtue^{-ut}f(X_t)$$

since the interior integral is uniformly bounded and vanishes off a set whose Ⓟ

measure is the same as the capacity of the support of f. (See SMP, (5.9).) ///

5.3. Definition. The perturbed form E^0 is defined on $f \in \underline{\underline{F}}$ by

$$E^0(f,f) = E(f,f) - \int dx\,\kappa(x) f(x) \mu M f(x)$$
$$= D(f,f) + \tfrac{1}{2} J\langle f,f\rangle + \tfrac{1}{2}\kappa\mu M\langle f,f\rangle$$
$$+ \int dx\,\kappa(x)\{1-\mu M1(x)\} f^2(x).$$

The perturbed space $\underline{\underline{F}}^0$ is the E_1^0 completion of $\underline{\underline{F}}$. The perturbed resolvent $\{G_u^0, u > 0\}$ is given by

(5.9) $$G_u^0 = \Sigma_{k=0}^{\infty} \{H_u M\}^k G_u.$$ ///

Of course

$$\kappa\mu M\langle f,f\rangle = \int dx\,\kappa(x) \int \mu M(x,dy)\{f(x)-f(y)\}^2.$$

Intuitively, the perturbed resolvent is defined by letting the process run until it hits the regular boundary Δ_r rather than the full boundary Δ. A special case of (5.8) is

(5.8') $$H_u Mf = G_u \kappa\mu Mf$$

and therefore (5.9) can also be written

(5.9') $$G_u^0 = \Sigma_{k=0}^{\infty} \{G_u \kappa\mu M\}^k G_u.$$

Theorem 5.4. The perturbed space $(\underline{\underline{F}}^0, E^0)$ is a regular Dirichlet space on $L^2(\underline{\underline{X}}, dx)$ and $\{G_u^0, u > 0\}$ is the associated resolvent. Moreover,

(5.10) $$E^{\sim}(f,f) = E^0(f,f)$$

for $f \in \underline{\underline{F}}$. ///

Proof. Again let $f = G_1 g$ with $g \in L^2(\underline{\underline{X}}, dx)$. By (5.1) such $f \in \underline{\underline{F}}^{\sim}$ and

$$E^{\sim}(f,f) = E(f,f) - \int \nu(dy) R_{(\infty)} \pi_1 g(y) \pi_1 g(y)$$

which by Lemma 4.6

$$= E(f,f) - \int \nu(dy) \pi_1 g(y) Mf(y)$$

$$= E(f,f) - \int dx g(x) H_1 Mf(x)$$

which by (5.8')

$$= E(f,f) - \int dx g(x) G_1 \kappa \mu Mf(x)$$

$$= E^0(f,f)$$

and (5.10) is proved. The first sentence of the theorem follows from SMP, Theorem 18.2. ///

5.4. Definition. The regular hitting operators are defined for $\varphi \geq 0$ on Δ_r and for $y \in \Delta$ by

$$K\varphi(y) = E^{\sim}_y[\sigma(\Delta_r) < +\infty ; \varphi(Y_{\sigma(\Delta_r)})]$$

$$K_u \varphi(y) = E^{\sim}_y e^{-ub(\sigma(\Delta_r))} \varphi(Y_{\sigma(\Delta_r)}). \qquad ///$$

It is understood in Definition 5.4 that $\sigma(\Delta_r)$ is the hitting time for Δ_r relative to the original $\{Y_\tau\}$ process. Of course $b(\sigma(\Delta_r))$ is the corresponding hitting time for the process $\{X^{\sim}_t\}$.

Lemma 5.5. For φ bounded on Δ

$$D^n\varphi \to K\varphi\,;\ (D_u)^n\varphi \to K_u\varphi$$

quasi-everywhere on Δ. Also for $u > 0$ actually $(D_u)^n \to K_u$ strongly on $\underline{\underline{H}}$ and K_u is a contraction relative to the $Q_{(u)}$ norm. ///

Proof. The results relative to $(\underline{\underline{H}}, Q_{(u)})$ will follow from Lemma 4.2 once we establish pointwise convergence. The operator D corresponds to $a(0)$, and it is clear from the very definition of Δ_r that the stopping times for the iterations D^n must increase to a stopping time $\sigma \leq \sigma(\Delta\ )$. On the set where this increase is strict σ is predictable, and so $Y_\sigma = Y_{\sigma-0}$. Thus the lemma will follow if we show that actually $\sigma = \sigma(\Delta_r \cup \{\delta\})$ or, equivalently, that $Y_\sigma \in \Delta_s$ is impossible. But this is a consequence of Theorem 4.3(i). It is easy to see that $\{Y_\tau,\ 0 \leq \tau \leq \sigma\}$ cannot stay within any set on which $R_{(\infty)}1$ is bounded from below. By Theorem 4.3(i) the set Δ_s is the union of an increasing sequence of such sets and the hitting times for the complements must increase to $\sigma(\Delta_r \cup \{\delta\})$. ///

Before proving an analogue to Theorem 4.1 we introduce some dual operators. For $u > 0$ let $N_{(u)}$ be the potential operator for the Dirichlet space $(\underline{\underline{H}}, Q_{(u)})$ and let $\textcircled{M}_{(+)}$ be the set of measures on Δ (necessarily Radon) having finite $Q_{(u)}$ energy. (As the notation indicates, $\textcircled{M}_{(+)}$ is independent of $u > 0$.) The set $\textcircled{M}_{(+)}$ is complete relative to the energy norm

$$Q_{(u)}(\lambda) = \{Q_{(u)}(N_{(u)}\lambda, N_{(u)}\lambda)\}^{\frac{1}{2}}$$

and $N_{(u)}\textcircled{M}_{(+)}$, the set of potentials in $(\underline{\underline{H}}, Q_{(u)})$ is closed. (See SMP, Section 3, for terminology and relevant results.) From the identity $R_{(v)}U_{u,v}R_{(u)} = R_{(u)}U_{u,v}R_{(v)}$ it follows that for $v > u$ the operator $R_{(v)}U_{u,v}$ preserves $N_{(u)}\textcircled{M}_{(+)}$ and therefore by Lemma 4.2 so does D_u. It then follows that there exists a unique transformation $\textcircled{D}_u$ on $\textcircled{M}_{(+)}$ which is linear in the obvious sense and also contractive with respect to the energy norm $Q_{(u)}(\lambda)$ such that

(5.11) $$D_u N_{(u)}\lambda = N_{(u)}ⒹD_u\lambda.$$

We will be interested only in the special case when $\lambda = \varphi\cdot\nu$ with $\varphi \geq 0$ in $L^2(\Delta,\nu)$ and the energy norm reduces to

$$Q_{(u)}(\varphi\cdot\nu) = \{\int\nu(dy)\varphi(y)R_{(u)}\varphi(y)\}^{\frac{1}{2}}.$$

Therefore we can tidy up the notation by working instead with the dual operator D_u^* mapping functions $\varphi \in L^2(\Delta,\nu)$ into measures on Δ and determined by

$$D_u^*\varphi = ⒹD_u(\varphi\cdot\nu)$$

and then extended to all of $L^2(\Delta,\nu)$ by linearity. The contractivity of ⒹD$_u$ is equivalent to the estimate

(5.12) $$Q_{(u)}(D_u^*\varphi) \leq \{\int\nu(dy)\varphi(y)R_{(u)}\varphi(y)\}^{\frac{1}{2}}$$

and (5.11) is equivalent to

(5.11') $$D_u R_{(u)}\varphi = N_{(u)}D_u^*\varphi.$$

Similarly the dual operator K_u^* is defined by

(5.13) $$K_u R_{(u)}\varphi = N_{(u)}K_u^*\varphi.$$

In fact K_u^* is the restriction to $L^2(\Delta,\nu)$ of the balayage operator defined in SMP, Section 7. In order to make the analogy with the case $\Delta_s = \phi$ as clear as possible we introduce the balayaged measure

$$\nu^0 = K_1^*1.$$

This is a Radon measure on Δ_r which dominates the restriction of ν to Δ_r. The point is that $K_u^* \varphi$ is absolutely continuous with respect to ν^0 (but not necessarily ν) and therefore we can for $u \geq 0$ define operators k_u^* from $L^2(\Delta, \nu)$ to $L^2(\Delta_r, \nu^0)$ by

(5.14) $$K_u^* \varphi = (k_u^* \varphi) \cdot \nu^0 .$$

Also we define the Green's operator $R^0_{(u)}$ by

(5.15) $$R^0_{(u)} \varphi = N_{(u)} (\varphi \cdot \nu^0)$$

from which follows for $u > 0$ the identity

(5.16) $$N_{(u)} K_u^* \varphi = R^0_{(u)} k_u^* \varphi .$$

From Lemma 5.5 follows

(5.17) $$K_u^* = \mathrm{Lim}_{n \uparrow \infty} (D_u^*)^n$$

with convergence being in the $Q_{(u)}$ energy norm. By (5.17), (5.11'), Lemma 5.5 and (5.16) we can extend (5.13) to

(5.18) $$K_u R_{(u)} \varphi = N_{(u)} K_u^* \varphi = K_u N_{(u)} K_u^* \varphi = K_u R^0_{(u)} k_u^* \varphi .$$

We will also need

Lemma 5.6. *For $u \geq 0$, for $f \geq 0$ on $\underline{\underline{X}}$ and for quasi-every $y \in \Delta$*

$$D_u Mf(y) = MH_u Mf(y).$$ ///

Proof. The left side equals

(5.19) $$\text{Ⓔ}_y \tilde{e}^{-ub[a(0)]} Mf(Y_{a(0)})$$

which is nonzero only when $Y_{a(0)} \in \Delta_s$. This is true only when $b[a(0)] > 0$ and so

(5.19)

$$= \tilde{E}_y[b[a(0)] > 0; e^{-ub[a(0)]} Mf(Y_{a(0)})]$$

which by (5. 9)

$$= R_{(\infty)} U_{u,\infty} Mf(y)$$

$$= \text{Lim}_{v \uparrow \infty} R_{(\infty)} v \pi_v H_u Mf(y)$$

which by Lemma 5.6

$$= \text{Lim}_{v \uparrow \infty} Mv G_v H_u Mf(y)$$

$$= MH_u Mf(y). \qquad ///$$

Now we are ready to establish a variant of (3. 41) which depends on decomposing according to the first hitting time for Δ_r rather than for Δ.

Theorem 5.7. For $u > 0$ and $g \in L^2(\underline{X}, dx)$

(5.20)
$$\tilde{G}_u g = G_u^0 g + H_u K_u R_{(u)} \pi_u g$$

$$= G_u^0 g + H_u K_u R_{(u)}^0 k_u^* \pi_u g. \qquad ///$$

Proof. It suffices to show that for $n \geq 1$

(5.21)
$$\tilde{G}_u g = \Sigma_{k=0}^n \{H_u M\}^k G_u g + H_u D_u^n R_{(u)} \pi_u g$$

since by Lemma 5.5 and (5.18)

$$\text{Lim}_{n \uparrow \infty} H_u D_u^n R_{(u)} \pi_u g = H_u K_u R_{(u)} \pi_u g$$

$$= H_u K_u N_{(u)} K_u^* \pi_u g.$$

For $n = 1$ the identity (5.21) follows since by (4.8) and Lemma 4.6

$$\begin{aligned} G_u^{\sim} g &= G_u g + H_u R_{(u)} \pi_u g \\ &= G_u g + H_u R_{(\infty)} \pi_u g + H_u D_u R_{(u)} \pi_u g \\ &= G_u g + H_u M G_u g + H_u D_u R_{(u)} \pi_u g. \end{aligned}$$

The inductive step depends on (4.8), Lemma 4.6 and Lemma 5.6,

$$\begin{aligned} &H_u D_u^n R_{(u)} \pi_u g \\ &= H_u D_u^n \{R_{(\infty)} + D_u R_{(u)}\} \pi_u g \\ &= H_u D_u^n M G_u g + H_u D_u^{n+1} R_{(u)} \pi_u g \\ &= H_u (M H_u)^n M G_u g + H_u D_u^{n+1} R_{(u)} \pi_u g. \end{aligned}$$ ///

5.5. Notation. γ_r is restriction to the regular boundary Δ_r. In particular $\gamma_r \underline{\underline{H}}$ is the restriction of $\underline{\underline{H}}$. ///

Since Δ_r is finely closed in Δ, the operator γ_r is exactly analogous to γ in SMP, paragraph 8.1.

The following theorem is an analogue to Theorem 5.1 for the case $\Delta_s \neq \phi$.

Theorem 5.8. (i) The perturbed space $\underline{\underline{F}}^0$ is a closed subspace of $\underline{\underline{F}}^{\sim}$ and (5.5) is valid for $f \in \underline{\underline{F}}^0$.

(ii) For f in the extended space $\underline{\underline{F}}^{\sim}_{(e)}$ there is a unique function $\gamma_r f$ in the restricted extended space $\gamma_r \underline{\underline{H}}_{(e)}$ such that the difference $f\text{-}HK\gamma_r f$ belongs to the extended perturbed space $\underline{\underline{F}}^0_{(e)}$ and

(5.22) $$E^{\sim}(f,f) = E^0(f-HK\gamma_r f, f-HK\gamma_r f) + Q(K\gamma_r f, K\gamma_r f).$$

Also if $\varphi \in \gamma_r \underline{\underline{H}}_{(e)}$ then $HK\varphi \in \underline{\underline{F}}^{\sim}_{(e)}$ and $\gamma_r HK\varphi = \varphi$ and of course

(5.23) $$Q(K\varphi, K\varphi) = E^{\sim}(HK\varphi, HK\varphi).$$

(iii) Let $u > 0$. For $f \in \underline{\underline{F}}^{\sim}$ the restriction $\gamma_r f \in \gamma_r \underline{\underline{H}}$, the difference $f - H_u K_u \gamma_r f \in \underline{\underline{F}}^0$ and

(5.22') $$E^{\sim}_u(f,f) = E^0(f-H_uK_u\gamma_r f, f-H_uK_u\gamma_r f) + Q_{(u)}(K_u\gamma_r f, K_u\gamma_r f).$$

In particular, $H_u K_u \gamma_r$ projects $\underline{\underline{F}}^{\sim}$ onto the $E^{\sim}_u$ orthogonal complement of $\underline{\underline{F}}$. Also if $\varphi \in \gamma\, \underline{\underline{H}}$, then $H_u K_u \varphi \in \underline{\underline{F}}^{\sim}$ and $\gamma H_u K_u \varphi = \varphi$ and

(5.23') $$Q_{(u)}(\varphi, \varphi) = E^{\sim}_u(H_uK_u\varphi, H_uK_u\varphi).$$ ///

Proof. (i) is an immediate consequence of Theorem 5.4. To prove (iii) we first define

$$\gamma_r G_u g = \gamma_r R_{(u)} \pi_u g$$

for $g \in L^2(\underline{\underline{X}}, dx)$. This makes sense since $\pi_u g \in L^2(\Delta, \nu)$ and therefore $R_{(u)}\pi_u g \in \underline{\underline{H}}$ by Corollary 3.9. The identity (5.22') for the special case $f = G_u g$ follows from the two calculations

(5.24) $$\begin{aligned} &E^{\sim}_u(G^0_u g, H_uK_uR_{(u)}\pi_u g) \\ &= E^{\sim}_u(G^0_u g, G^{\sim}_u g - G^0_u g) \\ &= 0, \end{aligned}$$

(5.25) $$\begin{aligned} &E^{\sim}_u(H_uK_uR_{(u)}\pi_u g, H_uK_uR_{(u)}\pi_u g) \\ &= E^{\sim}_u(G^{\sim}_u g, H_uK_uR_{(u)}\pi_u g) \\ &= \int dx\, g(x) H_uK_uR_{(u)}\pi_u g \end{aligned}$$

$$= \int \nu(dy)\pi_u g(y) K_u R_{(u)}\pi_u g(u)$$

$$= Q_{(u)}(R_{(u)}\pi_u g, K_u R_{(u)}\pi_u g)$$

which by SMP, Theorem 7.3(i),

$$= Q_{(u)}(K_u R_{(u)}\pi_u g, K_u R_{(u)}\pi_u g).$$

Clearly such functions $f = G_u^{\sim} g$ are dense in $\underline{\underline{F}}^{\sim}$ and so (ii) will follow if we show that functions $\gamma_r R_{(u)}\pi_u g$ are dense in $\gamma_r \underline{\underline{H}}$. If this is false, then there exists nontrivial $\varphi \in \gamma_r \underline{\underline{H}}$ such that $H_u K_u \varphi = 0$ [a.e. dx] on $\underline{\underline{X}}$. Then also $H_v K_u \varphi = 0$ for $v > u$ and therefore $U_{0,v} K_u \varphi = 0$ which implies that

$$\int \nu(dy) U_{0,v} 1(y)\{K_u \varphi\}^2(y) = \tfrac{1}{2} U_{0,v} \langle K_u \varphi, K_u \varphi \rangle \le Q(K_u \varphi, K_u \varphi)$$

independent of v and therefore $K_u \varphi \in L^2(\Delta, U_{0,\infty} 1 \cdot \nu)$ which contradicts Theorem 4.3(ii). This proves (iii), and (ii) follows upon passage to the limit $u \downarrow 0$ as in the proof of SMP, Lemma 8.3. ///

In order to establish the appropriate generalization of Theorem 5.2, we proceed somewhat indirectly. First we introduce what we will call the <u>expanded process</u> by including ν as part of the speed measure and thus expanding the time scale while the process is on Δ. To be precise we let

$$b_e(\tau) = b(\tau) + \tau$$

$$a_e(t) = \inf\{\tau > 0 : b_e(\tau) > t\}$$

$$E_e = \bigcup_{\tau \in S} [b_e(\tau - 0), b_e(\tau))$$

$$F_e = [0, b_e(\zeta^{\Delta})) - E_e$$

and then define

$$X^e_{b_e(\tau-0)} = X^{ex}_{\zeta^*}(w_\tau) \quad \text{for } \tau \in S$$

$$X^e_t = X_{\zeta^*+t-b_e(\tau-0)}(w_\tau) \quad \text{for } b_e(\tau-0) < t < b_e(\tau),\ \tau \in S$$

$$X^e_t = Y_{a_e(t)} \quad \text{for } t \in F_e$$

$$X^e_t = \delta \quad \text{for } t \geq b_e(\zeta^\Delta).$$

This defines the expanded process $\{X^e_t\}$ with initial point $y \in \Delta$. For initial point $x \in \underline{\underline{X}}$ we modify in the obvious way paragraphs 3.7.1 through 3.7.3. Let $\{G^e_u, u > 0\}$ be the associated resolvent operators. For a starting point $y \in \Delta$ and for $g \geq 0$ on $\underline{\underline{X}} \cup \Delta$

$$G^e_u g(y) = \tilde{E}_y \Sigma_{\tau \in S} e^{-ub(\tau-0)-u\tau} \int_{\tau+b(\tau-0)}^{\tau+b(\tau)} dt e^{-u\{t-\tau-b(\tau-0)\}} g(X^e_t)$$

$$+ \tilde{E}_y \int_0^{\zeta^\Delta} d\tau e^{-u\tau-ub(\tau)} g(Y_\tau)$$

which by (3.28) and Lemma 3.4

$$= R_{(u)u}\{g+\pi_u g\}(y)$$

and there follows

$$(5.26) \qquad G^e_u g = G_u g + H_u R_{(u)u}\{g+\pi_u g\}.$$

It follows directly that each G^e_u is symmetric on $L^2(\underline{\underline{X}} \cup \Delta, dx+\nu)$. The sub-markov property

$$(5.27) \qquad uG^e_u 1 \leq 1$$

follows directly from the definition of the G^e_u. Also it can be verified using (5.26) as follows. The operator $R_{(u)u}$ is the Green's operator on $L^2(\Delta, \nu)$ relative to the Dirichlet form $Q_{(u)u}$ and by the proof of Corollary 3.9

$$Q_{(u)}(\varphi,\varphi)-\int \nu(dy)\varphi^2(y)\{u\pi_u 1+u1\}$$

is contractive and therefore by SMP, Theorem 12.1,

$$R_{(u)u}\{u\pi_u 1+u1\} \leq 1$$

which directly implies (5.27). The resolvent identity

(5.28) $$G^e_u - G^e_v = (v-u)G^e_v G^e_u$$

follows from

$$G^e_u - G^e_v = G_u - G_v + H_u R_{(u)u}(\pi_u+1) - H_v R_{(v)v}(\pi_v+1)$$

$$= (v-u)G_v G_u + H_u(R_{(u)u} - R_{(v)v})(\pi_v+1)$$

$$+H_u R_{(u)u}(v-u)\pi_u G_v + (v-u)G_u H_v R_{(v)v}(\pi_v+1)$$

$$= (v-u)G_u G_v + (v-u)H_u R_{(u)u}\pi_u G_v + (v-u)G_u H_v R_{(v)v}(\pi_v+1)$$

$$+(v-u)H_u\{R_{(u)u}\pi_u H_v R_{(v)u} + R_{(v)u}R_{(v)v}\}(\pi_v+1)$$

since the last term

$$= (v-u)H_u R_{(u)u}\pi_u H_v R_{(v)v}(\pi_v+1)$$

$$+(v-u)^2 H_u R_{(u)u}\pi_u H_v R_{(v)u} R_{(v)v}(\pi_v+1)$$

$$+(v-u)H_u R_{(u)u} R_{(v)v}(\pi_v+1)$$

$$-(v-u)^2 H_u R_{(u)u}\pi_u H_v R_{(v)u} R_{(v)v}(\pi_v+1)$$

$$= (v-u)H_u R_{(u)u}(\pi_u+1)H_v R_{(v)v}(\pi_v+1)$$

and since $(v-u)H_u R_{(u)u}(\pi_u+1)G_v = (v-u)H_u R_{(u)u}\pi_u G_v$. Thus $\{G^e_u, u>0\}$ is a symmetric resolvent on $L^2(\underline{\underline{X}}\cup\Delta, dx+\nu)$. We analyze next the associated Dirichlet space $(\underline{\underline{F}}^e, E^e)$ on $L^2(\underline{\underline{X}}\cup\Delta, dx+\nu)$. Since the G^e_u dominate G_u, the function space $\underline{\underline{F}}$, regarded in the obvious way as a subspace of $L^2(\underline{\underline{X}}\cup\Delta, dx+\nu)$, is contained in $\underline{\underline{F}}^e$. (See SMP, (20.43).) We show that actually

(5.29) $$E^e(f,f) = E(f,f)$$

for $f \in \underline{\underline{F}}$. It suffices to consider $f = G_u g$ with $g \in L^2(\underline{\underline{X}}, dx)$ and then

$$\begin{aligned} &v\int dx\{G_u g(x) - vG^e_{u+v}G_u g(x)\}G_u g(x)\\ &= v\int dx\{G_u g(x) - vG_{u+v}G_u g(x)\}G_u g(x)\\ &\quad -v^2\int dx H_{u+v}R_{(u+v),u+v}(1+\pi_{u+v})G_u g(x)G_u g(x). \end{aligned}$$

As $v\uparrow\infty$ the first term converges to

$$\int dx g(x)G_u g(x) = E_u(G_u g, G_u g).$$

The second term

$$= -\int dx H_u R_{(u+v),u+v}\pi_u vG_{u+v}g(x)vG_{u+v}g(x)$$

which converges to 0 since clearly $R_{(v)v}\to 0$ and (5.29) follows. (The point here is that $R_{(v)v}\to 0$ even when $R_{(v)}\to R_{(\infty)}\neq 0$.)

Since $(\underline{\underline{F}}^e, E^e)$ is a Dirichlet space on $L^2(\underline{\underline{X}}\cup\Delta, dx+\nu)$, any $f\in\underline{\underline{F}}^e$ automatically has a restriction γf belonging to $L^2(\Delta,\nu)$ and if f belongs to the extended space $\underline{\underline{F}}^e_{(e)}$, then by SMP, Section 1, this restriction γf is still well defined modulo ν-null sets. In fact the proof of Theorem 5.7 can be adapted in an obvious way to show that γf actually has a refinement belonging to the extended space $\underline{\underline{H}}_{(e)}$, that $f-H\gamma f$ belongs to the extended space $\underline{\underline{F}}_{(e)}$ and that

(5.30) $$E^e(f,f) = E(f-H\gamma f, f-H\gamma f) + Q(\gamma f, \gamma f).$$

Also for $u > 0$

(5.30') $$E_u^e(f,f) = E_u(f-H_u\gamma f, f-H_u\gamma f) + Q_{(u)u}(\gamma f, \gamma f).$$

Exactly as for (5.3) the formula (3.10) can be used to establish for general $f \in \underline{\underline{F}}^e_{(e)}$

(5.31) $$E^e(f,f) = D(f,f) + \tfrac{1}{2}J\langle f,f\rangle + \int \kappa(dx) r(x) f^2(x)$$
$$+ \int \kappa(dx) \int \mu(x,dy)\{f(x) - \gamma f(y)\}^2 + \{Q-N\}(\gamma f, \gamma f).$$

At least when $\underline{\underline{X}} \cup \Delta$ is locally compact and $(\underline{\underline{F}}^e, E^e)$ is regular we can apply the results on random time change SMP, Section 8, to (5.31) and get a formula for $E^{\sim}$. The process $\{X_t^{\sim}\}$ can be recovered from $\{X_t^e\}$ by random time change with dx playing the role of ν in SMP, Section 8. By the results of that section for $f \in \underline{\underline{F}}^{\sim}_{(e)}$

(5.32) $$E^{\sim}(f,f) = E^e(H^{\sim}f, H^{\sim}f)$$

where $H^{\sim}$ is the hitting operator for $\underline{\underline{X}} \cup \Delta_r$ and the function $H^{\sim}f$ depends only on f restricted to $\underline{\underline{X}}$ and on the regular restriction $\gamma_r f$ defined in Theorem 5.7. Indeed it is easy to see that

(5.33) $$H^{\sim}f = f \text{ on } \underline{\underline{X}}$$
$$H^{\sim}f = \gamma_r f \text{ on } \Delta_r$$
$$H^{\sim}f = Mf + D_\infty \gamma_r f \text{ on } \Delta_s$$

where D_∞ is defined for $\varphi \geq 0$ on Δ_r and $y \in \Delta$ by

(5.34) $$D_\infty \varphi(y) = E_y[\sigma(\Delta_r) < +\infty; b(\sigma(\Delta_r)) = 0; \varphi(Y_{\sigma(\Delta_r)})]$$

$$= \text{Ⓔ}^{\sim}_{y}[a(0) < +\infty ; b[a(0)] = 0 ; \varphi(Y_{a(0)})].$$

In particular, $D_\infty \varphi = \varphi$ on Δ_r. In fact (5.32) is valid in general, and we refer to the appendix at the end of this chapter for a proof. Applying the results of SMP, Part II, to the original process $\{Y_\tau\}$, we can write for $\varphi \in \underline{\underline{H}}$

(5.35) $$Q(\varphi, \varphi) = N(\varphi, \varphi) + D^r(\varphi, \varphi) + D^s(\varphi, \varphi) + \tfrac{1}{2}J^{r,0}<\varphi, \varphi>$$
$$+ \tfrac{1}{2}J^{s,0}<\varphi, \varphi> + J^{r,s,0}<\varphi, \varphi> + \int \kappa^{r,0}(dy)\varphi^2(y) + \int \kappa^{s,0}(dy)\varphi^2(y)$$

where D^r, D^s and $\kappa^{r,0}, \kappa^{s,0}$ are defined by restricting the diffusion form D^Δ and the partial killing measure $\kappa^{\Delta,0}$ for $(\underline{\underline{H}}, Q)$ to Δ_r and Δ_s and where $J^{r,0}$, $J^{s,0}$, $J^{r,s,0}$ are defined by restricting the partial Lèvy kernel $J^{\Delta,0}$ to $\Delta_r \times \Delta_r$, $\Delta_s \times \Delta_s$ and $\Delta_r \times \Delta_s$. (See (3.20) for the meaning of $\kappa^{\Delta,0}$ and $J^{\Delta,0}$.) For example

$$J^{r,s,0}<\varphi, \varphi> = \iint_{\Delta_r \times \Delta_s} J^{\Delta,0}(dy, dz)\{\varphi(y) - \varphi(z)\}^2.$$

Therefore we can write for $f \in \underline{\underline{F}}^{\sim}_{(e)}$

(5.36) $$E^{\sim}(f, f) = D(f, f) + \tfrac{1}{2}J<f, f> + \int \kappa(dx) r(x) f^2(x)$$
$$+ \iint_{\underline{\underline{X}} \times \Delta_r} \kappa\mu(dx, dy)\{f(x) - \gamma_r f(y)\}^2 + D^r(\gamma_r f, \gamma_r f)$$
$$+ \tfrac{1}{2}J^r<\gamma_r f, \gamma_r f> + \int \kappa^r(dy)\{\gamma_r f(y)\}^2 + B(f, f)$$

where now we will focus our attention on

(5.37) $$B(f, f) = \iint_{\underline{\underline{X}} \times \Delta_s} \kappa\mu(dx, dy)\{f(x) - Mf(y) - D_\infty \gamma_r f(y)\}^2$$
$$+ D^s(Mf + D_\infty \gamma_r f, Mf + D_\infty \gamma_r f)$$
$$+ \tfrac{1}{2}J^{s,0}<Mf + D_\infty \gamma_r f, Mf + D_\infty \gamma_r f>$$

$$+ \int \kappa^s(dy)\{Mf(y)+D_\infty \gamma_r f(y)\}^2$$

$$+ \tfrac{1}{2}J^{r,s,0}<Mf+D_\infty \gamma_r f, Mf+D_\infty \gamma_r f>.$$

The singular boundary Δ_s is finely open relative to the Dirichlet space $(\underline{\underline{F}}^e, E^e)$ and therefore Theorem 3.2 and in particular (3.12') can be applied with the absorbed space for Δ_s playing the role of $(\underline{\underline{F}}, E)$ there. The second, third and fourth terms in (5.37) correspond to the first three terms on the right in (3.12'), and the first and fifth terms together in (5.37) correspond to the fourth term in (3.12'). The application of Theorem 3.2 gives

(5.38) $$B(f,f) = \tfrac{1}{2}\kappa\mu M<f,f> + \tfrac{1}{2}\Theta^r<\gamma_r f, \gamma_r f>$$

$$+ \iint_{\underline{\underline{X}}\times\Delta_r} \kappa\mu 1_{\Delta_s} D_\infty(dx,dy)\{f(x)-\gamma_r f(y)\}^2$$

$$+ \int dx\, \kappa(x)f^2(x)\mu\{1-M1-D_\infty 1\}(x)$$

$$+ \int_{\Delta_r} \nu(dy)q(y)\{\gamma_r f(y)\}^2.$$

The first term accounts for excursions from $\underline{\underline{X}}$ to $\underline{\underline{X}}$, the second for excursions from Δ_r to Δ_r and the third for excursions from $\underline{\underline{X}}$ to Δ_r. The indicator 1_{Δ_s} is inserted in the third term because only "honest excursions" which hit Δ_s contribute. The fourth term accounts for incomplete excursions from $\underline{\underline{X}}$ and the fifth for incomplete excursions from Δ_r. All excursions from $\underline{\underline{X}}$ enter by jumping into the interior of Δ_s. Excursions from Δ_r can either jump into the interior of Δ_s or enter at the boundary. The bilinear form $\Theta^r<\gamma^r f, \gamma^r f>$ is the restriction to $\Delta_r \times \Delta_r$ of the analogue of $U_{0,\infty}<f,f>$ in Section 3 with $\underline{\underline{X}}\cup\Delta_r$ here playing the role of Δ there. We simply state here that such a form exists. There seems to be little point in being more explicit. The first and third terms are the restrictions to $\underline{\underline{X}} \times \underline{\underline{X}}$ and $\underline{\underline{X}} \times \Delta_r$. The function $q(y)$ in the fifth term is the restriction to Δ_r of the analogue to $\pi r\kappa + \pi_* p$ in (3.19). The fourth term accounts for the corresponding restriction to $\underline{\underline{X}}$.

Plugging (5.38) into (5.36), we get

(5.39)
$$\begin{aligned} E^{\sim}(f,f) &= D(f,f) + \tfrac{1}{2}J\langle f,f\rangle + \int dx\kappa(x)r(x)f^2(x) \\ &+ \tfrac{1}{2}\kappa\mu M\langle f,f\rangle + \int dx\kappa(x)\mu\{1 - M1 - D_\infty 1\}(x)f^2(x) \\ &+ \iint_{\underline{\underline{X}}\times\Delta_r} \kappa\mu D_\infty(dx,dy)\{f(x) - \gamma_r f(y)\}^2 \\ &+ D^r(\gamma_r f, \gamma_r f) + \tfrac{1}{2}J^r\langle\gamma_r f, \gamma_r f\rangle + \int_{\Delta_r}\kappa^r(dy)\{\gamma_r f(y)\}^2 \\ &+ \tfrac{1}{2}\Theta^r\langle\gamma^r f, \gamma^r f\rangle + \int_{\Delta_r}\nu(dy)q(y)\{\gamma_r f(y)\}^2. \end{aligned}$$

The sixth term combines the third term in (5.38) with the fourth term in (5.36). To simplify the notation we define for $\varphi \in \gamma_r\underline{\underline{H}}_{(e)}$

(5.40)
$$\begin{aligned} G(\varphi,\varphi) &= D^r(\varphi,\varphi) + \tfrac{1}{2}J^r\langle\varphi,\varphi\rangle + \int_{\Delta_r}\kappa^r(dy)\varphi^2(y) \\ &+ \tfrac{1}{2}\Theta^r\langle\varphi,\varphi\rangle + \int_{\Delta_r}\nu(dy)q(y)\varphi^2(y) \end{aligned}$$

so that (5.39) can be written

(5.39')
$$\begin{aligned} E^{\sim}(f,f) &= D(f,f) + \tfrac{1}{2}J\langle f,f\rangle + \int dx\kappa(x)r(x)f^2(x) \\ &+ \tfrac{1}{2}\kappa\mu M\langle f,f\rangle + \int dx\kappa(x)\mu\{1 - M1 - D_\infty 1\}(x)f^2(x) \\ &+ \iint_{\underline{\underline{X}}\times\Delta_r} \kappa\mu D_\infty(dx,dy)\{f(x) - \gamma_r f(y)\}^2 + G(\gamma_r f, \gamma_r f). \end{aligned}$$

Under the special restriction mentioned above, we have used the expanded Dirichlet space $(\underline{\underline{F}}^e, E^e)$ to establish (5.39) for any $f \in \underline{\underline{F}}_{(e)}$ with $\gamma_r f$

as determined in Theorem 5.8. In fact (5.39) is valid without this special restriction, and we will verify it in the appendix at the end of the chapter. For the present we drop the restriction and continue as if (5.39) had been established in general. The desired analogue to Theorem 5.2 is

Theorem 5.9. *If* $f \in \tilde{\mathbf{F}}_{(e)}$, *then* f *satisfies*

$$D(f,f) + \tfrac{1}{2} J\langle f,f\rangle + \int dx\,\kappa(x)\{r(x) + \mu(1 - M1 - D_\infty 1)(x)\} f^2(x) + \tfrac{1}{2}\kappa\mu M\langle f,f\rangle < +\infty \tag{5.41}$$

and $\gamma_r f$ *is the unique function* $\varphi \in \gamma_r \mathbf{H}_{(e)}$ *satisfying* (5.4) *and such that*

$$\iint_{\mathbf{X} \times \Delta_r} \kappa\mu D_\infty(dx, dy)\{f(x) - \varphi(y)\}^2 < +\infty. \tag{5.42}$$

Moreover, the identity (5.39) *is valid. Conversely, if locally controlled* f *satisfies* (5.41) *and if there exists* $\varphi \in \gamma_r \mathbf{H}_{(e)}$ *satisfying* (5.4) *and* (5.42), *then* $f \in \tilde{\mathbf{F}}_{(e)}$ *and* $\varphi = \gamma_r f$. ///

Proof. Consider locally controlled f satisfying (5.41) for which there exists $\varphi \in \gamma_r \mathbf{H}_{(e)}$ satisfying (5.4) and (5.37). Then just as for Theorem 5.2, $g = f - HK\varphi$ satisfies $E^0(g,g) < +\infty$ and also (5.4) with $\varphi = 0$. By SMP, Theorem 14.5 applied to $(\mathbf{F}^0, E^0)$ such $g \in \mathbf{F}^0$ and therefore, by Theorem 5.8, $f \in \tilde{\mathbf{F}}_{(e)}$. Everything is proved except that the two conditions (5.42) and (5.4) for $\varphi \in \gamma_r \mathbf{H}_{(e)}$ imply $\varphi = \gamma_r f$. The difference $\psi = \varphi - \gamma_r f$ satisfies

$$\iint_{\mathbf{X} \times \Delta_r} \kappa\mu D_\infty(dx, dy)\{\psi(y)\}^2 < +\infty \tag{5.43}$$

$$\psi(X^{ex}_{\zeta-0}) = 0 \quad \text{when} \quad X^{ex}_{\zeta-0} \in \Delta. \tag{5.44}$$

Therefore we must prove that there is no nontrivial $\psi \in \gamma_r \mathbf{H}_{(e)}$ satisfying (5.43) and (5.44). Now (5.44) guarantees

$$H_u K_u \psi^2 = N_u \kappa \mu K_u \psi^2 \tag{5.45}$$

but we cannot combine (5.43) directly with Theorem 4.3(ii) to conclude that $\psi = 0$. Instead we apply the proof of Theorem 4.3(ii) but use instead the appropriate modified operators. For $0 \leq u < v$ we have

$$R^0_{(u)} = R^0_{(v)} + (v-u)R^0_{(u)} k^*_u \pi_u H_v K_v R^0_{(v)}$$

and so the role of the Feller operator $U_{u,v}$ is played here by $(v-u)k^*_u \pi_u H_v K_v$ and therefore the hypothesis in Theorem 4.3(ii) corresponds here to the condition

$$\mathrm{Lim}_{v \uparrow \infty} v \int_{\Delta_r} \nu^0(dy) k^* \pi H_v K_v \psi^2(y) < +\infty$$

or equivalently to

$$\mathrm{Lim}_{v \uparrow \infty} \int_\Delta \nu(dy) \pi v H_v K_v \psi^2(y) < +\infty$$

which by (5.45) is equivalent to

$$\mathrm{Lim}_{v \uparrow \infty} \int dx H1(x) v G_v \kappa \mu K_v \psi^2(y) < +\infty. \tag{5.46}$$

As $v \uparrow \infty$ clearly $K_v \psi^2 \to D_\infty \psi^2$ on Δ and since $vG_v H1 \uparrow H1 \leq 1$, the estimate (5.46) is implied by (5.43). Since the operators $R^0_{(v)} \downarrow 0$ on Δ_s, the proof of Theorem 4.3(ii) does guarantee that $\psi = 0$. ///

6. The Local Generator.

6.1. Definition. The local generator $\textcircled{A}$ is defined by

$$E(f, g) = -\int dx \textcircled{A} f(x) g(x) \quad \text{for } g \in \underline{\underline{F}} \cap C_{com}(\underline{\underline{X}}). \tag{6.1}$$ ///

It is understood in (6.1) that f is locally controlled so that the left side is well defined. The statement "f ∊ domain $\textcircled{A}$" is generally meant to imply that $\textcircled{A} f$ is locally integrable.

The main result in this section is

Theorem 6.1. Let f be bounded and locally controlled. Then for any $u \geq 0$ the following two conditions are equivalent.

(i) f is u-harmonic. (See Definition 14.3 in SMP.)

(ii) $\textcircled{A} f = uf$. ///

Proof. We can always replace E by E_u and therefore it suffices to consider the case $u = 0$. Fix D open with compact closure in $\underline{\underline{X}}$ and let $M = \underline{\underline{X}} \setminus D$. To show that (i) implies (ii), it suffices to check that $E(f, g) = 0$ for g belonging to the absorbed space $\underline{\underline{F}}^D$. But this follows directly from the formulae (13.7') and (16.1') in SMP. The converse follows from the same formulae after we invoke Lemma 14.2 in SMP to get a decomposition $f = f_0 + H^M f$. ///

Remark. The side condition that f be bounded is needed only to show that (ii) implies (i) and then only when E has a nonlocal part. It would certainly be of theoretical interest to remove this condition entirely. However, at least from the point of view of characterizing generators, this would have little consequence since, except for extremely simple situations where anyway the theorem is not needed, we have "a priori" control only for bounded harmonic functions. ///

7. The Normal Derivative.

7.1. Notation. $\{D_n\}_{n\geq 1}$ is an increasing sequence of open sets with compact closure such that $D_n \uparrow \underline{\underline{X}}$. Generally n and ~n will be used for superscripts and subscripts in place of the more clumsy D_n and $\underline{\underline{X}}\backslash D_n$. ///

In (3.8) we defined the terminal variable $\gamma_T f$ for $f \in \underline{\underline{F}}^{env}$ and so for $f \in \underline{\underline{\tilde{F}}}_{(e)}$. This is a random variable defined at the same time on the standard sample space Ω and on inverse limit spaces Ω_∞ and Ω^{ex}. Also for general φ on the boundary Δ we define the terminal variable $\gamma_T\varphi$ by

$$\gamma_T\varphi = \varphi(X^{ex}_{\zeta-0})I(X^{ex}_{\zeta-0} \in \Delta). \tag{7.1}$$

This is entirely consistent with (3.8) and indeed for $f \in \underline{\underline{\tilde{F}}}_{(e)}$

$$\gamma_T f = \gamma_T\gamma_r f = \gamma_T \gamma f. \tag{7.2}$$

In practice it is often convenient to view $\gamma_T f$ as a functional acting on bounded functions ψ defined on Δ. For $w(x) \geq 0$ let λ_w be the measure defined on Δ by

$$\int_\Delta \lambda_w(dy)\varphi(y) = \int dx\, Ⓔ_x \gamma_T\varphi. \tag{7.3}$$

Then at least when $w(x)dx$ has finite energy and when $\int dxw(x)Ⓔ_x\{\gamma_T f\}^2 < +\infty$, we have for bounded ψ on Δ

$$\int_\Delta \lambda_w(dy)\gamma_T f(y)\psi(y) = \mathrm{Lim}_{n\uparrow\infty}\int dxw(x)H^{\sim n} f Ⓔ_\cdot \gamma_T\psi(x). \tag{7.4}$$

The reason is that f has a representation

$$f = Ⓔ_\cdot \gamma_T f + N\kappa\mu f + f_0$$

and $\int dxw(x)H^{\sim n}|f_0(x)| \to 0$ by SMP, Theorem 7.3, while by the martingale and supermartingale convergence theorems, $\textcircled{E}_{X_t}\gamma_T f \to \gamma_T f$ and $N\kappa\mu\gamma f(X_t) \to 0$ in the appropriate sense as $t \uparrow \zeta$ on the set $[X^{ex}_{\zeta-0} \in \Delta]$.

At least for bounded f we can replace (7.4) by a limiting procedure which involves the <u>conditional</u> (<u>terminal</u>) <u>equilibrium distribution</u>. This is defined for $\varphi \geq 0$ on Δ and for each n and is the Radon measure ℓ^{φ}_n determined by

(7.5)
$$\int \ell^{\varphi}_n(dx)g(x) = \textcircled{E}[\sigma(D_n) < +\infty; \{\gamma_T\varphi\cdot\rho\}g(X_{\sigma(D_n)})].$$

If $g = Nv$ with $v \geq 0$ then (7.5) becomes

$$\begin{aligned}\int dxv(x)N\ell^{\varphi}_n(x) &= \textcircled{E}\{\gamma_T\varphi\cdot\rho\}\int^{\zeta}_{\sigma(D_n)} dtv(X_t)\\ &= \textcircled{E}\int^{\zeta}_{\zeta^*} dtv(X_t)I(\sigma^*(D_n) > t)\gamma_T\varphi\\ &= \int dxv(x)H^n\textcircled{E}\,\gamma_T\varphi\end{aligned}$$

and it follows that

(7.6)
$$N\ell^{\varphi}_n = H^n\textcircled{E}\,\gamma_T\varphi.$$

Thus ℓ^{φ}_n is determined by the condition that ℓ^{φ}_n is concentrated on the fine closure of D_n and that $N\ell^{\varphi}_n = \textcircled{E}\,\gamma_T\varphi$ on D_n. Often this characterization will permit us to compute ℓ^{φ}_n explicitly. Notice that if the original process has continuous trajectories, then ℓ^{φ}_n must be concentrated on the boundary of D_n. Now it is easy to check that if $f \in \tilde{\underline{\underline{F}}}_{(e)}$ is bounded, then

(7.7)
$$\int_{\Delta}\lambda_w(dy)\psi(y)\gamma_T f(y) = \mathrm{Lim}_{n\uparrow\infty}\int \ell^{\psi}_n(dx)f(x)Nw(x).$$

In the special case when $\psi = e_{\alpha}$, the indicator of singleton $\alpha \in \Delta$, we replace (7.4) and (7.7) by

(7.4')
$$\gamma_T f(\alpha) = \lambda_w(\alpha)^{-1} \operatorname{Lim}_{n\uparrow\infty} \int dx\, w(x) H^{\sim n} f \textcircled{E}\, \gamma_T e_\alpha(x)$$

(7.7')
$$\gamma_T f(\alpha) = \lambda_w(\alpha)^{-1} \operatorname{Lim}_{n\uparrow\infty} \int \ell_n^\alpha(dx) f(x) N w(x).$$

Notice that we are using α rather than e_α as a label for the equilibrium distribution.

In applications it is usually possible to replace the function $w(x)$ by a reference point 0 and then (7.3), (7.4) and (7.7) are replaced by

(7.3')
$$\int_\Delta \lambda_0(dy)\varphi(y) = E_0 \gamma_T \varphi$$

(7.4'')
$$\int_\Delta \lambda_0(dy)\psi(y)\gamma_T f(y) = \operatorname{Lim}_{n\uparrow\infty} H^{\sim n} f \textcircled{E}\, \gamma_T \psi(0)$$

(7.7'')
$$\int_\Delta \lambda_0(dy)\psi(y)\gamma_T f(y) = \operatorname{Lim}_{n\uparrow\infty} \int \ell_n^\psi(dx) f(x) N(x,0).$$

Our main purpose in this section is to establish a useful generalization of the classical Green's formula

(7.8)
$$\tfrac{1}{2}\int_D dx \operatorname{grad} f(x)\cdot \operatorname{grad} g(x) + \tfrac{1}{2}\int_D dx \Delta f(x) g(x) = -\tfrac{1}{2}\int_{\partial D} d\sigma(\partial f/\partial n)(\sigma) g(\sigma).$$

We first define the (inner) normal derivative as a linear functional.

7.2. Definition. Let $f \in \underline{\underline{F}}^e_{(e)}$, the expanded Dirichlet space. A function ψ defined on Δ belongs to the domain of the normal derivative $\partial f/\partial n$ if the following limit exists.

(7.9)
$$(\partial f/\partial n)(\psi) = \operatorname{Lim}_{n\uparrow\infty} \textcircled{E}^{ex}[\sigma(D_n) < +\infty ; \psi(X^{ex}_{\zeta^*-0})\{f(X_{\sigma(D_n)}) - \gamma f(X^{ex}_{\zeta^*-0})\}]. \qquad ///$$

Recall that for $f \in \underline{\underline{F}}^e_{(e)}$ the boundary function γf is the ordinary restriction to Δ. We agree that γf is represented by the refinement belonging to $\underline{\underline{H}}_{(e)}$. In the special case when the singular boundary $\Delta_s = \phi$, the boundary

function γf is determined by the restriction of f to $\underline{\underline{X}}$. (See Theorem 5.1.)

Theorem 7.1. Let $f \in \underline{\underline{F}}^{e}_{(e)}$ belong to the domain of the local generator Ⓐ and assume that

(7.11) $$\text{Ⓐ}f(x) + \kappa(x)\mu\gamma f(x)$$

is integrable. Then every bounded $\psi \in \underline{\underline{H}}_{(e)}$ belongs to the domain of the normal derivative $\partial f/\partial n$ and

(7.12) $$-(\partial f/\partial n)(\psi) = D(f, H\psi) + \tfrac{1}{2}J\langle f, H\psi\rangle$$
$$+ [\int dx\{\text{Ⓐ}f(x)+\kappa(x)f(x)\}H\psi(x) - \int dx\kappa(x)\int\mu(x,dy)\{f(x)-\gamma f(y)\}\psi(y)]. \qquad ///$$

Remark 1. The expression inside the square bracket in (7.12) is well defined since

(7.13) $$\int dx\kappa(x)\{\text{Ⓐ}f(x)+\kappa(x)f(x)\}H\psi(x) - \int dx\kappa(x)\int\mu(x,dy)\{f(x)-\gamma f(y)\}\psi(y)$$
$$= \int dx\{\text{Ⓐ}f(x)+\kappa\mu\gamma f(x)\}H\psi(x) + \int dx\kappa(x)r(x)f(x)H\psi(x)$$
$$+ \int dx\kappa(x)\int\mu(x,dy)\{f(x)-\gamma f(y)\}\{H\psi(x)-\psi(y)\}.$$

The first term on the right converges by hypothesis, and the other two terms converge since both f and $H\psi$ belong to $\underline{\underline{F}}^{e}_{(e)}$. The reason for the bracket in (7.12) is that in general the integrals inside do not converge separately. ///

Remark 2. The hypothesis that (7.11) be integrable is a natural one. We will see in Section 10 that if f belongs to the domain of the L^2 generator (or even belongs "modulo boundary conditions"), then (7.11) is square integrable and therefore integrable when dx is bounded. This requirement that dx be bounded could be suppressed if we were willing to work with a $u > 0$ version of (7.12). ///

Remark 3. Notice that (7.12) and (6.1) can be combined to give for bounded $g \in \underline{\underline{F}}^e_{(e)}$

$$-(\partial f/\partial n)(\gamma g) = D(f,g) + \tfrac{1}{2}J<f,g>$$
$$+ [\int dx\{\textcircled{A}f(x)+\kappa(x)f(x)\}g(x) - \int dx\kappa(x)\int\mu(x,dy)\{f(x)-\gamma f(y)\}\gamma g(y)]. \quad ///$$

<u>Proof.</u> For each n there exists a function in $\underline{\underline{F}}$ which is identically one quasi-everywhere on the closure of D_n and so by Theorem 7.3(i) in SMP the function $H^n 1$ belongs to $\underline{\underline{F}}$ and therefore is in $L^2(\kappa)$. It follows by the Cauchy-Schwarz inequality that for ψ bounded

$$\int dx\kappa(x)\int\mu(x,dy)\{f(x)-\gamma f(y)\}\psi(y)H^n 1(x)$$

converges. Similarly, $\int dx\kappa(x)\{f(x)-\mu\gamma f(x)\}H^n 1(x)$ converges, and so by integrability of (7.11)

(7.14) $$\int dx\{\textcircled{A}f(x)+\kappa(x)f(x)\}H^n 1(x)$$

converges and therefore also

(7.15) $$\int dx\{\textcircled{A}f(x)+\kappa(x)f(x)\}H^n H\psi(x)$$

converges. It is easy to see that

(7.16) $$\int dx\kappa(x)\int\mu(x,dy)\{f(x)-\gamma f(y)\}\psi(y)H^n 1(x)$$
$$= \textcircled{E}^{ex}[X_{\zeta^*} \in \underline{\underline{X}}; \sigma(D_n) < +\infty; \{f(X_{\zeta^*}) - \gamma f(X^{ex}_{\zeta^*-0})\}\psi(X^{ex}_{\zeta^*-0})].$$

It is also true that

(7.17) $$\int dx\{\textcircled{A}f(x)+\kappa(x)f(x)\}H^n H\psi(x)$$
$$= \textcircled{E}^{ex}[\sigma(D_n) < +\infty; \psi(X^{ex}_{\zeta^*-0})\{\gamma f(X^{ex}_{\zeta^*-0}) - f(X_{\sigma(D_n)})\}]$$

$$+ \text{Ⓔ}^{ex}[\sigma(D_n) < +\infty\,;\, \psi(X^{ex}_{\zeta^*-0})\{\gamma f(X^{ex}_{\zeta}) - \gamma f(X^{ex}_{\zeta^*-0})\}]$$

$$+ \text{Ⓔ}^{ex}[\sigma(D_n) < +\infty\,;\, X_{\zeta-0} \in \underline{\underline{X}}\,;\, \psi(X^{ex}_{\zeta^*-0})\{f(X_{\zeta-0}) - \gamma f(X^{ex})\}]\,,$$

but this requires some proof. Notice first that convergence in (7.14) guarantees that the potential $N\{\text{Ⓐ}f+\kappa f\}$ converges quasi-everywhere, since every nonpolar set carries a nontrivial measure whose potential is dominated by $H^n 1$ for some n. For fixed n clearly $\tilde{H}^n f + N^n \kappa\mu\gamma f$ is the E^e orthogonal projection of f onto the complement of the absorbed space for D_n. Thus for $g \in \underline{\underline{F}} \cap C_{com}(D_n)$ we have

$$\begin{aligned} &E(f-\tilde{H}^n f - N^n \kappa\mu\gamma f, g) \\ &= E^e(f,g) \\ &= D(f,g) + \tfrac{1}{2}J\langle f,g\rangle + \int dx\,\kappa(x)\{f(x)-\mu\gamma f(x)\}g(x) \\ &= -\int_{D_n} dx\{\text{Ⓐ}f(x)+\kappa f(x)\}g(x) + \int_{D_n} dx\,\kappa(x)\{f(x)-\mu\gamma f(x)\}g(x) \end{aligned}$$

and it follows that

$$-N^n \kappa\mu\gamma f = -N^n\{\text{Ⓐ}f+\kappa f\} + N^n \kappa\{f-\mu\gamma f\}$$

or equivalently

$$f = \tilde{H}^n f - N^n \text{Ⓐ} f.$$

Passing to the limit $n\uparrow\infty$ we get

$$f = \text{Ⓔ}_{\cdot}\, \gamma_T f - N\text{Ⓐ}f$$

or equivalently

$$\begin{aligned} N\{\text{Ⓐ}f+\kappa f\}(x) &= \text{Ⓔ}_x[I(X_{\zeta-0} \in \underline{\underline{X}})f(X_{\zeta-0}) + I(X_{\zeta-0} = \partial)\gamma f(X^{ex}_{\zeta}) - f(X_0)] \\ &= \text{Ⓔ}_x[\gamma f(X^{ex}_{\zeta}) + I(X_{\zeta-0} \in \underline{\underline{X}})\{f(X_{\zeta-0}) - \gamma f(X^{ex}_{\zeta})\} - f(X_0)] \end{aligned}$$

with the variable X_{ζ}^{ex} adjoined according to an obvious analogue of paragraph 3.6.2. Applying both sides to $X_{\sigma(D_n)}$, multiplying by $\psi(X^{ex}_{\zeta^*-0})I(\sigma(D_n)<+\infty)$, and then integrating with respect to the excursion measure $\textcircled{P}^{ex}$, we get

$$\textcircled{E}^{ex}[\psi(X^{ex}_{\zeta^*-0});\sigma(D_n)<+\infty;\int_{\sigma(D_n)}^{\zeta}dt\{\textcircled{A}f(X_t)+\kappa f(X_t)\}]$$

$$=\textcircled{E}^{ex}[\sigma(D_n)<+\infty;\psi(X^{ex}_{\zeta^*-0})\{\gamma f(X^{ex}_{\zeta})-f(X_{\sigma(D_n)})\}]$$

$$+\textcircled{E}^{ex}[\sigma(D_n)<+\infty;X_{\zeta-0}\in \underline{\underline{X}};\psi(X^{ex}_{\zeta^*-0})\{f(X_{\zeta-0})-\gamma f(X^{ex}_{\zeta})\}]$$

$$=\textcircled{E}^{ex}[\sigma(D_n)<+\infty;\psi(X^{ex}_{\zeta^*-0})\{\gamma f(X^{ex}_{\zeta^*-0})-f(X_{\sigma(D_n)})\}]$$

$$+\textcircled{E}^{ex}[\sigma(D_n)<+\infty;\psi(X^{ex}_{\zeta^*-0})\{\gamma f(X^{ex}_{\zeta})-\gamma f(X^{ex}_{\zeta^*-0})\}]$$

$$+\textcircled{E}^{ex}[\sigma(D_n)<+\infty;X_{\zeta-0}\in \underline{\underline{X}};\psi(X^{ex}_{\zeta^*-0})\{f(X_{\zeta-0})-\gamma f(X^{ex}_{\zeta})\}].$$

Applying time reversal invariance to the left side we get

$$\textcircled{E}^{ex}[\sigma(D_n)<+\infty;\int_{\zeta^*}^{\sigma^*(D_n)}dt\{\textcircled{A}f(X_t)+\kappa f(X_t)\}\psi(X^{ex}_{\zeta})]$$

$$=\textcircled{E}^{ex}\int_{\zeta^*}^{\zeta}dt\{\textcircled{A}f(X_t)+\kappa f(X_t)\}I(\theta_t\sigma(D_n)<+\infty)\psi(X^{ex}_{\zeta})$$

$$=\textcircled{E}^{ex}\int_{\zeta^*}^{\zeta}dt\{\textcircled{A}f(X_t)+\kappa f(X_t)\}H^nH\psi(X_t)$$

$$=\int dx\{\textcircled{A}f(x)+\kappa f(x)\}H^nH\psi(x),$$

and (7.17) is finally proved. On the other hand

(7.18) $$\int dx\kappa(x)r(x)f(x)H\psi(x)+\int dx\kappa(x)\int\mu(x,dy)\{f(x)-\gamma f(y)\}\{H\psi(x)-\psi(y)\}$$

$$=\textcircled{E}^{ex}[X_{\zeta-0}\in \underline{\underline{X}};\{f(X_{\zeta-0})-\gamma f(X^{ex}_{\zeta})\}\{H\psi(X_{\zeta-0})-\psi(X^{ex}_{\zeta})\}],$$

and so (3.2'') implies

$$(7.19)\qquad D(f,H\psi)+\tfrac{1}{2}J\langle f,H\psi\rangle$$
$$=\tfrac{1}{2}Ⓔ^{ex}\{\psi(X_\zeta^{ex})-\psi(X_{\zeta^*-0}^{ex})\}\{\gamma f(X_\zeta^{ex})-\gamma f(X_{\zeta^*-0}^{ex})\}$$
$$-Ⓔ^{ex}[X_{\zeta-0}\in \underline{\underline{X}};\{f(X_{\zeta-0})-\gamma f(X_\zeta^{ex})\}\{H\psi(X_{\zeta-0})-\psi(X_\zeta^{ex})\}].$$

As $n\uparrow\infty$ the second term on the right in (7.17) converges to the negative of the first term on the right in (7.19). The third term on the right in (7.17) minus the right side of (7.16) is the same as

$$(7.20)\quad Ⓔ^{ex}[\sigma(D_n)<+\infty;X_{\zeta-0}\in\underline{\underline{X}};\{\psi(X_{\zeta^*-0}^{ex})-\psi(X_\zeta^{ex})\}\{f(X_{\zeta-0})-\gamma f(X_\zeta^{ex})\}].$$

Since

$$Ⓔ^{ex}[X_{\zeta-0}\in\underline{\underline{X}};\{f(X_{\zeta-0})-\gamma f(X_\zeta^{ex})\}^2]$$
$$=\int dx\kappa(x)r(x)f^2(x)+\int dx\kappa(x)\int\mu(x,dy)\{f(x)-\gamma f(y)\}^2$$

converges and since

$$Ⓔ^{ex}[X_{\zeta-0}\in\underline{\underline{X}};\{\psi(X_{\zeta^*-0}^{ex})-\psi(X_\zeta^{ex})\}^2]$$
$$=\int dx\kappa(x)\int\mu(x,dy)\int H(x,dz)\{\psi(y)-\psi(z)\}^2\le U_{0,\infty}\langle\psi,\psi\rangle$$

converges, we can pass to the limit $n\uparrow\infty$ in (7.20) to get a term which exactly cancels the second term on the right in (7.19). Thus the theorem follows after subtracting (7.16) from (7.17), adding (7.19) and then passing to the limit $n\uparrow\infty$ using integrability of (7.15). ///

The <u>terminal normal derivative</u> $(\partial f/\partial n)_T(\psi)$ is defined by

(7.21) $(\partial f/\partial n)_T(\psi)$

$$= \text{Lim}_{n\uparrow\infty} \textcircled{E}^{ex}[\sigma(D_n) < +\infty ; X^{ex}_{\zeta^*} \in \Delta ; \psi(X^{ex}_{\zeta^*})\{f(X_{\sigma(D_n)}) - \gamma f(X^{ex}_{\zeta^*})\}].$$

We will see below that it is often possible to decompose the normal derivative into

(7.22) $$(\partial f/\partial n)(\psi) = (\partial f/\partial n)_T(\psi) + \text{Lim}_{n\uparrow\infty}\int_{D_n} dx\kappa(x)\int\mu(x, dy)\{f(x)-\gamma f(y)\}\psi(y)$$

or even

(7.23) $$(\partial f/\partial n)(\psi) = (\partial f/\partial n)_T(\psi) + \int dx\kappa(x)\int\mu(x, dy)\{f(x)-\gamma f(y)\}\psi(y)$$

when the last integral converges. If f satisfies the hypotheses of Theorem 7.1 and (7.22) or (7.23) is valid, then the Green's formula (7.12) can be replaced respectively by

(7.12') $$-(\partial f/\partial n)_T(\psi) = D(f, H\psi) + \tfrac{1}{2}J<f, H\psi> + \text{Lim}_{n\uparrow\infty}\int_{D_n} dx\{\textcircled{A}f(x)+\kappa f(x)\}H\psi(x)$$

or by

(7.12'') $$-(\partial f/\partial n)_T(\psi) = D(f, H\psi) + \tfrac{1}{2}J<f, H\psi> + \int dx\{\textcircled{A}f(x)+\kappa f(x)\}H\psi(x).$$

These formulae are closer analogues than (7.12) to the classical Green's formula.

We finish this section by considering a special situation where (7.23) and (7.12'') are valid. As in Section 6, we use the notation $(\partial f/\partial n)(\alpha)$, $(\partial f/\partial n)_T(\alpha)$ when $\psi = e_\alpha$ so that in particular

(7.24) $$(\partial f/\partial n)_T(\alpha) = \text{Lim}_{n\uparrow\infty}\int\ell^n(dx)\{f(x)-\gamma f(\alpha)\}.$$

Also we use the special notation

(7.25) $$\mu_\alpha(x) = \mu(x, \{\alpha\}).$$

<u>Theorem 7.2. Let $f \in \underline{\underline{F}}^{e}_{(e)}$ satisfy the hypotheses of Theorem</u> 7.1. <u>Also assume that γf is bounded and let $\alpha \in \Delta$ be such that</u>

(7.26) $$e_\alpha \in \underline{\underline{H}}\ ;\ 1-e_\alpha \in \underline{\underline{H}}.$$

<u>Then the limit</u> (7.17) <u>exists, $\{Ⓐf(x)+\kappa f(x)\}e_\alpha(x)$ is integrable, and the following two formulae are valid:</u>

(7.23') $$(\partial f/\partial n)(\alpha) = (\partial f/\partial n)_T(\alpha) + \int dx\kappa(x)\mu_\alpha(x)\{f(x)-\gamma f(\alpha)\}$$

(7.12''') $$-(\partial f/\partial n)(\alpha) = D(f, He_\alpha) + \tfrac{1}{2}J\langle f, He_\alpha\rangle + \int dx\{Ⓐf(x)+\kappa f(x)\}He_\alpha(x).$$ ///

<u>Proof.</u> It suffices to prove that

(7.27) $$\lim_{n\uparrow\infty} Ⓔ^{ex}[\sigma(D_n) < +\infty\ ;\ X_{\zeta^*} \in \underline{\underline{X}}\ ;\ X_{\zeta^*-0} = \alpha\ ;\ f(X_{\sigma(D_n)})-\gamma f(\alpha)]$$
$$= \int dx\kappa(x)\mu_\alpha(x)\{f(x)-\gamma f(\alpha)\}.$$

This will follow with the help of the dominated convergence theorem if we can establish convergence for the right side in (7.27) and also

(7.28) $$\lim_{n\uparrow\infty}\int dx\kappa(x)\mu_\alpha(x)\{H^n f(x)-f(x)H^n 1(x)\} = 0.$$

We note first that by the proof of Theorem 7.1 the function f has a representation

$$f = Ⓔ_{.}\ \gamma_T f - NⒶf$$
$$= H\gamma f - N\{Ⓐf+\kappa\mu f\}$$

and so $f-H\gamma f$ is dominated by the potential of an integrable function. But for $v(x)$ an integrable function

$$\int dx\kappa(x)\mu_\alpha(x)Nv(x) \le \int dx v(x)$$

and so we need only consider the special case when $f = H\gamma f$. For $\gamma f(\alpha) = 0$ both (7.28) and convergence on the right in (7.27) follow (since γf is bounded) from the estimate

$$\begin{aligned}\int dx\kappa(x)\mu_\alpha(x)H(1-e_\alpha)(x) &\leq \int\int U_{0,\infty}(dy,dz)e_\alpha(y)(1-e_\alpha)(z)\\ &= -\tfrac{1}{2}U_{0,\infty}\langle e_\alpha, 1-e_\alpha\rangle\end{aligned}$$

which is finite by (7.26) and (3.19). It only remains for us to consider the function $f = He_\alpha$. The right side of (7.27)

$$= -\int dx\kappa(x)\mu_\alpha(x)\{H(1-e_\alpha)(x) + N\kappa r(x)+p(x)\},$$

and convergence will follow from convergence of

$$\int dx\kappa(x)\mu_\alpha(x)N\kappa r(x) + \int dx\kappa(x)\mu_\alpha(x)p(x).$$

But this also follows from (3.19) since $e_\alpha = e_\alpha^2$ and since

$$\begin{aligned}\int\pi^* p(y)e_\alpha(y) &= \operatorname{Lim}_{v\uparrow\infty}\int dxp(x)vH_v e_\alpha(x)\\ &\geq \operatorname{Lim}_{v\uparrow\infty}\int dxp(x)vG_v\kappa\mu_\alpha(x)\\ &= \operatorname{Lim}_{v\uparrow\infty}\int dx\kappa(x)\mu_\alpha(x)vG_v p(x)\\ &= \int dx\kappa(x)\mu_\alpha(x)p(x).\end{aligned} \tag{7.29}$$

Finally, the exact same argument takes care of (7.28) since $H^n f(x)-f(x)H^n 1(x) = H^n(f-1)(x)-(f(x)-1)H^n 1(x)$. ///

8. L^2 Generator for $(\underline{\underline{F}}, E)$.

The transition operators P_t, $t > 0$, defined by

$$P_t f(x) = \textcircled{E}_x f(X_t)$$

form a strongly continuous contractive semigroup of self-adjoint operators on the Hilbert space $L^2(dx)$. The L^2 generator A is defined by

$$Af = \text{Lim}_{t \downarrow 0} (1/t)\{P_t f - f\}$$

with the understanding that this limit must converge strongly. It is an easy consequence of the spectral theory that

$$\text{domain } A \subset \underline{\underline{F}} \tag{8.1}$$

and A is determined by

$$E(f, g) = -\int dx A f(x) g(x) \quad \text{for} \quad g \in \underline{\underline{F}}. \tag{8.2}$$

It is understood in (8.2) that $f \in$ domain A if and only if there exists a function in $L^2(dx)$, necessarily unique, such that (8.2) is valid. We will not bother to point this out in analogous situations below. This is clearly equivalent to

__Theorem 8.1.__ _The L^2 generator A for (F, E) is defined by_

$$Af = \textcircled{A} f \tag{8.3}$$

supplemented by the following two conditions.

(i) $f \in \underline{\underline{F}}^{ref}$,

(ii) $\gamma_T f = 0$. ///

Here and below we use the terminology of Theorem 8.1 in place of the following more precise but clumsier statement. "$f \in L^2(dx)$ belongs to domain A if and only if Ⓐ$f \in L^2(dx)$ and f satisfies (i) and (ii). In this case (8.3) is valid."

8.1. Convention. In the remainder of this chapter the reference measure dx is bounded. ///

In general, the explicit requirement $f \in \underline{F}^{ref}$ is hard to check. The following technical result justifies suppressing it for a special case.

Lemma 8.2. *Let f be bounded and locally controlled with Ⓐf bounded.*

(i) $f = f_0+h$ with $f_0 \in \underline{F}$ and h bounded and 1-harmonic. Therefore $\gamma_T f$ is well defined by (3.8).

(ii) If $\gamma_T f = 0$, then $f \in \underline{F}$.

Proof. The point is that $h = f-G_1(f-af)$ is bounded and Ⓐ$h = h$. The lemma follows after applying Theorem 6.1. ///

Now it is easy to prove

Theorem 8.3. *Let f be bounded and locally controlled and let Ⓐf be bounded.*

(i) The terminal variable $\gamma_T f$ is well defined by (3.8).

(ii) If $\gamma_T f = 0$, then $f \in$ domain A. ///

Remark. Let $(\underline{F}, E)$ correspond to absorbed Brownian motion on the half line $(0,\infty)$ but with reference measure $m(x)dx$ where $m(x)$ has a finite second moment. Then $f(x) = x$ satisfies all of the conditions of Theorem 8.3 except that it is not bounded and clearly f is not in domain A since it is not in

$\underline{\underline{F}}$. The explicit requirement $f \in \underline{\underline{F}}^{ref}$ is needed to eliminate f as a candidate for domain A. A more subtle example will be given in Section 19. In general, we can eliminate the requirement in Theorem 8.3 that f and Ⓐf be bounded only when we can show directly that Ⓐ-1 has no annihilators in $L^2(dx)$ except for those which can be eliminated by the condition $\gamma_T f = 0$. If the killing density $\kappa(x)$ is bounded away from 0, then the potential operator N is bounded on $L^2(dx)$ and we can work with annihilators of Ⓐ rather than of Ⓐ-1. ///

Sometimes we can establish additional regularity properties for the transition operators P_t.

Theorem 8.4. *Assume that the potential operator N maps bounded functions into $C_0(\underline{\underline{X}})$, the Banach space of continuous functions vanishing at infinity.*

(i) *Each resolvent operator G_u, $u > 0$, maps bounded functions into $C_0(\underline{\underline{X}})$.*

(ii) *Let $C^{\#}$ be the uniform closure in $C_0(\underline{\underline{X}})$ of the set of functions Nf with f bounded. Then P_t, $t > 0$, is a strongly continuous semigroup on $C^{\#}$.* ///

Conclusion (i) is clear since $G_u f = N\{f - uG_u f\}$ and then (ii) follows from the Hille-Yosida theorem since for f bounded, $uG_u Nf = Nf - G_u f \to Nf$ uniformly as $u \uparrow \infty$. In general, it may be necessary to replace the operators G_u and P_t by appropriate refinements before Theorem 8.4 is valid. Also it is legitimate to replace N in the hypothesis by any one of the resolvent operators G_u.

9. Boundary Conditions when $\mu = 0$.

In this and the next two sections we study the $L^2(dx)$ generators $A^\sim$ for the process constructed in Section 3. The definition is the same as in Section 8 except that P_t is replaced by

$$P_t^\sim f(x) = \textcircled{E}_x f(X_t)$$

and (8.1), (8.2) are replaced by

(9.1) $$\text{domain } A^\sim \subset \underline{\underline{F}}^\sim$$

(9.2) $$E^\sim(f,g) = -\int dx A^\sim f(x) g(x) \quad \text{for } g \in \underline{\underline{F}}^\sim .$$

In this section we restrict attention to the case $\mu = 0$. Thus there is no auxiliary set $\underline{\underline{M}}$ in Section 3, and certainly $\Delta_s = \phi$:

Theorem 9.1. *The L^2 generator $A^\sim$ for $(\underline{\underline{F}}^\sim, E^\sim)$ is defined by*

(9.3) $$A^\sim f = \textcircled{A} f$$

supplemented by the following three conditions.

(i) $f \in \underline{\underline{F}}^{ref}$.

(ii) *There exists* $\gamma f \in \underline{\underline{H}}$, *necessarily unique, such that* $\gamma_T f = \gamma_T \gamma f$.

(iii) *Every bounded* $\psi \in \underline{\underline{H}}$ *belongs to the domain of the terminal normal derivative* $(\partial f/\partial n)_T$ *and*

(9.4) $$(\partial f/\partial n)_T(\psi) = \{Q-N\}(\gamma f, \psi).$$ ///

Proof. If f is in domain $A^\sim$, then certainly $f \in \underline{\underline{F}}^\sim$ and by Theorem 5.2 this implies (i) and (ii). Also by (5.3) the identity (9.2) is equivalent to

(9.5) $$-\int dx A^\sim f(x) g(x) = D(f,g) + \tfrac{1}{2} J\langle f, g\rangle + \int dx \kappa(x) f(x) g(x) + \{Q-N\}(\gamma f, \gamma g)$$

for general $g \in \underline{\underline{F}}$. Applying this with $g \in \underline{\underline{F}} \cap C_{com}(\underline{\underline{X}})$ we get (9.3). Applying this with $g = H\psi$ where $\psi \in \underline{\underline{H}}$ is bounded and rearranging terms, we get

$$(9.6) \qquad -D(f, H\psi) - \tfrac{1}{2}J\langle f, H\psi\rangle - \int dx\{(A)f(x)+\kappa f(x)\}H\psi(x) = \{Q-N\}(\gamma f, \psi)$$

and (iii) follows from the Green's formula (7.12) since $(\partial f/\partial n)_T = (\partial f/\partial n)$. Conversely, consider $f \in L^2(dx)$ such that $(A)f \in L^2(dx)$ and f satisfies (i) through (iii). Conditions (i), (ii) guarantee that $f \in \underline{\underline{F}}^{\sim}$ and then $(A)f \in L^2(dx)$ implies (9.2) for $g \in \underline{\underline{F}} \cap C_{com}(\underline{\underline{X}})$ and (iii) guarantees it for $g = H\psi$ with $\psi \in \underline{\underline{H}}$ bounded and this is sufficient to prove $f \in$ domain $A^{\sim}$. ///

Remark 1. The Dirichlet spaces $(\underline{\underline{F}}^{\sim}, E^{\sim})$ which occur here are precisely those covered by the First Structure Theorem (Theorem 15.2 in SMP). Indeed, it follows from the results in that section that we obtain all possible Dirichlet spaces $(\underline{\underline{F}}^{\sim}, E^{\sim})$ on $L^2(dx)$ for which (9.3) is valid. It is true that the local generator is defined differently in SMP, but it follows from Theorem 6.1 that the definitions agree for bounded functions for which $(A)f$ is bounded, and this is enough for carrying over the results. ///

Remark 2. The requirement $\gamma_T f = \gamma_T \gamma f$ is essential in (ii). In general our regularization at the beginning of Section 3 may map part of the terminal set $\underline{\underline{T}}$ into the dead point δ and then there can exist $f \in \underline{\underline{F}}^{ref}$ not belonging to $\underline{\underline{F}}_{(e)}$ such that $\gamma f = 0$. We will give a simple example in Section 12, Case 2. ///

Remark 3. It is also possible to formulate the boundary condition (9.4) in a different manner, ignoring the normal derivative and using instead the generator A_Δ for the Dirichlet space $(\underline{\underline{H}}, Q)$. Suppose for example that the operator π^* is bounded from $L^2(\underline{\underline{X}}, dx)$ to $L^2(\Delta, \nu)$. Then (9.4) can be replaced by the requirement γf is in domain A_Δ and

(9.4')
$$A_\Delta \gamma f = \pi^* \text{Ⓐ} f.$$

This is the approach used by H. Kunita[34]. (See also the example treated near the end of Section 15.) ///

Consider again f bounded and locally controlled with Ⓐf bounded so that by Lemma 8.2 $\gamma_T f$ is well defined. If there exists $\gamma f \in \underline{\underline{H}}$ such that $\gamma_T f = \gamma_T \gamma f$, then $f - N_1 \kappa \mu \gamma f \in \underline{\underline{F}}$, and this is enough to guarantee $f \in \underline{\underline{F}}^{\sim}$. Thus

Theorem 9.2. Let f be bounded and locally controlled and let Ⓐf be bounded.

(i) The terminal variable $\gamma_T f$ is well defined by (3.8).

(ii) If there exists $\gamma f \in \underline{\underline{H}}$ such that $\gamma_T f = \gamma_T \gamma f$, then $f \in \underline{\underline{F}}^{\sim}$.

(iii) If in addition γf satisfies the boundary condition (8.3), then $f \in$ domain $A^{\sim}$. ///

An analogue to Theorem 8.4 depends on regularity properties for the Green's operators associated with the Dirichlet space $(\underline{\underline{H}}, Q)$ and also for the hitting operator H. We refer to Section 2.1 where details are given for a non-trivial example. (In fact, $\mu \neq 0$, and the terminal boundary $\underline{\underline{T}} = \phi$ for the example in Section 21, but this doesn't really matter.)

10. Boundary Conditions when $\Delta_s = \phi$.

In this section we admit a nontrivial jumping measure μ, but we assume that the singular boundary Δ_s is not present. An analytic condition for this is given in Theorem 4.3. The Dirichlet spaces $(\underset{=}{F}{}^{\sim}, E^{\sim})$ are precisely those covered by the Second Structure Theorem (Theorem 20.2 in SMP). (However, see the remark following Corollary 4.4 concerning an error in SMP.)

Theorem 10.1. The L^2 generator $A^{\sim}$ for $(\underset{=}{F}{}^{\sim}, E^{\sim})$ is defined by

$$(10.1) \qquad A^{\sim}f(x) = \textcircled{A}f(x) + \kappa(x)\mu\gamma f(x)$$

supplemented by the following three conditions.

  (i) f is locally controlled and

$$(10.2) \qquad D(f, f) + \tfrac{1}{2}J<f, f> + \int dx\kappa(x)r(x)f^2(x) < +\infty.$$

  (ii) There exists $\gamma f \in \underset{=}{H}$, necessarily unique, such that $\gamma_T f = \gamma_T \gamma f$ and such that

$$(10.3) \qquad \int dx\kappa(x)\int \mu(x, dy)\{f(x) - \gamma f(y)\}^2 < +\infty.$$

  (iii) Every bounded $\psi \in \underset{=}{H}$ belongs to the domain of the normal derivative $\partial f/\partial n$ and

$$(10.4) \qquad (\partial f/\partial n)(\psi) = \{Q-N\}(\gamma f, \psi). \qquad\qquad ///$$

The proof goes exactly as in Section 8 except that we apply Theorem 5.2 in full strength. Suppose now that f is bounded and locally controlled and that there exists bounded $\varphi \in \underset{=}{H}$ such that $\textcircled{A}f + \kappa\mu\varphi$ is bounded. Then $\textcircled{A}\{f-H\varphi\} = \textcircled{A}f + \kappa\mu\varphi$ is bounded, and it follows from Lemma 8.2 that the terminal variable $\gamma_T f$ is well defined. If in addition $\gamma_T f = \gamma_T \varphi$, then $\gamma_T\{f-H\varphi\} = 0$ and it follows again from Lemma 8.2 that $f - H\varphi \in \underline{\underline{F}}$ and therefore $f \in \underline{\underline{F}}^{\sim}$ and $\gamma f = \varphi$. Thus

Theorem 10.2. Let f be bounded and locally controlled and suppose that there exists bounded $\varphi \in \underline{\underline{H}}$ such that $Ⓐf + \kappa\mu\varphi$ is bounded.

(i) The terminal variable $\gamma_T f$ is well defined by (3.8).

(ii) If $\gamma_T f = \gamma_T \varphi$, then $f \in \underline{\underline{F}}^{\sim}$ and $\varphi = \gamma f$.

(iii) If in addition f satisfies the boundary condition (10.4), then $f \in$ domain $A^{\sim}$. ///

Again we do not formulate an analogue to Theorem 8.4 but instead refer to Section 21 for detailed treatment of an example.

11. Boundary Conditions when $\Delta_s \neq \phi$.

We assume now that the singular boundary Δ_s is nonempty. The Dirichlet space $(\underline{\underline{F}}^{\sim}, E^{\sim})$ is never an extension in the language of SMP, Section 20, but it is covered by the Third Structure Theorem in SMP, Section 21. However, we cannot by the construction in Section 3 obtain all of the Dirichlet spaces covered there. (See Case 6 in Section 12 below.)

We first examine the L^2 generator A^e for the expanded Dirichlet space $(\underline{\underline{F}}^e, E^e)$.

Theorem 11.1. *The $L^2(\underline{\underline{X}} \cup \Delta, dx+\nu)$ generator A^e for $(\underline{\underline{F}}, E)$ is defined by*

$$A^e f(x) = \textcircled{A} f(x) + \kappa(x)\mu\gamma f(x) \quad \text{on } \underline{\underline{X}} \tag{11.1}$$

$$\int \nu(dy) A^e f(y)\psi(y) = (\partial f/\partial n)(\psi) - \{Q-N\}(\gamma f, \psi) \quad \text{for bounded } \psi \in \underline{\underline{H}} \tag{11.2}$$

supplemented by the following two conditions.

(i) The restriction of f to $\underline{\underline{X}}$ is locally controlled and satisfies (10.2).

(ii) The restriction γf of f to Δ belongs to $\underline{\underline{H}}$. Moreover, $\gamma_T f = \gamma_T \gamma f$ and (10.3) is valid. ///

Proof. Any $f \in \underline{\underline{F}}^e$ certainly satisfies (i) and (ii). If also $f \in$ domain A^e, then for any $g \in \underline{\underline{F}}^e$

$$\begin{aligned} -\int dx A^e f(x) g(x) - \int \nu(dy) A^e f(y) g(y) &= D(f_0, g_0) + \tfrac{1}{2} J\langle f_0, g_0 \rangle \\ &+ \int dx \kappa(x) r(x) f_0(x) g_0(x) + \int dx \kappa(x) \int \mu(x, dy)\{f_0(x) - \gamma f(y)\}\{g_0(x) - \gamma g(y)\} \\ &+ \{Q-N\}(\gamma f, \gamma g). \end{aligned} \tag{11.3}$$

Taking $g \in \underline{\underline{F}}^e \cap C_{com}(\underline{\underline{X}}) = \underline{\underline{F}} \cap C_{com}(\underline{\underline{X}})$ so that $\gamma g = 0$ we get

$$-\int dx A^{e} f(x) g(x) = D(f_0, g_0) + \tfrac{1}{2} J\langle f_0, g_0\rangle + \int dx\, \kappa(x)\{f_0(x) - \mu\gamma f(y)\} g_0(x)$$

and (11.1) follows. Taking $g = H\psi$ with ψ bounded we get

$$-\int_{\underline{\underline{X}}} dx \{A f_0(x) + \kappa\mu\gamma f(x)\} H\psi(x) - \int_{\Delta} \nu(dy) A^{e} f(y)\psi(y)$$

$$= D(f_0, H\psi) + \tfrac{1}{2} J\langle f_0, H\psi\rangle + \int dx\, \kappa(x) r(x) f_0(x) H\psi(x)$$

$$+ \int dx\, \kappa(x) \int \mu(x, dy)\{f_0(x) - \gamma f(y)\}\{H\psi(x) - \psi(y)\} + \{Q - N\}(\gamma f, \psi)$$

which after rearrangement and application of (7.12) implies (11.2). Conversely, if f satisfies (i) and (ii), then as in the proof of Theorem 5.2 the difference $f - H\gamma f$ belongs to $\underline{\underline{F}}$ and therefore f itself belongs to $\underline{\underline{F}}^{(e)}$. Finally, (11.1) and (11.2) (interpreted by our conventions) guarantee that $f \in$ domain A^{e}. ///

We will not bother here with the appropriate analogue to Theorem 10.2. Instead we concentrate our attention now on the generator $A^{\sim}$ for the Dirichlet space $(\underline{\underline{F}}^{\sim}, E^{\sim})$. At present we can obtain satisfactory results only for f bounded and only after imposing the integrability condition

(11.4)
$$\int dx\, \kappa(x) \int_{\Delta_s} \mu(x, dy) < +\infty .$$

We do not know of any examples where (11.4) fails. It follows from Theorem 4.3 (ii) that there exists $\psi \in \underline{\underline{H}}_{(e)}$ positive on Δ_s and vanishing on Δ_r such that $U_{0,\infty}(1, \psi^2) < +\infty$ which implies

(11.5)
$$\int_{\underline{\underline{X}}} dx\, \kappa(x) H1(x) \int_{\Delta_s} \mu(x, dy)\psi^2(y) < +\infty .$$

The argument at the end of the proof of Theorem 7.2 can be used to extend (11.5) to

(11.5')
$$\int_{\underline{\underline{X}}} dx\, \kappa(x) \int_{\Delta_s} \mu(x, dy)\psi^2(y) < +\infty .$$

Thus (11.4) willfollow if the function ψ^2 is bounded away from 0. This is the case for example if the singular boundary Δ_s is a finite set.

The restriction (11.4) has two effects. The most important is that if $\psi \in \gamma_r \underline{\underline{H}}$ is bounded, then the difference $HK\psi - H\psi$ is the potential of a measure with finite energy and therefore

$$HK\psi - H\psi \in \underline{\underline{F}} \; ; \; \psi \in \underline{\underline{H}}_b. \tag{11.6}$$

The second effect is that various terms which occur below converge separately and therefore can be easily manipulated.

__Theorem 11.2.__ _Assume that the integrability condition_ (11.4) _is satisfied. Then the_ L^2 _generator_ $A^\sim$ _for_ $(\underline{\underline{F}}^\sim, E^\sim)$ _is defined on bounded functions_ f _by_

$$A^\sim f(x) = Af(x) + \kappa(x)\mu Mf(x) + \kappa(x)\mu D_\infty \gamma_r f(x) \tag{11.7}$$

supplemented by the following three conditions.

(i) f _is locally controlled and satisfies_ (10.2).

(ii) _There exists_ $\gamma_r f \in \gamma_r \underline{\underline{H}}$, _necessarily unique, such that_ $\gamma_T f = \gamma_T \gamma_r f$ _and such that_

$$\int dx \kappa(x) \int_{\Delta_r} \mu(x, dy)\{f(x) - \gamma_r f(y)\}^2 < +\infty. \tag{11.8}$$

(iii) _Every bounded_ $\psi \in \gamma_r \underline{\underline{H}}$ _belongs to the domain of the normal derivative_ $\partial f/\partial n$ _and_

$$(\partial f/\partial n)(\psi) + \int dx \int \kappa(x) \int_{\Delta_s} \mu(x, dz) \int_{\Delta_r} D_\infty(z, dy)\{f(x) - \gamma_r f(y)\}\psi(y) = G(\gamma_r f, \psi). \tag{11.9}$$

///

The term $\mu D_\infty \gamma_r f$ in (11.7) is the appropriate analogue for $\mu\gamma f$ in (10.1). The term $\kappa\mu Mf(x)$ should be viewed as a contribution to the "jumping part" of the local generator. The right side of (11.9) is defined by (5.40). It is analogous to the right side of (10.4), but unlike the latter, it is not given directly in the prescription for the construction in Section 3. Instead it must be computed.

Proof. By Theorem 5.9 any $f \in \underline{\underline{F}}^{\sim}$ and therefore any $f \in$ domain $A^{\sim}$ satisfies (i) and (ii), and (5.39') together with (9.2) gives for general $g \in \underline{\underline{F}}^{\sim}$

(11.10)
$$-\int dx A^{\sim} f(x)g(x) = D(f,g) + \tfrac{1}{2}J\langle f,g\rangle + \int dx \kappa(x) r(x) f(x) g(x) + \tfrac{1}{2}\kappa\mu M\langle f,g\rangle + \int dx \kappa(x)\mu\{1 - M1 - D_{\infty}1\}(x) f(x) g(x) + \iint \kappa\mu D_{\infty}(dx,dy)\{f(x) - \gamma_r f(x)\}\{g(x) - \gamma_r g(x)\} + G(\gamma_r f, \gamma_r g).$$

For f, g bounded

(11.11)
$$\tfrac{1}{2}\kappa\mu M\langle f,g\rangle = \int dx \kappa(x)\{f(x)g(x)\mu M1(x) - \mu Mf(x)g(x)\},$$

and so (11.10) is equivalent to

(11.11')
$$-\int dx A^{\sim} f(x)g(x) = D(f,g) + \tfrac{1}{2}J\langle f,g\rangle - \int dx \kappa(x)\mu Mf(x)g(x) + \int dx \kappa(x)\{r(x) + \mu(1 - D_{\infty}1)(x)\} f(x) g(x) + \iint \kappa\mu D_{\infty}(dx,dy)\{f(x) - \gamma_r f(y)\}\{g(x) - \gamma_r g(y)\} + G(\gamma_r f, \gamma_r g).$$

(The identity (11.11') depends on the symmetry result, Lemma 5.3.) Taking $g \in \underline{\underline{F}} \cap C_{com}(\underline{\underline{X}})$ we get

$$-\int dx A^{\sim} f(x)g(x) = D(f,g) + \tfrac{1}{2}J\langle f,g\rangle - \int dx \kappa(x)\mu Mf(x)g(x) + \int dx \kappa(x)\{f(x) - \mu D_{\infty}\gamma_r f(x)\} g(x)$$

and (11.7) follows. Taking $g = H\psi$ with $\psi \in \gamma_r \underline{\underline{H}}$ bounded we get

$$-\int dx\{\textcircled{A}f(x)+\kappa\mu Mf(x)+\kappa\mu D_\infty\gamma_r f(x)\}H\psi(x)$$

$$= D(f, H\psi) + \tfrac{1}{2}J\langle f, H\psi\rangle - \int dx\,\kappa(x)\mu Mf(x)H\psi(x)$$

$$+ \int dx\,\kappa(x)\{r(x)+\mu(1-D_\infty 1)(x)\}f(x)H\psi(x)$$

$$+ \int\int \kappa\mu D_\infty(dx, dy)\{f(x)-\gamma_r f(y)\}\{H\psi(x)-\psi(y)\}$$

$$+ G(\gamma_r f, \psi)$$

and after rearrangement and cancellation

$$-[\int dx\{\textcircled{A}f(x)+\kappa f(x)\}H\psi(x)-\int dx\,\kappa(x)\int_{\Delta_r}\mu(x, dy)\{f(x)-\gamma_r f(y)\}\psi(y)]$$

$$-D(f, H\psi)-\tfrac{1}{2}J\langle f, H\psi\rangle$$

$$= -\int dx\,\kappa(x)\int_{\Delta_s}\mu(x, dz)\int_{\Delta_r}D_\infty(z, dy)\{f(x)-\gamma_r f(y)\}\psi(y)$$

$$+G(\gamma_r f, \psi),$$

and (11.9) follows by Theorem 7.1. The converse follows in the usual way with the help of Theorem 5.9. ///

Assume now that f is locally controlled and bounded and that there exists bounded $\varphi \in \gamma_r\underline{\underline{H}}$ such that $\textcircled{A}f+\kappa\mu Mf+\kappa\mu D_\infty\varphi$ is bounded. Then $\textcircled{A}\{f-N\kappa\mu Mf-HD_\infty\varphi\} = \textcircled{A}f-N\kappa\mu Mf+\kappa\mu D_\infty f$ is bounded and therefore $\gamma_T f$ is defined. Suppose now that $\gamma_T f = \gamma_T\varphi$. The condition (11.4) guarantees that $N\kappa\mu Mf \in \underline{\underline{F}}$ and since $\gamma_T\{f-N\kappa\mu Mf-HD_\infty\varphi\} = 0$ it follows that $f \in \underline{\underline{F}}^\sim$ and $\varphi = \gamma_r f$. Thus

<u>Theorem 11.3.</u> <u>Assume that the integrability condition</u> (11.4) <u>is satisfied. Let</u> f <u>be bounded and locally controlled, and suppose that there exists bounded</u> $\varphi \in \gamma_r\underline{\underline{H}}$ <u>such that</u> $\textcircled{A}f+\kappa\mu Mf+\kappa\mu D_\infty\varphi$ <u>is bounded.</u>

(i) <u>The terminal variable</u> $\gamma_T f$ <u>is well defined by</u> (3.8).

(ii) <u>If</u> $\gamma_T f = \gamma_T\varphi$, <u>then</u> $f \in \underline{\underline{F}}^\sim$ <u>and</u> $\varphi = \gamma_r f$.

(iii) *If in addition* f *satisfies the boundary condition* (11.9), *then* $f \in$ *domain* $A^{\sim}$. ///

Theorems 11.2 and 11.3 are awkward in that the conclusion involves the operators M and D_∞ and also the bilinear form G which are not given directly. This is avoided in the following reformulation. (However, see cases 4 and 5 in Section 12.)

Theorem 11.4. *Assume that the integrability condition* (11.4) *is satisfied and let* f *be bounded and locally controlled. Then* f *belongs to the domain of* $A^{\sim}$ *and* $A^{\sim}f$ *is bounded if and only if there exists bounded* $\varphi \in \underline{\underline{H}}$ *satisfying the following conditions.*

(i) $Ⓐf+\kappa\mu\varphi$ *is bounded.*

(ii) $\gamma_T f = \gamma_T \varphi$.

(iii) *Every bounded* $\psi \in \underline{\underline{H}}$ *belongs to the domain of the normal derivative* $\partial f/\partial n$ *and*

(11.12)
$$(\partial f/\partial n)(\psi) = \{Q-N\}(\varphi,\psi).$$

If f *satisfies these conditions, then*

(11.13)
$$A^{\sim}f = Ⓐf+\kappa\mu\varphi. \qquad ///$$

Proof. If $f \in$ domain $A^{\sim}$ with $A^{\sim}f$ bounded, then Theorem 11.2 implies (i) and (ii) with $\varphi = Mf+D_\infty\gamma_r f$. Conclusion (iii) follows from (11.2) since by the results in SMP, Sections 7, 8, the pair $(f,\varphi) \in L^2(\underline{\underline{X}}\cup\Delta, dx+\nu)$ belongs to domain A^e and $A^e(f,\varphi) = 0$, a.e. ν on Δ. Conversely, (i) and (ii) guarantee by the usual argument that $(f,\varphi) \in \underline{\underline{F}}^e$ and therefore $f \in \underline{\underline{F}}^{\sim}$ and then with the help of (iii) and the results in SMP, Sections 7 and 8, it is easy to check that $f \in$ domain $A^{\sim}$. ///

In Theorem 11.4 the boundary conditions involve the entire boundary Δ and not just the regular part Δ_r as was the case in Theorem 11.2 and Theorem 11.3. The only difference with the situation in Section 10 is that in general there is no intrinsic connection between f and the values of the boundary function φ on Δ_s. Even this distinction disappears if certain regularity conditions are satisfied. Then if $f \in$ domain $\tilde{A}$ and if both f and $\tilde{A} f$ are bounded, f has a unique continuous extension to $\underline{\underline{X}} \cup \Delta$ and φ is just the usual restriction of f to φ. This explains why in [52] there is no difference between the cases $0 < \alpha < 1$ and $1 < \alpha < 2$. We refer to Section 19 below for a discussion of the situation in [52] and to Section 21 for a discussion in a more complicated context.

Appendix A. Regularity of the Process.

In this appendix we prove that the process $X_t^{\sim}$ constructed in Section 3 is a "Hunt process modulo a polar set"--that is, it has the properties prescribed in SMP, Theorems 4.8 through 4.17. At the same time we justify (5.32) in the general case. The point is that the results in SMP, Chapter I, are directly applicable only when the relevant Dirichlet spaces are regular.

To achieve complete generality we begin by studying the expanded process $\{X_t^e\}$. We then verify that our results on random time change in SMP, Section 8, are applicable and in particular that (5.32) is valid. Finally, we establish the desired properties for the process $\{X_t^{\sim}\}$. In the special case when the singular boundary $\Delta_s = \phi$ it would be possible, and certainly more efficient to work directly with the process $\{X_t^{\sim}\}$.

The reference Hilbert space for the expanded Dirichlet space is $L^2(\underline{\underline{X}} \cup \Delta, dx+\nu)$. In general $(\underline{\underline{F}}^e, E^e)$ is not transient and so we use the Dirichlet form

$$E_1^e(f,f) = E^e(f,f) + \int_{\underline{\underline{X}}} dx f^2(x) + \int_{\Delta} \nu(dy) f^2(y)$$

in place of E^e.

We begin by choosing a particular regularizing space for $(\underline{\underline{F}}^e, E_1^e)$ in the sense of SMP, Section 19. Let $\underline{\underline{B}}_0^e$ be a subset of $\underline{\underline{F}}^e$ satisfying the conditions 2.2.1 through 2.2.4 in SMP except that $\underline{\underline{F}}$ is replaced by $\underline{\underline{F}}^e$. With $\underline{\underline{H}}_0$ as in paragraph 3.5 we assume further that $\underline{\underline{B}}_0^e$ contains $H_1\underline{\underline{H}}_0$ and also a countable dense subset of $\underline{\underline{F}} \cap C_{com}(\underline{\underline{X}})$ and is contained in the algebra generated by the vector space sum $\underline{\underline{F}} \cap C_{com}(\underline{\underline{X}}) + H_1\underline{\underline{H}}_0$. It is easy to see that such a choice is possible. Our regularizing space $\underline{\underline{Y}}$ is the maximal ideal space for $\underline{\underline{B}}_0^e$. At the same time we identify δ with the trivial homomorphism so that $\underline{\underline{Y}} \cup \{\delta\}$ is compact. The space $\underline{\underline{X}}^{\sim} = \underline{\underline{X}} \cup \Delta$ can be identified with a subset of $\underline{\underline{Y}}$ in an obvious way, and the subspace topology on $\underline{\underline{X}} \cup \Delta \cup \{\delta\}$ is identical with the topology prescribed in Section 3. Furthermore $\underline{\underline{X}}$ is open and

$\Delta \cup \{\delta\}$ is compact in $\underline{\underline{Y}} \cup \{\delta\}$ and therefore Δ is closed in $\underline{\underline{Y}}$. In particular $\underline{\underline{X}}^{\sim}$ is Borel in $\underline{\underline{Y}}$. Thus $dx+\nu$ can be viewed also as a measure on $\underline{\underline{Y}}$ which must be Radon and $(\underline{\underline{F}}^e, E^e)$ becomes a regular Dirichlet space on $(\underline{\underline{Y}}, dx+\nu)$. The crucial tool for establishing regularity of the process $\{X_t^e\}$ is

<u>Theorem A.1.</u> (i) <u>The complement $\underline{\underline{Y}} \backslash \underline{\underline{X}}^{\sim}$ is polar for $(\underline{\underline{F}}^e, E_1^e)$.</u>

(ii) <u>A subset of $\underline{\underline{X}}$ is polar for $(\underline{\underline{F}}^e, E_1^e)$ if and only if it is polar for</u> $(\underline{\underline{F}}, E)$.

(iii) <u>The capacities for $(\underline{\underline{F}}^e, E_1^e)$ and $(\underline{\underline{H}}, Q_{(1)1})$ are identical on subsets of</u> Δ.

<u>Proof.</u> (iii) is a direct consequence of (5.30') with $u = 1$ and (ii) follows as in the proof of SMP, Lemma 10.2(ii). For (i) it suffices by SMP, Corollary 3.13 to show that if f is an $(\underline{\underline{F}}^e, E_1^e)$ potential then there exist Radon measures μ_0, μ_1 on $\underline{\underline{X}}$ and Δ such that

$$E_1^e(f, g) = \int_{\underline{\underline{X}}} \mu_0(dx) g(x) \quad \text{for } g \in \underline{\underline{F}} \cap C_{com}(\underline{\underline{X}}) \tag{A.1}$$

$$E_1^e(f, H_1\varphi) = \int_{\underline{\underline{X}}} \mu_0(dx) H_1\varphi(x) + \int_{\Delta} \mu_1(dy)\varphi(y) \quad \text{for } \varphi \in \underline{\underline{H}} \cap C_{com}(\Delta). \tag{A.2}$$

The existence of μ_0 follows directly from the proof of Lemma 3.1 in SMP applied to $(\underline{\underline{F}}, E_1)$ and then (A.1) extends to general $g \in \underline{\underline{F}}$. For $\varphi \geq 0$ in $\underline{\underline{H}} \cap C_{com}(\Delta)$ we have

$$E_1^e(f, H_1\varphi) = \operatorname{Lim}_{v \uparrow \infty} E_1^e(f, vG_{v+1}^e H_1\varphi)$$

which by (5.26) and (5.30') for $u = 1$

$$= \operatorname{Lim}_{v \uparrow \infty} \{E_1^e(f, vG_{v+1} H_1\varphi) + Q_{(1)1}(\gamma f, vR_{(v+1), v+1}(\pi_{v+1}+1) H_1\varphi)\}$$

with both terms inside { } being nonnegative. Since $vG_{v+1}H_1\varphi \uparrow H_1\varphi$ quasi-everywhere and therefore [a. e. μ_0] on $\underline{\underline{X}}$ it follows that

$$E_1^e(f, H_1\varphi) \geq \int_{\underline{\underline{X}}}\mu_0(dx)H_1\varphi(x).$$

This together with another application of the proof in SMP, this time to $(\underline{\underline{H}}, Q_{(1)1})$, establishes the existence of μ_1, and the theorem is proved. ///

By the results in SMP, Sections 3 and 4, there exists a process $\{Y_t\}$ with sample space probabilities Q_x taking values in $\underline{\underline{Y}} \cup \{\delta\}$ which corresponds to the Dirichlet space $(\underline{\underline{F}}^e, E^e)$ and which is regular in the sense of the first paragraph. Since $\underline{\underline{Y}} \setminus \underline{\underline{X}}^{\sim}$ is polar, it follows from SMP, Proposition 4.15, together with an argument like the one preceding SMP, Lemma 4.1, that the process $\{Y_t\}$ can be viewed as taking values in $\underline{\underline{X}}^{\sim} \cup \{\delta\}$. To establish regularity for $\{X_t^e\}$ it suffices to show that for quasi-every $x \in \underline{\underline{X}}^{\sim}$ the trajectories $X_{\cdot}^e$ and $Y_{\cdot}$ are identically distributed relative to P_x^e and Q_x respectively. Since $\{X_t^e\}$ and $\{Y_t\}$ both have right continuous trajectories, it suffices to show that for each $t > 0$ and for quasi-every $x \in X$ the variables X_t^e and Y_t have the same distribution relative to P_x^e and Q_x. But this is clear since by (5. 31) the resolvent operators agree modulo polar sets. (Notice that the right side of (5. 26) belongs to $\underline{\underline{F}}^{(e)}$ and therefore is unique modulo a polar set. Also it is important for the above argument that the subspace topology on $\underline{\underline{X}}^{\sim}$ agrees with the given one.) The results on random time change in SMP, Section 8, are certainly applicable to the process $\{Y_t\}$ and therefore also to $\{X_t^e\}$, which suffices to justify (5. 32). It only remains to establish regularity for the process $\{X_t^{\sim}\}$. If $\underline{\underline{X}} \cup \Delta_r$ is closed in $\underline{\underline{Y}}$ then $(\underline{\underline{F}}^{\sim}, E^{\sim})$ is regular and we can argue as above for $\{X_t^e\}$ to establish regularity. Finally, the general case follows since we can always consider instead an equivalent representation for $(\underline{\underline{F}}^e, E^e)$ such that the associated quasi-homeomorphism in the sense of SMP, Section 19, maps $\underline{\underline{X}} \cup \Delta_r$ into a closed set (modulo a polar set).

Chapter III. Examples Related to Brownian Motion

For all the examples in this chapter, the original Dirichlet space $(\underline{\underline{F}}, E)$ corresponds to absorbing barrier Brownian motion on an appropriate space $\underline{\underline{X}}$, sometimes with killing in the interior. The only exception to this pattern is Section 16 where we prove a representation theorem of A. Beurling and J. Deny for Dirichlet spaces on Euclidean space.

In Sections 12 and 13 the space $\underline{\underline{X}}$ is one-dimensional, and our intention is to illustrate as much as possible in a context where the technicalities are minimal. We consider seven cases in succession in Section 12 with $\underline{\underline{X}} = (0,1)$. The first three cases are very similar to the situation studied by Feller [14] except that we allow jumps to and from the boundary. In Cases 4 and 5 an extra piece is added to the natural 2 point boundary. For Case 4 it is a singleton, and for Case 5 a line segment carrying a second Brownian motion. In both cases the extra piece is necessarily singular, and the net effect both at the sample space level and the Dirichlet space level is essentially the same. Also in both cases the descriptions of the generator given by Theorems 11.2 and 11.3 seem to be the natural ones. (This contrasts with the situation in Chapter IV.) The two descriptions are more or less the same, but there is an interesting technical distinction. For Case 4 the description given in Theorem 11.4 can be transformed into the other one by a direct substitution. For Case 5 this would not be so easy, and it seems more natural instead to apply the technique of Theorems 11.2 and 11.3. In Cases 6 and 7 we look briefly at the case when the killing intensity κ is not integrable. Perhaps the most interesting point is that for Case 7 the terminal boundary is empty, and therefore the ordinary derivatives necessarily vanish at the endpoints (when f is a "candidate" for belonging to the domain of the generator) and therefore do not appear in the actual boundary conditions. In Section 13 we let $\underline{\underline{X}} = [0, \infty)$. The main new feature is that if κ is not integrable, then the regular boundary can be quite complicated. In fact, we construct an example where the regular boundary is a full interval carrying a second Brownian motion. This should be contrasted with Case 5 in Section 12 where the adjoined interval is singular. Also we study in some detail the much less interesting situation where κ is integrable. First we wish to compare our present techniques with the rather pedestrian approach used in SMP, Section 24. Second, we show that the topology prescribed at the beginning of Section 3 can be rather artificial in practice. (The appendix at the end of Chapter II guarantees that this makes no difference.)

In Section 14 we look briefly at standard Brownian motion on a full Euclidean space. The novelty here is the notion of a "normal derivative at infinity" and its connection with $\int dx \Delta f(x)$.

In Section 15 we study Brownian motion in the upper half plane. This is a good example to work out in detail since the boundary theory is nontrivial, and yet it is relatively easy to make calculations. In particular, we derive an explicit formula for the normal derivative. Also at the end of the section we formulate in precise terms the technical problem which must be resolved before one can give a complete classification of extensions.

Section 16 contains a proof of the representation theorem of Beurling and Deny. This states that every Dirichlet norm on Euclidean space must be of the form (0.8). Little seems to be known about which expressions can actually occur as Dirichlet norms. In Section 17 we show that a special case of this problem is equivalent to the technical problem which is formulated at the end of Section 15.

12. Brownian Motion on (0, 1).

Throughout this section $\underline{\underline{X}} = (0,1)$, the open unit interval and dx is normalized Lebesgue measure. The diffusion form is the classical one

$$D(f,f) = \frac{1}{2}\int_0^1 dx \, \{f'(x)\}^2 \tag{12.1}$$

where $f'(x)$ denotes the ordinary derivative. The Levy measure $J = 0$.

Our purpose in this section is to illustrate as much as possible in this relatively simple context. After developing some general theory we will consider in succession seven different cases. We recommend Feller's paper [14] as a general reference for analysis of the differential equations which occur below.

12.1. Notation. The Sobolev space $\underline{\underline{W}}$ is the collection of functions F defined and absolutely continuous on the closed interval [0.1] such that $f'(x) \in L^2(\underline{\underline{X}}, dx)$. ///

The reflected space

$$\underline{\underline{F}}^{ref} = \underline{\underline{W}} \cap L^2(\kappa) .$$

and the Dirichlet norm E is defined on $f \in \underline{\underline{F}}^{ref}$ by

$$E(f,f) = \frac{1}{2}\int_0^1 dx \, \{f'(x)\}^2 + \int_0^1 dx \, \kappa(x) f^2(x) . \tag{12.2}$$

The Dirichlet space $\underline{\underline{F}}$ is the E-closure of $\underline{\underline{W}} \cap C_{com}(\underline{\underline{X}})$. From the formula

$$f(x) = \int_0^x dt \, f'(t) = - \int_x^1 dt \, f'(t)$$

valid for $f \in \underline{\underline{W}} \cap C_{com}(\underline{\underline{X}})$ follows the estimate

(12.3) $$|f(x)| \leq E^{\frac{1}{2}}(f,f)$$

for $f \in \underline{\underline{F}}$ which guarantees on the one hand that singletons are nonpolar for $(\underline{\underline{F}}, E)$ and guarantees on the other hand for $f \in \underline{\underline{F}}$ the boundary conditions

(12.4) $$f(0) = 0 \; ; \; f(1) = 0 \; .$$

By partitioning and applying convolutions it is easy to show that (12.4) characterizes those $f \in \underline{\underline{F}}^{ref}$ which belong to $\underline{\underline{F}}$. However it is also true that (12.4) is valid for all $f \in \underline{\underline{F}}^{ref}$ and therefore irrelevant for characterizing $\underline{\underline{F}}$ at an end point near which $\kappa(x)$ is not integrable. In particular $\underline{\underline{F}} = \underline{\underline{F}}^{ref}$ when κ is integrable near neither end point. Since every $f \in \underline{\underline{F}}^{ref}$ is continuous

$$\underline{\underline{F}}_{(e)} = \underline{\underline{F}} \; ; \; \underline{\underline{F}}^{ref} = \underline{\underline{F}}^{ref}_a \; .$$

Clearly f is "locally controlled" and satisfies (3.7) if and only if $f \in \underline{\underline{W}}$. Therefore all of the Dirichlet spaces considered in this section will be contained in $\underline{\underline{W}}$.

Since $J = 0$ the associated process has continuous trajectories. (See SMP, Theorem 11.10.) Indeed it is easy to check that, in the language of [32], the process is the absorbing barrier diffusion on (0,1) with standard scale and speed measure and with killing measure $\kappa(x)\,dx$.

Next we identify the harmonic functions as defined in SMP, Section 14. Let $D = (a,b)$ with $0 < a < b < 1$ and $M = \underline{\underline{X}} \setminus D$. By continuity of trajectories any harmonic h can be represented on D

(12.5) $$h(x) = H^M h(x) = h(a) p_{ab}(x) + h(b) p_{ba}(x)$$

where

$$p_{ab}(x) = \text{Ⓔ}_x[\sigma_a<\sigma_b] \; ; \; p_{ba}(x) = \text{Ⓔ}_x[\sigma_b<\sigma_a] \, .$$

(In defining hitting times we are using a, b rather than the more clumsy $\{a\}, \{b\}$ as labels.) In particular there exists $h^D \in \underline{\underline{F}}$ such that $H^M h = H^M h^D$ on D and so by SMP, Theorem 7.3

(12.6) $$\frac{1}{2}\int_0^1 dx\, h'(x)g'(x) + \int_0^1 dx\, \kappa(x)h(x)g(x) = 0$$

whenever $g \in \underline{\underline{W}} \cap C_{com}(D)$ and therefore, since D is arbitrary, whenever $g \in \underline{\underline{W}} \cap C_{com}(\underline{\underline{X}})$. But it is easy to see that (12.6) is equivalent to the equation

(12.7) $$\frac{1}{2}h''(x) - \kappa(x)h(x) = 0$$

interpreted in the distribution sense. This means that h is continuously differentiable, h' is absolutely continuous, and (12.7) is valid almost everywhere. Thus every harmonic function satisfies (12.7) on $\underline{\underline{X}}$. Similarly the functions p_{ab}, p_{ba} satisfy (12.7) on (a, b). They also satisfy the boundary conditions

(12.8) $$p_{ab}(a+0) = p_{ba}(b-0) = 1 \; ; \; p_{ab}(b-0) = p_{ba}(a+0) = 0 \, ,$$

which is easily proved using the martingale convergence theorem. For example, for any initial point $x \in D$ the process $\{p_{ab}(X_t)\}_{t<\sigma(M)}$ is a bounded martingale and therefore $p_{ab}(X_t) \to 1$ as $t \uparrow \sigma_a$ on the set $[\sigma_a<\sigma_b]$ which clearly implies $p_{ab}(a+0) = 1$. By the theory of the equation (12.7), every solution h of (12.7) is a linear combination of p_{ab} and p_{ba} on (a, b) and therefore $h = H^M h$. Thus (12.7) actually characterizes harmonic functions for $(\underline{\underline{F}}, E)$. It is also clear from the theory of (12.7) that p_{ab}, p_{ba} are the unique solutions of (12.7) which satisfy the boundary conditions (12.8). By (12.2) we have for $f \in \underline{\underline{F}}$ and $x \in \underline{\underline{X}}$

$$\frac{1}{2}\int_0^1 dt\, f'(t)\, \partial_t N(t,x) + \int_0^1 dt\, \kappa(t) f(t) N(t,x) = f(x)$$

with ∂_t denoting the partial derivative with respect to t. In particular this is valid for $f \in C^{\infty}_{com}(\underline{\underline{X}})$ and therefore in the distribution sense

(12.9) $$\frac{1}{2}\partial_t \partial_t N(t,x) - \kappa(t) N(t,x) = -\epsilon_x$$

Thus $N(x,t)$ is the classical Green function for (12.7) with boundary condition (12.4). It is well-known that $N(t,x)$ can be written out explicitly in terms of solutions of the homogeneous equation (12.7). We give now a derivation of this result, at the same time proving some related well-known results which will be used below.

Let ψ be any solution of (12.7) which is positive and bounded away from 0 on $\underline{\underline{X}}$. Such ψ always exists since $\kappa \geq 0$. Define

(12.10) $$\eta_0(x) = \psi(x) \int_x^1 dt\, \psi^{-2}(t)$$

$$\eta_1(x) = \psi(x) \int_0^x dt\, \psi^{-2}(t)$$

$$C_\psi = \{ \int_0^1 dt\, \psi^{-2}(t) \}^{-1} .$$

It is easy to check that η_0, η_1 satisfy the homogeneous equation (12.7) and the boundary conditions

(12.11) $$\eta_0(1) = \eta_1(0) = 0 .$$

Also it is not hard to see that η_0, η_1 are determined by this up to constant multiples and therefore varying ψ only replaces η_0, η_1 by constant multiples.

By direct calculation

$$C_\psi = \{\eta_0(x)\eta_1'(x) - \eta_0'(x)\eta_1(x)\}^{-1} \tag{12.12}$$

for all $x \in \underline{\underline{X}}$. The potential kernel can be represented

$$N(t,x) = 2C_\psi \eta_1(t\wedge x)\eta_0(t\vee x). \tag{12.13}$$

To verify this fix x and note first that

$$\begin{aligned} \partial_t N(t,x) &= 2C_\psi \eta_1'(t)\eta_0(x) && \text{for } t < x \\ \partial_t N(t,x) &= 2C_\psi \eta_0'(t)\eta_1(x) && \text{for } t > x\,. \end{aligned} \tag{12.14}$$

By convexity $\eta_1'(t)$ is bounded near 0 and $\eta_0'(t)$ is bounded near 1 and therefore $N(\cdot,x) \in \underline{\underline{W}}$. For the same reason

$$\int_0^x dt\, \kappa(t)\{\eta_1(t)\}^2 = \int_0^x dt\, \eta_1''(t)\eta_1(t)\ ;\ \int_x^1 dt\, \kappa(t)\{\eta_0(t)\}^2 = \int_x^1 dt\, \eta_0''(t)\eta_0(t)$$

both converge and therefore $N(\cdot,x) \in \underline{\underline{F}}^{\mathrm{ref}}$. The boundary conditions (12.4) are clear and so actually $N(\cdot,x) \in \underline{\underline{F}}$. Finally we differentiate (12.14) in the distribution sense to get

$$\frac{1}{2}\partial_t\partial_t N(t,x) - \kappa(t)N(t,x) = C_\psi\{\eta_0'(x)\eta_1(x) - \eta_0(x)\eta_1'(x)\}\epsilon_x$$

and (12.9) follows.

Next we note the equivalences

$$\begin{aligned} \eta_0'(0) &> -\infty && \text{iff} && \int_0^{\frac{1}{2}} dt\, \kappa(t) < +\infty \\ \eta_1'(1) &< +\infty && \text{iff} && \int_{\frac{1}{2}}^1 dt\, \kappa(t) < +\infty \end{aligned} \tag{12.15}$$

$$(12.16)\qquad \eta_0(0) < +\infty \quad \text{iff} \quad \int_0^{\frac{1}{2}} dt\, \kappa(t)t < +\infty$$

$$\eta_1(1) < +\infty \quad \text{iff} \quad \int_{\frac{1}{2}}^{1} dt\, \kappa(t)(1-t) < +\infty$$

To our knowledge it was Feller [14] who first discovered (12.16) and understood the significance of (12.15) and (12.16) for boundary value problems. We refer either to [14] or to SMP, page 23.3 for a proof.

Suppose for the moment

$$(12.17)\qquad \int_0^{\frac{1}{2}} dt\, \kappa(t)t < +\infty \quad ; \quad \int_{\frac{1}{2}}^{1} dt\, \kappa(t)(1-t) < +\infty$$

and therefore the following functions are well defined

$$(12.18)\qquad p_0(x) = \eta_0(x)/\eta_0(0) \quad ; \quad p_1(x) = \eta_1(x)/\eta_1(1)$$

and are the unique solutions of the equation (12.7) satisfying the boundary conditions

$$(12.19)\qquad p_0(1) = p_1(0) = 0 \quad ; \quad p_0(0) = p_1(1) = 1\,.$$

Clearly (12.13) can be replaced by

$$(12.13')\qquad N(t,x) = 2Cp_1(t\wedge x)p_0(t\vee x)$$

where now

$$(12.12')\qquad C = \{p_0(x)p_1'(x) - p_0'(x)p_1(x)\}^{-1}$$

Moreover $C = \eta_0(0)\eta_1(1)C_\psi$ and it follows that

$$(12.20)\qquad C = \psi(0)\psi(1)\int_0^1 dt\, \psi^{-2}(t)$$

and in particular the right side of (12.20) is independent of the original choice of ψ. When either condition in (12.17) is violated we agree that the corresponding $p_i(x)$ vanishes identically and therefore (12.18) is still valid.

Now we are ready to consider the seven cases. In Cases 1 through 5 we assume that $\kappa(x)$ is integrable near both endpoints. Case 1 accounts for the "generic" two point boundary conditions. Case 2 corresponds to one absorbing boundary and Case 3 to periodic boundary conditions. Also under Case 1 we show that the terminal normal derivative of Section 5 corresponds to the usual derivative at the end point. For Cases 4 and 5 the singular boundary Δ_s is nonempty. We consider the two extremes when Δ_s is a singleton and when Δ_s is a separate interval carrying another Brownian motion. The net effect on the process $\{\tilde{X}_t\}$ is essentially the same. We also look at the corresponding expanded process. For Case 5 this illustrates in a nontrivial context a general procedure for "piecing together" processes.

In Case 6 we drop the condition that κ be integrable but retain the moment condition. The main difference is that restrictions must be placed on the jumping measure to avoid trivialities. In particular if the jumping measure is null, then nothing new can happen. Also we correct an error in SMP, Section 23 concerning the existence of normal derivatives.

In Case 7 we also drop the moment condition on κ. The new feature is that for all candidates f for the domain of the generator $\tilde{A}$ the terminal normal derivative must be 0 and therefore the ordinary derivative at the end point not only exists but must be 0. The boundary conditions simplify accordingly. Before proceeding we introduce a convenient notation.

12.3. Notation. $X^{(R)}_{\zeta=0}$ is the left-hand limit

$$X^{(R)}_{\zeta-0} = \mathrm{Lim}_{t\uparrow\zeta}\, X_t$$

with limits defined in the real line $\underline{\underline{R}}$ (and indeed in the closed interval $[0,1]$) rather than in the one point compactification of $\underline{\underline{X}} = (0,1)$. ///

The terminal set $\underline{\underline{T}}$ always splits into the two atoms $[X^{(R)}_{\zeta-0}=0], [X^{(R)}_{\zeta-0}=1]$ which may or may not be null.

Notice that the notation 12.3 is consistent with Section 1.

Case 1. We assume that $\kappa(x)$ is integrable on $\underline{\underline{X}}$. Then (12.17) is certainly satisfied and both p_0 and p_1 are nontrivial. By the arguments used above to analyze p_{ab} and p_{ba} the terminal set $\underline{\underline{T}}$ consists of the two non-null atom $[X^{(R)}_{\zeta-0}=0], [X^{(R)}_{\zeta-0}=1]$ and

(12.21) $$p_0(x) = Ⓟ_x[X^{(R)}_{\zeta-0}=0] \;;\; p_1(x) = Ⓟ_x[X^{(R)}_{\zeta-0}=1].$$

We take the auxiliary space $\underline{\underline{M}}$ to be empty and we let μ and r be arbitrary. As in Section 7 we use the notation

(12.22) $$\mu_0(x) = \mu(x, [X^{(R)}_{\zeta-0}=0]) \;;\; \mu_1(x) = \mu(x, [X^{(R)}_{\zeta-0}=1]).$$

For harmonic measure on $\underline{\underline{T}}$ we use counting measure rather than the measure defined by (3.1). The excursion form (3.2) becomes

(12.23) $$\begin{aligned} N(\Phi,\Phi) = {} & P(X^{(R)}_{\zeta^*}=0; X^{(R)}_{\zeta-0}=1)\{\Phi(0)-\Phi(1)\}^2 \\ & + (\mu_0\kappa N\kappa\mu_1)\{\Phi(0)-\Phi(1)\}^2 \\ & + (\mu_0\kappa p_1 + \mu_1\kappa p_0)\{\Phi(0)-\Phi(1)\}^2 \\ & + (r\kappa p_0)\Phi^2(0) + (r\kappa p_1)\Phi^2(1) \end{aligned}$$

(12.23 continued) $$+ (r\kappa N\kappa\mu_0)\Phi^2(0) + (r\kappa N\kappa\mu_1)\Phi^2(1) .$$

Here Φ is a terminal variable and $\Phi(i)$ is its value on the terminal atom $[X^{(R)}_{\zeta-0}=i]$. In (12.23) we are using notations such as

$$\mu_0\kappa N\kappa\mu_1 = \int dx \int dt\, \kappa(x)\mu_0(x)N(x,t)\kappa(t)\mu_1(t)$$

$$\mu_0\kappa p_1 = \int dx\, \kappa(x)\mu_0(x)p_1(x) .$$

Before continuing we look more carefully at the first term in (12.23). In the notation of paragraph 7.1 we put

(12.24) $$D_n = (\frac{1}{n}, 1-\frac{1}{n}) , \qquad n \geq 3.$$

The conditional equillibrium distributions $\ell^0_n(dx), \ell^1_n(dy)$ corresponding to the atoms $[X^{(R)}_{\zeta-0}=0]$ and $(X^{(R)}_{\zeta-0}=1]$ are concentrated respectively at $\frac{1}{n}$ and $1-\frac{1}{n}$. For notational convenience we define constants ℓ^0_n, ℓ^1_n by

$$\ell^0_n(dx) = \ell^0_n\, \epsilon_{1/n} ; \quad \ell^1_n(dx) = \ell^1_n\, \epsilon_{1-1/n} .$$

These constants are determined by

(12.25) $$p_0(1/n) = N(1/n, 1/n)\ell^0_n$$
$$p_1(1-1/n) = N(1-1/n, 1-1/n)\ell^1_n .$$

which by (12.13') can be simplified to

(12.25') $$2Cp_1(1/n)\ell^0_n = 1 \ ; \ 2Cp_0(1-1/n)\ell^1_n = 1 .$$

Now for each $n \geq 3$ clearly

$$\textcircled{P}[X^{(R)}_{\zeta^*}=0;X^{(R)}_{\zeta-0}=1] = \int \ell^0_n(dx)p_1(x) = \ell^0_n p_1(1/n) = p_0(1/n)p_1(1/n)N^{-1}(1/n,1/n)$$

and therefore by (12.13')

$$P[X^{(R)}_{\zeta^*}=0;X^{(R)}_{\zeta-0}=1] = (2C)^{-1} . \tag{12.26}$$

In particular (12.23) always converges and the excursion space $\underline{\underline{N}}$ is the full two dimensional function space on $\underline{\underline{T}}$. (See the end of Section 25 for an interesting situation where the analogous expression does not converge.) For the Dirichlet space on the boundary, we take

$$\underline{\underline{H}} = \underline{\underline{N}} \tag{12.27}$$

and it is clear that the regularized boundary Δ is the two point set

$$\Delta = \{0,1\} \tag{12.28}$$

adjoined in the usual way to $\underline{\underline{X}} = (0,1)$, that $X^{ex}_{\zeta-0} = X^{(R)}_{\zeta-0}$, and that the singular boundary $\Delta_s = \phi$. For general φ on Δ we have

$$H\varphi(x) = \{p_0(x)+N\kappa\mu_0(x)\}\varphi(0) + \{p_1(x)+N\kappa\mu_1(x)\}\varphi(1)$$

The expression (12.23) can now be rewritten

$$\begin{aligned} N(\varphi,\varphi) = {} & (1/2C)\{\varphi(0)-\varphi(1)\}^2 + (\mu_0\kappa N\kappa\mu_1)\{\varphi(0)-\varphi(1)\}^2 \\ & + (\mu_0\kappa p_1+\mu_1\kappa p_0)\{\varphi(0)-\varphi(1)\}^2 + (r\kappa p_0)\varphi^2(0) + (r\kappa p_1)\varphi^2(1) \\ & + (r\kappa N\kappa\mu_0)\varphi^2(0) + (r\kappa N\kappa\mu_1)\varphi^2(1) . \end{aligned} \tag{12.29}$$

The formula (3.10) is applicable to general $f \in \underline{\underline{W}}$ and becomes

(12.30)
$$\frac{1}{2}\int_0^1 dx\,\{f'(x)\}^2 + \int_0^1 dx\,\kappa(x)r(x)f^2(x)$$
$$+ \int_0^1 dx\,\kappa(x)[\mu_0(x)\{f(x)-f(0)\}^2 + \mu_1(x)\{f(x)-f(1)\}^2]$$
$$= \frac{1}{2}\int_0^1 dx\,\{(f-H\gamma f)'(x)\}^2 + \int_0^1 dx\,\kappa(x)\{(f-H\gamma f)(x)\}^2$$
$$+ N(\gamma f, \gamma f)\ .$$

Of course γf is the "ordinary" restriction of f to the boundary $\Delta = \{0,1\}$. The formulas(12.29) and (12.30) together yield for general $f \in \underline{\underline{W}}$ the curious identity

(12.31)
$$\frac{1}{2}\int_0^1 dx\,\{f'(x)\}^2 + \int_0^1 dx\,\kappa(x)r(x)f^2(x)$$
$$+ \int_0^1 dx\,\kappa(x)[\mu_0(x)\{f(x)-f(0)\}^2 + \mu_1(x)\{f(x)-f(1)\}^2]$$
$$= \frac{1}{2}\int_0^1 dx\,\{f'(x)-Hf'(x)\}^2 + \int_0^1 dx\,\kappa(x)\{f(x)-Hf(x)\}^2$$
$$+ \{(1/2C) + \mu_0\kappa N\kappa\mu_1 + \mu_0\kappa p_1 + \mu_1\kappa p_0\}\{f(0)-f(1)\}^2$$
$$+ \{r\kappa p_0 + r\kappa N\kappa\mu_0\}f^2(0) + \{r\kappa p_1 + r\kappa N\kappa\mu_1\}f^2(1)\ .$$

Clearly the local generator Ⓐ is defined by

(12.32)
$$Ⓐf(x) = \frac{1}{2}f''(x) - \kappa(x)f(x)$$

in the distribution sense. The restriction in Lemma 8.2 that Ⓐf be bounded can easily be removed and since every $f \in \underline{\underline{W}}$ is certainly bounded, it follows from Theorems 8.1 and 8.2 that the generator A for $(\underline{\underline{F}}, E)$ is defined by

(12.33)
$$Af(x) = \frac{1}{2}f''(x) - \kappa(x)f(x)$$

supplemented by the requirement that f be bounded and satisfy the boundary conditions

$$f(0) = f(1) = 0. \tag{12.34}$$

In fact the requirement that f be bounded can be omitted and Theorem 8.4 is applicable with $C^{\#} = C_0(0,1)$.

We look next at the normal derivative. Theorem 7.2 is applicable and therefore

$$(\partial f/\partial n)(0) = (\partial f/2n)_T(0) + \int_0^1 dx\, \kappa(x)\mu_0(x)\{f(x)-f(0)\} \tag{12.35}$$

$$(\partial f/\partial n)(1) = (\partial f/2n)_T(1) + \int_0^1 dx\, \kappa(x)\mu_1(x)\{f(x)-f(1)\}$$

whenever $\frac{1}{2}f''(x) - \kappa(x)\{f(x)-f(0)\mu_0(x) - f(1)\mu_1(x)\}$ is integrable or, equivalently, whenever $f''(x)$ is integrable. Also

$$(\partial f/\partial n)_T(0) = \operatorname{Lim}_{n\uparrow\infty} \ell_n^0\{f(1/n) - f(0)\} \tag{12.36}$$

$$(\partial f/\partial n)_T(1) = \operatorname{Lim}_{n\uparrow\infty} \ell_n^1\{f(1-1/n) - f(1)\}$$

and since (12.25') is equivalent to

$$2\ell_n^0\psi(0)\psi(1/n) \int_0^{(1/n)} dt\, \psi^{-2}(t) = 1 \tag{12.37}$$

$$2\ell_n^1\psi(1)\psi(1-1/n) \int_{1-(1/n)}^1 dt\, \psi^{-2}(t) = 1$$

there follows

$$\ell_n^0 \sim n/2 \;;\; \ell_n^1 \sim n/2 \tag{12.38}$$

and therefore (12.36), becomes

(12.39) $$(\partial f/\partial n)_T(0) = \frac{1}{2}f'(0) \; ; \; (\partial f/\partial n)_T(1) = -\frac{1}{2}f'(1)$$

and (12.35)

(12.35') $$(\partial f/\partial n)(0) = \frac{1}{2}f'(0) + \int_0^1 dx\, \kappa(x)\mu_0(x)\{f(x) - f(0)\}$$

$$(\partial f/\partial n)(1) = -\frac{1}{2}f'(1) + \int_0^1 dx\, \kappa(x)\mu_1(x)\{f(x) - f(1)\} .$$

The Green's identity (7.12'') is just

(12.40) $$\frac{1}{2}\int_0^1 dx\, f'(x)h'(x) + \frac{1}{2}\int_0^1 dx\, f''(x)h(x) = \frac{1}{2}f'(1)h(1) - \frac{1}{2}f'(0)h(0)$$

for h satisfying (12.7). For this case the results in Section 7 are no deeper than an elementary integration by parts.

We look now at the possibilities for $(\underline{\underline{F}}^{\sim}, E^{\sim})$. The most general Q satisfying paragraph 3.4.1 has the form

(12.41) $$Q(\varphi,\varphi) = N(\varphi,\varphi) + a\varphi^2(0) + b\varphi^2(1) + c\{\varphi(0) - \varphi(1)\}^2$$

with $a, b, c \geq 0$. (Of course 3.4.2 is vacuous when Δ is finite.) In every case $\underline{\underline{F}}^{\sim} = \underline{\underline{W}}$ and

$$E^{\sim}(f,f) = \frac{1}{2}\int_0^1 dx\, \{f'(x)\}^2 + \int_0^1 dx\, \kappa(x)r(x)f^2(x)$$

$$+ \int_0^1 dx\, \kappa(x)[\mu_0(x)\{f(x) - f(0)\}^2 + \mu_1(x)\{f(x) - f(1)\}^2]$$

$$+ af^2(0) + bf^2(1) + c\{f(0) - f(1)\}^2 .$$

Theorems 9.1 and 9.2 imply that the generators $A^{\sim}$ is defined by

(12.42) $$A^{\sim}f(x) = \frac{1}{2} f''(x) - \kappa(x)\{f(x) - f(0)\mu_0(x) - f(1)\mu_1(x)\}$$

supplemented by the boundary conditions

(12.43) $$\frac{1}{2} f'(0) + \int dx\, \kappa(x)\mu_0(x)\{f(x) - f(0)\} = af(0) + c\{f(0) - f(1)\}$$

(12.44) $$-\frac{1}{2} f'(0) + \int dx\, \kappa(x)\mu_1(x)\{f(x) - f(1)\} = bf(1) + c\{f(1) - f(0)\} .$$

Again the requirement that f be bounded can be omitted.

Case 2. The assumption on κ and M are as in Case 1. The difference is that now we take

(12.27) $$\underline{\underline{H}} = \underline{\underline{N}}_0$$

the subspace of terminal variables Φ which vanish on the atom $[X^{(R)}_{\zeta-0}=1]$. Then Δ reduces to the singleton

(12.28') $$\Delta = \{0\}$$

and $\underline{\underline{X}} \cup \Delta = [0,1)$ with the usual topology. Also

$$X^{ex}_{\zeta-0} = 0 \quad \text{when} \quad X^{(R)}_{\zeta-0} = 0$$

$$X^{ex}_{\zeta-0} = \delta \quad \text{when} \quad X^{(R)}_{\zeta-0} = 1 .$$

The sum $\mu_1(x) + r(x)$ plays the role of $r(x)$ in Case 1. The condition $\gamma_T f = \gamma_T \gamma f$ in Theorem 9.1 and 10.1 are equivalent to the null boundary condition

(12.44) $$f(1) = 0.$$

The Dirichlet space $\underline{\underline{F}}^{\sim}$ is always the set of $f \in \underline{\underline{W}}$ satisfying (12.44) and the boundary condition (12.43) at the endpoint 1 is replaced by (12.44). Otherwise everything goes as in Case 1 with some simplification since Δ is a singleton.

Case 3. κ and M are as in Case 1 but now

$$\underline{\underline{H}} = \underline{\underline{N}}_p \tag{12.27''}$$

the subspace of constant terminal variables. Again Δ reduces to a singleton (12.28') but now $\underline{\underline{X}} \cup \Delta = [0,1]$ with the endpoints 0 and 1 identified and

$$X^{ex}_{\zeta-0} = 0 \text{ when } X^{(R)}_{\zeta-0} = 0 \text{ or } 1 .$$

Also $\mu_0(x) + \mu_1(x)$ plays the role of $\mu_0(x)$ in Case 2, and $p_0(x) + p_1(x)$ plays the role of $p_0(x)$. The condition $\gamma_T f = \gamma_T \gamma f$ is equivalent to the periodic boundary condition

$$f(0) = f(1) . \tag{12.45}$$

The Dirichlet space $\underline{\underline{F}}^{\sim}$ is the set of $f \in \underline{\underline{W}}$ satisfying (12.45) and the boundary conditions (12.42), (12.43) collapse into one boundary condition

$$\frac{1}{2}\{f'(0) - f'(1)\} + \int_0^1 dx\, \kappa(x)\{\mu_0(x) + \mu_1(x)\}\{f(x) - f(0)\} = af(0) . \tag{12.46}$$

We have already mentioned in Remark 2 following Corollary 4.4 that Cases 1 and 3 together give an example of Dirichlet spaces $(\underline{\underline{F}}_1, E_1)$ and $(\underline{\underline{F}}_2, E_2)$ such that $\underline{\underline{F}}_1$ is contained in $\underline{\underline{F}}_2$ and E_2 reduces to E_1 on $\underline{\underline{F}}_1$ but $(\underline{\underline{F}}_2, E_2)$ is not an extension of $(\underline{\underline{F}}_1, E_1)$ and indeed does not dominate it in the sense of SMP, Section 21. It suffices to let $\underline{\underline{F}}_1 = \underline{\underline{F}}^{\sim}$ in Case 3 and $\underline{\underline{F}}_2 = \underline{\underline{F}}^{\sim}$ in Case 1 and to let a in Case 3 agree with a + b in Case 1. Observe that $\underline{\underline{F}}_1$ is not an ideal in $\underline{\underline{F}}_2$ and therefore SMP, Theorem 20.1 is not applicable.

Case 4. κ is the same as in Case 1 but now the auxiliary space $\underline{\underline{M}}$ is a singleton

(12.47) $$\underline{\underline{M}} = \{m\} .$$

To simplify the presentation we will assume for all x

(12.48) $$\mu_0(x) = \mu_1(x) = r(x) = 0 ; \mu_m(x) = 1 .$$

In the preliminary definition of the excursion form we must consider pairs (Φ,φ) on $\underline{\underline{T}} \cup M$ and (12.23) becomes

(12.49) $$N((\Phi,\varphi),(\Phi,\varphi)) = (1/2C)\{\Phi(0) - \Phi(1)\}^2 + (\kappa p_0)\{\Phi(0) - \varphi(m)\}^2 + (\kappa p_1)\{\Phi(1) - \varphi(m)\}^2.$$

We take $\underline{\underline{H}}$ to be the full excursion space $\underline{\underline{N}}$ so that the boundary

(12.50) $$\Delta = \{0,1,m\} .$$

For general φ on Δ we have

$$H\varphi(x) = p_0(x)\varphi(0) + p_1(x)\varphi(1) + N\kappa(x)\varphi(m) .$$

The boundary points 0 and 1 are adjoined so that [0,1] has the usual topology. Since $G\kappa(x) = 1 - p_0(x) - p_1(x)$ is bounded away from 1, the boundary point m is adjoined as an isolated point. The indicator e_m belongs to $\underline{\underline{H}}$ and

$$U_{0,\infty}(e_m^2, 1) = \operatorname{Lim}_{v\uparrow\infty} v \int dx\, H1(x)H_v e_m^2(x)$$

$$= \operatorname{Lim}_{v\uparrow\infty} \int dx\, vG_v\kappa(x)$$

$$= \int_0^1 dx\, \kappa(x)$$

which is finite. Thus by Theorem 4.3 or Corollary 4.4

$$\Delta_r = \{0,1\} \; ; \; \Delta_s = \{m\} \; .$$

The most general Q has the form

$$Q(\varphi,\varphi) = a_0\varphi^2(0) + a_1\varphi^2(1) + a_m\varphi^2(m)$$
$$+ a_{01}\{\varphi(0) - \varphi(1)\}^2 + a_{0m}\{\varphi(0) - \varphi(m)\}^2$$
$$+ a_{1m}\{\varphi(1) - \varphi(m)\}^2 + N(\varphi,\varphi)$$

with all of the constants $a_i, a_{ij} \geq 0$. This can be written also

(12.51)
$$Q(\varphi,\varphi) = a_0\varphi^2(0) + a_1\varphi^2(1) + a_m\varphi^2(m)$$
$$+ \{a_{01} + (1/2C)\}\{\varphi(0) - \varphi(1)\}^2$$
$$+ \{a_{0m} + \kappa p_0\}\{\varphi(0) - \varphi(m)\}^2$$
$$+ \{a_{1m} + \kappa p_1\}\{\varphi(1) - \varphi(m)\}^2 \; .$$

We consider first the expanded Dirichlet space $(\underline{\underline{F}}^e, E^e)$ introduced in Section 4. The reference Hilbert space consists of functions f defined on $[0,1] \cup \{m\}$ with inner product

(12.52)
$$\int_0^1 dx\, f^2(x) + f^2(0) + f^2(1) + f^2(m) \; .$$

The Dirichlet space $\underline{\underline{F}}^e$ consists precisely of those f whose restriction to $[0,1]$ belongs to $\underline{\underline{W}}$ and

(12.53)
$$E^e(f,f) = \frac{1}{2}\int_0^1 dx\, \{f'(x)\}^2 + \int_0^1 dx\, \kappa(x)\{f(x) - f(m)\}^2$$

$$+ \Sigma_i a_i f^2(i) + \frac{1}{2} \Sigma_{i,j} a_{ij} \{f(i) - f(j)\}^2$$

where the indices i, j run over $0, 1, m$ and $a_{ij} = a_{ji}$. (Keep in mind that $f(0)$ and $f(1)$ are defined by continuity for $f \in \underline{\underline{F}}^e$.) Theorem 7.2 can be applied to give $(\partial f/\partial n)(m) = \int \kappa(dx)\{f(x) - f(m)\}$ and by the calculations for Case 1

$$(\partial f/\partial n)(0) = (\partial f/\partial n)_T(0) = \frac{1}{2} f'(0)$$

$$(\partial f/\partial n)(1) = (\partial f/\partial n)_T(1) = -\frac{1}{2} f'(1) .$$

It follows with the help of Theorem 11.1 that the expanded generator A^e is defined by

(12.54) $$A^e f(x) = \frac{1}{2} f''(x) - \kappa(x)\{f(x) - f(m)\} \quad \text{on } (0,1)$$

$$A^e f(0) = \frac{1}{2} f'(0) - a_0 f(0) - a_{01}\{f(0) - f(1)\} - a_{0m}\{f(0) - f(m)\}$$

$$A^e f(1) = -\frac{1}{2} f'(1) - a_1 f(1) - a_{01}\{f(1) - f(0)\} - a_{1m}\{f(1) - f(m)\}$$

$$A^e f(m) = \int_0^1 dx\, \kappa(x)\{f(x) - f(m)\} - a_m f(m) - a_{0m}\{f(m) - f(0)\} - a_{1m}\{f(m) - f(1)\} .$$

We can immediately apply Theorem 11.4 to get a characterization of the generator $A^\sim$ for $(\underline{\underline{F}}^\sim, E^\sim)$. This is defined by

(12.55) $$A^\sim f(x) = \frac{1}{2} f''(x) - \kappa(x)\{f(x) - f(m)\}$$

supplemented by the two boundary conditions

(12.56) $$\frac{1}{2} f'(0) = a_0 f(0) + a_{01}\{f(0) - f(1)\} + a_{0m}\{f(0) - f(m)\}$$

$$-\frac{1}{2} f'(1) = a_1 f(1) + a_{01}\{f(1) - f(0)\} + a_{1m}\{f(1) - f(m)\} ,$$

where $f(m)$ is the constant defined by

(12.57) $$f(m) = \{\int_0^1 dx\, \kappa(x)f(x)+a_{0m}f(0)+a_{1m}f(1)\}/\{\int_0^1 dx\, \kappa(x)+a_{0m}+a_{1m}+a_m\} .$$

Before describing the Dirichlet space $(\underline{\underline{F}}^{\sim}, E^{\sim})$ we must do some preliminary work. The singleton $\{m\}$ is nearly open and it is clear from (12.53) that

$$\kappa^e(\{m\}) = a_m$$

$$J^e(\{m\}), \{0\}) = a_{0m} \quad ; \quad J^e(\{m\}, \{1\}) = a_{1m}$$

$$J^e(\{m\}, dx) = \kappa(x)\, dx$$

where κ^e, J^e denote the killing measure and Levy measure for $(\underline{\underline{F}}^e, E^e)$. It follows from the probabilistic interpretation of κ and J given in SMP, Part II that if we let

$$\sigma = \inf \{t > 0 : X_t^e \neq m\} ,$$

the time of first exit from $\{m\}$, then

(12.58) $$Ⓟ_m^e(\sigma>t) = e^{-tq_m}$$

$$Ⓟ_m^e(X_\sigma = \delta) = P(m, \delta)$$

$$Ⓟ_m^e(X_\sigma = 0) = P(m, 0) \; ; Ⓟ_m^e(X_\sigma = 1) = P(m, 1)$$

$$Ⓔ_m^e[X_\sigma \in (0,1); g(X_\sigma)] = \int_0^1 P(m, dx)g(x)$$

where

(12.59) $$q_m = a_m + a_{0m} + a_{1m} + \int_0^1 dx\, \kappa(x)$$

(12.59 continued) $$P(m,\delta) = a_m/q_m$$

$$P(m,0) = a_{0m}/q_m \;;\; P(m,1) = a_{1m}/q_m$$

$$P(m,dx) = (1/q_m)\kappa(x)\,dx \quad \text{on} \quad (0,1)\ .$$

That is, the singleton $\{m\}$ is stable and the first exit from m is exponentially distributed at the rate q_m. The position of the first jump has distribution $P(m,\cdot)$. It is clear then that

(12.60) $$Mf(m) = (1/q_m)\int_0^1 dx\,\kappa(x)f(x)$$

$$D_\infty f(m) = (a_{0m}/q_m)f(0) + (a_{1m}/q_m)f(1)\ .$$

The perturbed space $\underline{\underline{F}}^o = \underline{\underline{F}}$ and the perturbed form E^o is defined on $\underline{\underline{F}}$ by

$$E^o(f,f) = \frac{1}{2}\int_0^1 dx\{f'(x)\}^2 + (1/2q_m)\int_0^1 dx\int_0^1 dy\,\kappa(x)\kappa(y)\{f(x)-f(y)\}^2$$

$$+\{(a_m+a_{0m}+a_{1m})/q_m\}\int_0^1 dx\,\kappa(x)f^2(x)\ .$$

The full Dirichlet space $\underline{\underline{F}}^\sim = \underline{\underline{W}}$ and for $f \in \underline{\underline{F}}^\sim$ we have, in the notation of (5.36) and (5.37),

(12.61) $$E^\sim(f,f) = \frac{1}{2}\int_0^1 dx\,\{f'(x)\}^2 + a_0 f^2(0) + a_1 f^2(1) + a_{01}\{f(0)-f(1)\}^2 + B(f,f)$$

where

(12.62) $$B(f,f) = \int_0^1 dx\,\kappa(x)\{f(x)-Mf(m)-D_\infty f(m)\}^2$$

$$+ a_m\{Mf(m)+D_\infty f(m)\}^2$$

$$+ a_{0m}\{f(0)-Mf(m)-D_\infty f(m)\}^2 - a_{1m}\{f(1)-Mf(m)-D_\infty f(m)\}^2$$

In the notation of (5.38) it is clear that

$$\frac{1}{2}\Theta^r\langle\gamma_r f, \gamma_r f\rangle = P(m,0)J^e(m,\{1\})\{f(0)-f(1)\}^2$$
$$+P(m,1)J^e(m,\{0\})\{f(0)-f(1)\}^2$$
$$+\{a_{0m}a_{1m}/q_m\}\{f(0)-f(1)\}^2$$
$$+\{a_{0m}a_{1m}/q_m\}\{f(0)-f(1)\}^2$$

$$\iint_{X\times\Delta_r} \kappa\mu 1_{\Delta_s} D_\infty(dx,dy)\{f(x)-\gamma_r f(y)\}^2 = (a_{0m}/q_m)\int_0^1 dx\,\kappa(x)\{f(x)-f(0)\}^2$$
$$+(a_{1m}/q_m)\int_0^1 dx\,\kappa(x)\{f(x)-f(1)\}^2$$

$$\int \kappa(dx)f^2(x)\mu\{1-M1-D_\infty 1\}(x) = (a_m/q_m)\int_0^1 dx\,\kappa(x)f^2(x)$$

$$\int_{\Delta_r} \nu(dy)g(y)\{\gamma_r f(y)\}^2 = \kappa^e(\{m\})P(m,0)f^2(0) + \kappa^e(\{m\})P(m,1)f^2(1)$$

$$= \{a_{0m}a_m/q_m\}f^2(0) + \{a_{1m}a_m/q_m\}f^2(1)\,.$$

Thus

(12.63)
$$B(f,f) = (1/2q_m)\int_0^1 dx\int_0^1 dy\,\kappa(x)\kappa(y)\{f(x)-f(y)\}^2$$
$$+\{a_{0m}a_{1m}/q_m\}\{f(0)-f(1)\}^2 + (a_{0m}/q_m)\int_0^1 dx\,\kappa(x)\{f(x)-f(0)\}^2$$
$$+(a_{1m}/q_m)\int_0^1 dx\,\kappa(x)\{f(x)-f(1)\}^2 + (a_m/q_m)\int_0^1 dx\,\kappa(x)f^2(x)$$
$$+(a_{0m}a_m/q_m)f^2(0) + (a_{1m}a_m/q_m)f^2(1)\,.$$

and plugging this into (12.61) we get

$$(12.64)\quad E^{\sim}(f,f) = \frac{1}{2}\int_0^1 dx\,\{f'(x)\}^2 + (1/2q_m)\int_0^1 dx\int_0^1 dy\,\kappa(x)\kappa(y)\{f(x)-f(y)\}^2$$

$$+ (a_m/q_m)\int_0^1 dx\,\kappa(x)f^2(x) + (a_{0m}/q_m)\int_0^1 dx\,\kappa(x)\{f(x)-f(0)\}^2$$

$$+ (a_{1m}/q_m)\int_0^1 dx\,\kappa(x)\{f(x)-f(1)\}^2 + G(\gamma_r f, \gamma_r f)$$

where for φ defined on $\Delta_r = \{0,1\}$

$$(12.65)\qquad G(\varphi,\varphi) = \{a_{01}+(a_{0m}a_{1m}/q_m)\}\{\varphi(0)-\varphi(1)\}^2$$

$$+ \{a_0+(a_{0m}a_m/q_m)\}\varphi^2(0)$$

$$+ \{a_1+(a_{1m}a_m/q_m)\}\varphi^2(1)\ .$$

From the point of view of Theorems 11.2 and 11.3 the generator $A^{\sim}$ is defined by

$$(12.55')\quad A^{\sim}f(x) = \frac{1}{2}f''(x) - \kappa(x)\{f(x)-(1/q_m)[\int_0^1 dy\,\kappa(y)f(y)+a_{0m}f(0)+a_{1m}f(1)]\}$$

supplemented by the two boundary conditions

$$(12.56')\qquad \frac{1}{2}f'(0) + (a_{0m}/q_m)\int_0^1 dx\,\kappa(x)\{f(x)-f(0)\}$$

$$= \{a_0+(a_{0m}a_m/q_m)\}f(0) + \{a_{01}+(a_{0m}a_{1m}/q_m)\}\{f(0)-f(1)\}$$

$$-\frac{1}{2}f'(1) + (a_{1m}/q_m)\int_0^1 dx\,\kappa(x)\{f(x)-f(1)\}$$

$$= \{a_1+(a_{1m}a_m/q_m)\}f(1) + \{a_{01}+(a_{1m}a_{0m}/q_m)\}\{f(1)-f(0)\}\ .$$

This is consistent with our previous description and indeed can be derived from it by substituting (12.57) into (12.55) and (12.56). In this case the second description of $A^{\sim}$ seems to be the "natural one." Indeed the first description would never be considered if it were not for the specific construction involved.

Case 5. Again $\kappa(x)$ is integrable but now the auxiliary space $\underline{\underline{M}}$ is the open interval

(12.66) $$\underline{\underline{M}} = (2, 3)$$

We continue to use the special notation

$$\mu_0(x) = \mu(x, [X^{(R)}_{\zeta-0} = 0]) \; ; \; \mu_1(x) = \mu(x, [X^{(R)}_{\zeta-0} = 1])$$

for the restriction of the jumping measures $\mu(x, \cdot)$ to $\underline{\underline{T}}$. Also to simplify matters slightly we assume that the restrictions to $(2, 3)$ are all absolutely continuous and therefore can be represented

(12.67) $$\mu(x, ds) = \mu(x, s)\, ds \; .$$

The excursion form is given by

(12.68)
$$N((\Phi, \varphi), (\Phi, \varphi))$$
$$= \{(1/2C) + \mu_0 \kappa N \kappa \mu_1 + \mu_0 \kappa p_1 + \mu_1 \kappa p_0\}\{\Phi(0) - \Phi(1)\}^2$$
$$+ \int_0^1 dx\, \kappa(x)\{N\kappa\mu_0(x) + p_0(x)\} \int_2^3 ds\, \mu(x, s)\{\Phi(0) - \varphi(s)\}^2$$
$$+ \int_0^1 dx\, \kappa(x)\{N\kappa\mu_1(x) + p_1(x)\} \int_2^3 ds\, \mu(x, s)\{\Phi(1) - \varphi(s)\}^2$$
$$+ \int_0^1 dx\, \kappa(x) \int_0^1 dy\, \kappa(y) N(x, y) \int_2^3 ds \int_2^3 dt\, \mu(x, s)\mu(y, t)\{\varphi(s) - \varphi(t)\}^2$$
$$+ \{r\kappa N \kappa\mu_0 + r\kappa p_0\}\Phi^2(0) + \{r\kappa N\kappa\mu_1 + r\kappa p_1\}\Phi^2(1)$$
$$+ \int_0^1 dx\, \kappa(x) r(x) \int_2^3 ds\, \mu(x, s)\varphi^2(s) \; .$$

We take $\underline{\underline{H}}$ to be the set of pairs (Φ, φ) satisfying

(12.69) $$\varphi \in \underline{\underline{W}}((2,3)) \; ; \; \varphi(2) = \Phi(0) \; ; \; \varphi(3) = \Phi(1)$$

where $\underline{\underline{W}}((2,3))$ is defined as in paragraph 12.1 except that the interval $(0,1)$ is replaced by $(2,3)$. The matching conditions make sense since all such φ have continuous extensions to the closed interval. The boundary Δ can be identified with $[2,3]$ and $\underline{\underline{X}} \cup \Delta = [0,1] \cup [2,3]$ topologized in the usual way except that 0 is identified with 2 and 1 is identified with 3. We take harmonic measure to be normalized Lebesgue measure on the open interval $(2,3)$ and counting measure on the two point set $\{2,3\}$. For general $\varphi \in \underline{\underline{H}}$ we have

$$H\varphi(x) = \{p_0(x)+N\kappa\mu_0(x)\}\varphi(2) + \{p_1(x)+N\kappa\mu_1(x)\}\varphi(3)$$
$$+ \int_0^1 dy\, N(x,y)\kappa(y)\int_2^3 ds\, \mu(y,s)\varphi(s)$$

If $\varphi \in \underline{\underline{H}}$ satisfies $\varphi(2) = \varphi(3) = 0$ then

$$U_{0,\infty}(\varphi^2, 1) = \operatorname{Lim}_{v\uparrow\infty} v\int_0^1 dx\, H1(x)H_v\varphi^2(x)$$
$$\leq \operatorname{Lim}_{v\uparrow\infty} \int_0^1 dy\, v\bar{G}_v 1(y)\kappa(y)\int_2^3 ds\, \mu(y,s)\varphi^2(s)$$
$$= \int_0^1 dy\, \kappa(y)\int_0^1 ds\, \mu(y,s)\varphi^2(s)$$

is finite and it follows from Theorem 4.3 that

$$\Delta_r = \{2,3\} \; ; \; \Delta_s = (2,3) \; .$$

We assume that Q has the form

$$Q(\varphi,\varphi) = \frac{1}{2}\int_2^3 ds\,\{\varphi'(s)\}^2 + a\varphi^2(2) + b\varphi^2(3)$$

$$+ c\{\varphi(2)-\varphi(3)\}^2 + N(\varphi,\varphi)$$

with $a, b, c \geq 0$. As for Case 4 we begin by looking at the expanded Dirichlet space $(\underline{\underline{F}}^e, E^e)$. The reference Hilbert space consists of functions f defined on $[0,1] \cup [2,3]$ with inner product

$$\int_0^1 dx\, f^2(x) + f^2(0) + f^2(1) + \int_2^3 ds\, f^2(s).$$

The Dirichlet space $\underline{\underline{F}}^e$ is the subcollection satisfying

$$f \in \underline{\underline{W}}((0,1)\cup(2,3))\ ;\ f(0) = f(2)\ ;\ f(1) = f(3)$$

and the form E^e is given by

$$E^e(f,f) = \frac{1}{2}\int_0^1 dx\,\{f'(x)\}^2 + \int_0^1 dx\,\kappa(x)r(x)f^2(x)$$

$$+ \int_0^1 dx\,\kappa(x)\{\mu_0(x)(f(x)-f(0))^2 + \mu_1(x)(f(x)-f(1))^2\}$$

$$+ \int_0^1 dx\,\kappa(x)\int_2^3 ds\,\mu(x,s)\{f(x)-f(s)\}^2$$

$$+ \int_2^3 ds\,\{f'(s)\}^2 + af^2(0) + bf^2(1) + c\{f(0)-f(1)\}^2 .$$

If f belongs to the domain of the expanded generator A^e, then by Theorem 11.1

(12.70) $$A^e f(x) = \frac{1}{2}f''(x) - \kappa(x)\{f(x) - \mu_0(x)f(0) - \mu_1(x)f(1)$$

$$- \int_2^3 ds\,\mu(x,s)f(s)\} \qquad \text{on } (0,1)$$

and for bounded $\psi \in \underline{\underline{H}}$

$$(12.71)\quad A^e f(2)\psi(2) + A^e f(3)\psi(3) + \int_2^3 ds\, A^e f(s)\psi(s) = (\partial f_0/\partial n)(\psi) - \{Q-N\}(\gamma f, \psi) .$$

Also, Theorem 7.2 is applicable so that

$$(12.72)\quad (\partial f_0/\partial n)(\psi) = (\partial f_0/\partial n)_T(\psi) + \psi(0)\int_0^1 dx\, \kappa(x)\mu_0(x)\{f(x)-f(0)\}$$

$$+ \psi(1)\int_0^1 dx\, \kappa(x)\mu_1(x)\{f(x)-f(1)\}$$

$$+ \int_0^1 dx\, \kappa(x)\int_2^3 ds\, \mu(x,s)\{f(x)-f(s)\}\psi(s) .$$

Taking $\psi \in \underline{\underline{W}}([2,3]) \cap C_{com}((2,3))$ in (12.71) we get

$$\int_2^3 ds\, A^e f(s)\psi(s) = \int_0^1 dx\, \kappa(x)\int_2^3 ds\, \mu(x,s)\{f(x)-f(s)\}\psi(s)$$

$$- \frac{1}{2}\int_2^3 ds\, f'(s)\psi'(s)$$

and therefore

$$(12.73)\quad A^e f(s) = \frac{1}{2} f''(s) + \int_0^1 dx\, \kappa(x)\mu(x,s)\{f(x)-f(s)\} \quad \text{on } (2,3) .$$

Let p_0 be extended to $[2,3]$ so that it satisfies on $(2,3)$ the homogeneous equation

$$(12.74)\quad \frac{1}{2} h''(s) - \int_0^1 dx\, \kappa(x)\mu(x,s)h(s) = 0$$

and the boundary conditions $p_0(2) = 1$, $p_0(3) = 0$. Similarly let p_1 be extended so that it satisfies (11.74) on $(2,3)$ and the boundary conditions $p_1(2) = 0$, $p_1(3) = 1$. By the calculations for Case 1

$$(\partial f/\partial n)_T(\gamma p_0) = \frac{1}{2} f'(0) \; ; \; (\partial f/\partial n)_T(\gamma p_1) = -\frac{1}{2} f'(1) .$$

Therefore $\psi = \gamma p_0$ in 12.71 gives

$$A^e f(0) + \int_2^3 ds\, A^e f(s) p_0(s) = \frac{1}{2} f'(0)$$

$$+ \int_0^1 dx\, \kappa(x)\mu_0(x)\{f(x)-f(0)\} + \int_0^1 dx\, \kappa(x) \int_2^3 ds\, \mu(x,s)\{f(x)-f(s)\} p_0(s)$$

$$- \frac{1}{2} \int_2^3 ds\, A^e f'(s) p_0'(s) - af(0) - c\{f(0)-f(1)\} .$$

Either after integration by parts or again applying the results in Section 7 we get

$$\int_2^3 ds\, \{A^e f(s) - \int_0^1 dx\, \kappa(x)\mu(x,s)(f(x)-f(s))\} p_0(s) + \frac{1}{2} \int_2^3 ds\, A^e f'(s) p_0'(s)$$

$$= -\frac{1}{2} A^e f'(2)$$

and therefore

$$(12.75) \quad A^e f(0) = \frac{1}{2} f'(0) + \frac{1}{2} f'(2) + \int_0^1 dx\, \kappa(x)\mu_0(x)\{f(x)-f(0)\} - af(0) - c\{f(0)-f(1)\} .$$

Similarly

$$(12.76) \quad A^e f(1) = -\frac{1}{2} f'(1) - \frac{1}{2} f'(3) + \int_0^1 dx\, \kappa(x)\mu_1(x)\{f(x)-f(1)\} - bf(1) - c\{f(1)-f(0)\} .$$

Thus $A^e f$ is defined by (12.70, 12.73, 12.75 and 12.76). Notice that the sums $\frac{1}{2} f'(0) + \frac{1}{2} f'(2)$ and $-\frac{1}{2} f'(1) - \frac{1}{2} f'(3)$ actually correspond to a jump in the first derivative at the corresponding boundary.

Applying Theorem 11.4 we conclude immediately that the generator $A^\sim$ is defined by

$$(12.77) \qquad A^{\sim}f(x) = \frac{1}{2}f''(x) - \kappa(x)\{f(x) - \mu_0(x)f(0) - \mu_1(x)f(1) - \int_2^3 ds\, \mu(x,s)\varphi(s)\}$$

supplemented by the two boundary conditions

$$(12.78) \quad \frac{1}{2}f'(0) + \int_0^1 dx\, \kappa(x)\mu_0(x)\{f(x) - f(0)\} = -\frac{1}{2}\varphi'(2) - af(0) - c\{f(0) - f(1)\}$$

$$(12.79) \quad -\frac{1}{2}f'(1) + \int_0^1 dx\, \kappa(x)\mu_1(x)\{f(x) - f(1)\} = -\frac{1}{2}\varphi'(3) - bf(1) - c\{f(1) - f(0)\}$$

where φ defined on $[2,3]$ is the solution of

$$(12.80) \qquad \frac{1}{2}\varphi''(s) + \int_0^1 dx\, \kappa(x)\mu(x,s)\{f(x) - \varphi(s)\} = 0 .$$

supplemented by the matching conditions

$$(12.81) \qquad \varphi(2) = f(0)\ ;\ \varphi(3) = f(1) .$$

Of course the function φ is the analogue to the constant $f(m)$ in Case 4. For Case 4 it would have been trivial to eliminate $f(m)$ at this stage and obtain directly the second description for $A^{\sim}$ without first studying $(\underline{\underline{F}}^{\sim}, E^{\sim})$. This is not so easily done here.

We turn now to the Dirichlet space $(\underline{\underline{F}}^{\sim}, E^{\sim})$. It is clear that

$$D_\infty f(s) = p_0(s)f(0) + p_1(s)f(1)$$

$$Mf(s) = \int_2^3 dt\, N^s(s,t) \int_0^1 dx\, \kappa(x)\mu(x,t)f(x)$$

where now $N^s(s,t)$ is the Green function

$$N^s(s,t) = 2C^s p_1(t \wedge s) p_0(t \vee s)$$

defined for $2 \leq s,\ t \leq 3$ and

$$C^s = \psi^s(2)\psi^s(3)\int_2^3 dt\ \{\psi^s(t)\}^{-2}$$

with ψ^s any solution of (12.74) bounded away from 0. Again the perturbed space $\underline{\underline{F}}^o = \underline{\underline{F}}$ and the perturbed form E^o is defined by

(12.82) $$E^0(f,f) = \frac{1}{2}\int_0^1 dx\ \{f'(x)\}^2 + \frac{1}{2}\int_0^1 dx\ \kappa(x)\int_0^1 dy\ \kappa(y)\lambda(x,y)\{f(x)-f(y)\}^2$$
$$+ \int_0^1 dx\ \kappa(x)\{1-\lambda(x)\}f^2(x)$$

where for typographic convenience we have introduced the notations

$$\lambda(x,y) = \int_2^3 ds \int_2^3 dt\ N^s(s,t)\mu(x,s)\mu(y,t)$$
$$\lambda(x) = \int_0^1 dy\ \kappa(y)\lambda(x,y)\ .$$

Again $\underline{\underline{F}}^{\sim} = \underline{\underline{W}}$ and for $f \in \underline{\underline{F}}^{\sim}$ we have

(12.83) $$E^{\sim}(f,f) = \frac{1}{2}\int_0^1 dx\ \{f'(x)\}^2 + \int_0^1 dx\ \kappa(x)r(x)f^2(x)$$
$$+ \int_0^1 dx\ \kappa(x)[\mu_0(x)\{f(x)-f(0)\}^2 + \mu_1(x)\{f(x)-f(1)\}^2]$$
$$+ af^2(0) + bf^2(1) + c\{f(0)-f(1)\}^2$$
$$+ B(f,f)$$

where now

(12.84) $$B(f,f) = \int_0^1 dx\ \kappa(x)\int_2^3 ds\ \mu(x,s)\{f(x)-Mf(s)-D_\infty f(s)\}^2$$
$$+\frac{1}{2}\int_2^3 ds\ \{Mf'(s)+D_\infty f'(s)\}^2\ .$$

It is clear from the proof of (12.29) that in the notation of (5.38)

$$\frac{1}{2}\Theta^r\langle\varphi,\varphi\rangle = (1/2C^s)\{\varphi(0)-\varphi(1)\}^2$$

and therefore (5.38) becomes

$$(12.85)\quad B(f,f) = \frac{1}{2}\int_0^1 dx\,\kappa(x)\int_0^1 dy\,\kappa(y)\lambda(x,y)\{f(x)-f(y)\}^2 + (1/2C^s)\{f(0)-f(1)\}^2$$
$$+\int_0^1 dx\,\kappa(x)\int_2^3 ds\,\mu(x,s)p_0(s)\{f(x)-f(0)\}^2$$
$$+\int_0^1 dx\,\kappa(x)\int_2^3 ds\,\mu(x,s)p_1(s)\{f(x)-f(1)\}^2$$

and after plugging into (12.83) we get

$$(12.86)\qquad E^{\sim}(f,f) = \frac{1}{2}\int_0^1 dx\{f'(x)\}^2 + \int_0^1 dx\,\kappa(x)r(x)f^2(x)$$
$$+\int_0^1 dx\,\kappa(x)[\mu_0(x)\{f(x)-f(0)\}^2 + \mu_1(x)\{f(x)-f(1)\}^2]$$
$$+\frac{1}{2}\int_0^1 dx\,\kappa(x)\int_0^1 dy\,\kappa(y)\lambda(x,y)\{f(x)-f(y)\}^2$$
$$+\int_0^1 dx\,\kappa(x)[\mu p_0(x)\{f(x)-f(0)\}^2 + \mu p_1(x)\{f(x)-f(1)\}^2]$$
$$+G(\gamma_r f,\gamma_r f)$$

where for φ defined on Δ_r

$$G(\varphi,\varphi) = \{(1/2C^s)+c\}\{\varphi(0)-\varphi(1)\}^2 + af^2(0) + bf^2(1)\ .$$

In (12.86) we have introduced the shorthand

$$\mu p_0(x) = \int_2^3 ds\,\mu(x,s)p_0(s)\ ;\ \mu p_1(x) = \int_2^3 ds\,\mu(x,s)p_1(s)\ .$$

By Theorem 11.2

$$A^{\sim}f(x) = \frac{1}{2}f''(x) - \kappa(x)\{f(x) - \int_0^1 dy\ \lambda(x,y)\kappa(y)f(y) - (\mu_0(x)+\mu p_0(x))f(0) - (\mu_1(x)-\mu p_1(x))f(1)\} \tag{12.87}$$

which is supplemented by the two boundary conditions

$$(\partial f/\partial n)(e_0) + \int_0^1 dx\ \kappa(x)\mu p_0(x)\{f(x)-f(0)\} = af(0) + \{(1/2C_s)+c\}\{f(0)-f(1)\}$$

$$(\partial f/\partial n)(e_1) + \int_0^1 dx\ \kappa(x)\mu p_1(x)\{f(x)-f(1)\} = bf(1) + \{(1/2C_s)+c\}\{f(1)-f(0)\}\ .$$

Again by Theorem 7.2 and the calculations for Case 1

$$(\partial f/\partial n)(e_1) = \frac{1}{2}f'(0) + \int_0^1 dx\ \kappa(x)\mu_0(x)\{f(x)-f(0)\}$$

$$(\partial f/\partial n)(e_1) = -\frac{1}{2}f'(1) + \int_0^1 dx\ \kappa(x)\mu_1(x)\{f(x)-f(1)\}$$

and so the boundary conditions can be written

$$\frac{1}{2}f'(0) + \int_0^1 dx\ \kappa(x)\{\mu_0(x)+\mu p_0(x)\}\{f(x)-f(0)\} = af(0) + \{(1/2C^s)+c\}\{f(0)-f(1)\} \tag{12.88}$$

$$-\frac{1}{2}f'(1) + \int_0^1 dx\ \kappa(x)\{\mu_1(x)+\mu p_1(x)\}\{f(x)-f(1)\} = bf(1) + \{(1/2C^s)+c\}\{f(1)-f(0)\}\ .$$

Thus the final result is quite similar to what we got for Case 4.

Case 6. We assume that $\kappa(x)$ is not integrable near either endpoint

(12.89) $$\int_0^{\frac{1}{2}} dx\, \kappa(x) = +\infty \quad ; \quad \int_{\frac{1}{2}}^1 dx\, \kappa(x) = +\infty$$

but κ does satisfy the moment condition near both endpoints

(12.90) $$\int_0^{\frac{1}{2}} dx\, \kappa(x)x < +\infty \quad ; \quad \int_{\frac{1}{2}}^1 dx\, \kappa(x)(1-x) < +\infty$$

and therefore $p_0(x), p_1(x)$ are both nontrivial and also $C < +\infty$. We take the auxiliary space $\underline{\underline{M}}$ to be empty and using the notation (12.22) we assume

(12.91) $$\int_0^{\frac{1}{2}} dx\, \kappa(x)\{r(x)+\mu_1(x)\} < +\infty \quad ; \quad \int_{\frac{1}{2}}^1 dx\, \kappa(x)\{r(x)+\mu_0(x)\} < +\infty .$$

Again harmonic measure is counting measure and the excursion form N is given by (12.29). Notice that the restriction (12.91) is precisely what is needed to guarantee that $\underline{\underline{N}}$ is the full two dimensional function space on $\underline{\underline{T}}$. Again we take $\underline{\underline{H}} = \underline{\underline{N}}$ so that Δ and $\underline{\underline{X}} \cup \Delta$ and the possibilities for Q are as in Case 1. The Dirichlet space $\underline{\underline{F}}^{\sim}$ is the proper subset of $f \in \underline{\underline{W}}$ satisfying the integrability condition

(12.92) $$\int_0^1 dx\, \kappa(x)[\mu_0(x)\{f(x)-f(0)\}^2 + \mu_1(x)\{f(x)-f(1)\}^2] < +\infty .$$

Again the generator $A^{\sim}$ is defined by (12.42) supplemented by the two boundary conditions (12.43), (12.44). It was incorrectly stated in SMP on pages 23.7 and 23.8 that the derivatives at the boundary may exist only in a formal sense. Since every harmonic function is bounded, it is still true here that we can omit the reguirement that f be bounded when we are characterizing domain $A^{\sim}$.

Finally we point out that in the present case we cannot use the method of Section 3 to obtain the most general process which dominates the absorbing one in the sense of SMP, Section 21. In particular, we cannot get the reflected Brownian motion without killing. It is clear from the discussion in SMP, Section 21 that in general one has to proceed in two steps - first suppressing the relevant killing and then applying the method of Section 3.

Case 7. We assume now that $\kappa(x)$ fails to satisfy a moment condition near either endpoint

(12.93) $$\int_0^{\frac{1}{2}} dx\, \kappa(x)x = +\infty \quad ; \quad \int_{\frac{1}{2}}^{1} dx\, \kappa(x)(1-x) = +\infty$$

and therefore p_0 and p_1 both vanish identically. Now the terminal set $\underline{\underline{T}}$ is empty. We take for the auxiliary space

$$M = \{0, 1\}$$

and replace (12.22) by

(12.22') $$\mu_0(x) = \mu(x, \{0\}) \; ; \; \mu_1(x) = \mu(x, \{1\}) \, .$$

Again we assume (12.91) and let harmonic measure be counting measure. The excursion form is given by

(12.94) $$N(\varphi, \varphi) = (\mu_0 \kappa N \kappa \mu_1)\{\varphi(0) - \varphi(1)\}^2 + (r\kappa N \kappa \mu_0)\varphi^2(0) + (r\kappa N \kappa \mu_1)\varphi^2(1)$$

and again (12.91) guarantees that the excursion space $\underline{\underline{N}}$ is the full two dimensional function space on $\underline{\underline{M}}$. We take $\underline{\underline{H}} = \underline{\underline{N}}$ and then Δ and $\underline{\underline{X}} \cup \Delta$ are exactly as for Cases 1 and 6. The Dirichlet space $(\underline{\underline{F}}^{\sim}, E^{\sim})$ is exactly as for Case 6 but the boundary conditions are different. Since $\underline{\underline{T}} = \phi$, Theorem 7.2 now implies

$$(\partial f/\partial n)(0) = \int_0^1 dx\, \kappa(x)\mu_0(x)\{f(x)-f(0)\}$$

$$(\partial f/\partial n)(1) = \int_0^1 dx\, \kappa(x)\mu_1(x)\{f(x)-f(1)\}$$

and so (12.43), (12.44) simplify to

$$\int_0^1 dx\, \kappa(x)\mu_0(x)\{f(x)-f(0)\} = af(0) + c\{f(0)-f(1)\} \tag{12.95}$$

$$\int_0^1 dx\, \kappa(x)\mu_1(x)\{f(x)-f(1)\} = bf(1) + c\{f(1)-f(0)\} .$$

But now the condition that f be bounded cannot be omitted when we are characterizing domain $A^{\sim}$ or even domain A. For example if $\kappa(x) = (4/9)x^{-2}$ for x near 0, then there exists a harmonic function h such that $h(x) = x^{-\frac{1}{3}}$ near 0. Such h does not belong to $\underline{\underline{W}}$ and therefore h is not in domain A. But $h \in L^2(dx)$ and there is no condition "$\gamma_T h = 0$" to eliminate it from consideration.

It turns out that Theorem 7.2 gives some additional information in the present case. Suppose that $f \in \underline{\underline{F}}^{\sim}$ and also $\frac{1}{2}f'' + \kappa\mu f$ is integrable. Then Theorem 7.2 guarantees that $\frac{1}{2}f''(x)N\kappa\mu_0(x)$ and $\frac{1}{2}f''(x)N\kappa\mu_1(x)$ are integrable and that

$$\frac{1}{2}\int_0^1 dx\, f'(x)(N\kappa\mu_0)'(x) + \frac{1}{2}\int_0^1 dx\, f''(x)N\kappa\mu_0(x) = 0$$

$$\frac{1}{2}\int_0^1 dx\, f'(x)(N\kappa\mu_1)'(x) + \frac{1}{2}\int_0^1 dx\, f''(x)N\kappa\mu_1(x) = 0 .$$

After integration by parts we conclude

$$f'(0) = 0 \ ; \ f'(1) = 0 \ . \tag{12.96}$$

Unlike for Case 1 the results in Section 7 give us information beyond what we can get after integration by parts above. However, it is true that (12.96) could be deduced directly from (12.93). We emphasize that (12.96) is a consequence of (12.42) suitably supplemented and should not be viewed as a boundary condition. In Case 1 with $\mu_0 = \mu_1 = 0$ this would be a possible boundary condition corresponding to pure reflection. No reflection is possible here or for Case 6.

13. Brownian Motion on $(0,\infty)$.

In this section $\underline{\underline{X}} = (0,\infty)$, the open positive half line and the reference measure is $m(x)\,dx$ where $m(x)$ is everywhere positive and integrable on $\underline{\underline{X}}$. The diffusion form is given by (12.1) with $(0,1)$ replaced by $(0,\infty)$ and again the Levy measure $J = 0$. We assume throughout this section that the killing density $\kappa(x)$ is integrable near 0,

$$\int_0^1 dx\, \kappa(x) < +\infty. \tag{13.1}$$

The Sobolev space $\underline{\underline{W}}$ is defined as in Section 12 except that $[0,1]$ is replaced by $[0,\infty)$. That is, f belongs to $\underline{\underline{W}}$ if it is absolutely continuous on $[0,\infty)$ and if

$$\int_0^\infty dx\, \{f'(x)\}^2 < +\infty .$$

Again the reflected space $\underline{\underline{F}}^{ref} = \underline{\underline{W}} \cap L^2(\kappa)$ and the Dirichlet norm E is given by

$$E(f,f) = \frac{1}{2}\int_0^\infty dx\, \{f'(x)\}^2 + \int_0^\infty dx\, \kappa(x) f^2(x) . \tag{13.2}$$

The significant new feature here is that functions $f \in \underline{\underline{W}}$ can behave rather wildly near infinity. When $\kappa(x)$ is not integrable near infinity, this allows much more complicated boundary behavior than in Section 12. In fact we will construct an example where the regular boundary Δ_r is a full line segment carrying another Brownian motion. In Section 12 the regular boundary could have at most two points.

We will first consider the relatively uninteresting case when $\kappa(x)$ is integrable near infinity. There are two reasons for this. First we illustrate in a simple context that the topology prescribed in Section 3 for $\underline{\underline{X}} \cup \Delta$ can be rather arbitrary and in some cases even "silly." Fortunately the appendix at the end of Chapter II assures us that this doesn't matter. As a general rule we will not even attempt to examine the topology in more complicated situations. The second reason is to permit comparison with SMP, Section 24 where we used "ad hoc" methods to analyze $(\underline{\underline{F}}^{\sim}, E^{\sim})$.

Before imposing any conditions on $\kappa(x)$ near infinity we develop some general theory.

It is clear from considerations as at the beginning of Section 12 that every $f \in \underline{\underline{F}}_{(e)}$, the extended space, satisfies the boundary condition

$$f(0) = 0 \; . \tag{13.3}$$

It is slightly deeper that (13.3) is also sufficient for $f \in \underline{\underline{F}}^{ref}$ to belong to $\underline{\underline{F}}_{(e)}$. The real point is that for fixed $\varepsilon > 0$ the functions

$$\begin{aligned} \varphi_n(x) &= 0 && \text{for } 0 \leq x < \varepsilon \quad \text{and for } x \geq n+1 \\ &= (x-\varepsilon)/\varepsilon && \text{for } \varepsilon \leq x < 2\varepsilon \\ &= 1 && \text{for } 2\varepsilon \leq x < n \\ &= (n+1-x) && \text{for } n \leq x < n+1 \end{aligned} \tag{13.4}$$

belong to $\underline{\underline{F}}^{ref} \cap C_{com}(\underline{\underline{X}}) = \underline{\underline{F}} \cap C_{com}(\underline{\underline{X}})$, are uniformly bounded in the norm (13.2) with the second term suppressed (with the bound depending on ε), and converge pointwise to 1 on the interval $(2\varepsilon, \infty)$. Thus if $f \in \underline{\underline{F}}^{ref}$ satisfies $0 \leq f \leq 1$ and is supported by $[2\varepsilon, \infty)$, then $f \wedge \varphi_n \to f$ pointwise and is uniformly bounded in E norm. Since each $f \wedge \varphi_n$ belongs to $\underline{\underline{F}}^{ref}$ and has compact support, it must belong to $\underline{\underline{F}}$ by SMP, Theorem 14. But then by SMP, paragraph 1.6.1' such $f \in \underline{\underline{F}}_{(e)}$. The restriction $0 \leq f \leq 1$ can be removed by splitting into positive and negative parts and then considering truncations, which disposes of

$f \in \underline{\underline{F}}^{ref}$ vanishing in a neighborhood of 0. If f satisfies (13.3) and is supported by a bounded interval, then it can be approximated by the translated functions

$$f_\varepsilon(x) = 0 \text{ for } 0 \leq x < 2\varepsilon \;;\; f_\varepsilon(x) = f(x-2\varepsilon) \text{ for } x \geq 2\varepsilon$$

which belong to $\underline{\underline{F}}_{(e)}$ by the above argument. Finally it is easy to check that any $f \in \underline{\underline{F}}^{ref}$ is the sum of a function vanishing in a neighborhood of 0 and a function supported by a bounded interval and sufficiency of (13.3) is proved. The reference Hilbert space is $L^2(\underline{\underline{X}}, m)$ and therefore

$$\underline{\underline{F}} = \underline{\underline{F}}_{(e)} \cap L^2(\underline{\underline{X}}, m) \;;\; \underline{\underline{F}}_a^{ref} = \underline{\underline{F}}^{ref} \cap L^2(\underline{\underline{X}}, m) .$$

The containments $\underline{\underline{F}} \subset \underline{\underline{F}}_{(e)}$ and $\underline{\underline{F}}_a^{ref} \subset \underline{\underline{F}}^{ref}$ are always proper, which will be of little consequence since there is equality for bounded functions. In the language of [32] the process associated with $(\underline{\underline{F}}, E)$ is the absorbing barrier diffusion on $(0, \infty)$ with standard scale, with speed measure $m(x)\,dx$ and with killing measure $\kappa(x)\,dx$. As in Section 12 a function h is harmonic if and only if it satisfies (12.7) on the appropriate interval. (The speed function $m(x)$ plays no role in this.) Let $\psi(x)$ be any solution of (12.7) which is bounded away from 0 and which is strictly increasing for sufficiently large x so that

$$(13.5) \qquad \eta_0(x) = \psi(x) \int_x^\infty dt\, \psi^{-2}(t)$$

converges for $x > 0$. The existence of such $\psi(x)$ is clear and convergence in (13.5) follows because $\psi(x)$ is convex and therefore dominates a multiple of x near ∞. Furthermore the integrability condition (13.1) guarantees $\psi(0) < +\infty$ and therefore $\eta_0(0) < +\infty$. Every bounded harmonic function is a multiple of η_0 and in particular

(13.6) $$\eta_\infty(x) = \psi(x)\int_0^x dt\, \psi^{-2}(t)$$

is unbounded near ∞. (The behavior of $\kappa(x)$ near ∞ is irrelevant for this.) It follows that the terminal algebra Ⓣ consists of the single atom $[X^{(R)}_{\zeta-0} = 0]$ and that

(13.7) $$p_0(x) = Ⓟ_x(X^{(R)}_{\zeta-0} = 0)$$

where $p_0(x) = \eta_0(x)/\eta_0(0)$ is the unique bounded solution of (12.7) satisfying the boundary condition

(13.8) $$p_0(0) = 1 .$$

The potential kernel $N(t,x)$ can be represented

(13.9) $$N(t,x) = 2C_\psi \eta_\infty(t \wedge x) p_0(t \vee x)$$

where now

(13.10) $$C_\psi = \{p_0(x)\eta_\infty'(x) - p_0'(x)\eta_\infty(x)\}^{-1}$$

and taking $x = 0$ we get

(13.10') $$C_\psi = \{1/\eta_\infty'(0)\} = \psi(0) .$$

There is no analogue for the "universal constant" C in Section 12.

We assume at first that also $\kappa(x)$ is integrable near infinity,

(13.11) $$\int_1^\infty dx\, \kappa(x) < +\infty$$

and we let the auxiliary space $\underline{\underline{M}}$ be a singleton

(13.12) $$\underline{\underline{M}} = \{\infty\} .$$

As in Section 12 we use the notation $\mu_0(x), \mu_\infty(x)$ for the jumping measures. The excursion space $\underline{\underline{N}}$ is the two dimensional subspace of functions on $\underline{\underline{T}} \cup \underline{\underline{M}}$ and we make the choice $\underline{\underline{H}} = \underline{\underline{N}}$. The boundary $\Delta = \{0, \infty\}$ and 0 is always adjoined so that for any finite R the subspace topology on $[0, R)$ is the usual one. Also it is always true (assuming (13.11)) that

(13.13) $$\Delta_r = \{0\} ; \Delta_s = \{\infty\} .$$

It seems to have no intrinsic significance but it is true that the manner in which ∞ and δ (and also 0 "near infinity") are adjoined depends on $\mu_0(x)$ and $\mu_\infty(x)$ as well as on $\kappa(x)$ itself. We illustrate this by examining two cases.

Case 1. We assume that $\kappa(x)$ has bounded support $(0, R]$. As above $\psi(x)$ is a fixed positive solution of (12.7) which is bounded away from 0 and strictly increasing for large x. We normalize so that

(13.14) $$\psi(x) = x \quad \text{for} \quad x \geq R .$$

This is always possible if $\kappa(x)$ is nontrivial. Then it is easy to check that for $x \geq R$

$$\eta_0(x) = 1 \quad ; \quad \eta_\infty(x) = Ax - 1$$

where $A = \int_0^\infty dt\, \psi^{-2}(t)$. Also $\eta_0(0) = A\psi(0)$ and therefore

$$p_0(x) = \eta_0(x)/A\psi(0) = \eta_0(x)/AC_\psi$$

$$N(x,t) = 2C_\psi \eta_\infty(t \wedge x) p_0(t \vee x) = (2/A)\eta_\infty(t \wedge x)\eta_0(t \vee x) .$$

Furthermore $p_0(x), N\kappa\mu_0(x)$ and $N\kappa\mu_\infty(x)$ all have limits as $x \to +\infty$ and $p_0(\infty) = 1/A\psi(0) > 0$ while the limits $N\kappa\mu_0(\infty), N\kappa\mu_\infty(\infty), N\kappa r(\infty)$ can vanish only if the corresponding function is trivial. Thus δ and ∞ are always adjoined as isolated points and except for the special case $\mu_0 = 1$, the half line $[0,\infty)$ has the usual locally compact topology. When $\mu_0 = 1$ the point 0 is adjoined so that also it acts as a "point at infinity" and therefore $[0,\infty)$ is compact. ///

Case 2. We make the specific choice

$$\begin{aligned} \kappa(x) &= x^{-2} && \text{for} \quad x \geq 1 \\ &= 2 && \text{for} \quad 0 \leq x \leq 1 \end{aligned}$$

and then it is easy to check that we can take

$$\begin{aligned} \psi(x) &= x^2 && \text{for} \quad x \geq 1 \\ &= e^{2x-2} && \text{for} \quad 0 < x < 1 . \end{aligned}$$

Again putting $A = \int_0^\infty dt\, \psi^{-2}(t)$ we have for $x \geq 1$

$$\eta_0(x) = 1/3x \quad ; \quad \eta_\infty(x) = Ax^2 - 1/3x$$

$$p_0(x) = \eta_0(x)/\eta_0(0) = (3A\psi(0)x)^{-1} .$$

Now let

$$\mu_\infty(x) = 1 \text{ for } (n-1)! < x \leq (n+1)!$$

whenever n is even but not divisible by 4 and let

$$\mu_0(x) = 1 \text{ for } (n-1)! < x \leq (n+1)!$$

whenever n is divisible by 4. Since $\mu_0 + \mu_\infty + r = 1$ this completely prescribes

$\mu_0(x), \mu_\infty(x), r(x)$ for $x \geq 1$. The values on any bounded interval and in particular for $x \leq 1$ will be irrelevant for what follows. If n is even but not divisible by 4 then

$$N\kappa\mu_\infty(n!) \geq (2/A)\eta_0(n!)\int_{(n-1)!}^{n!} dt\, \kappa(t)\eta_\infty(t) + (2/A)\eta_\infty(n!)\int_{n!}^{(n+1)!} dt\, \kappa(t)\eta_0(t)$$

$$= (2/A)(1/3n!)\int_{(n-1)!}^{n!} dt\, \{A-(1/3)t^{-3}\}$$

$$+ (2/A)\{A(n!)^2-(1/3n!)\}\int_{n!}^{(n+1)!} dt\, (1/3t^3)$$

which converges to 1 as $n \to \infty$. Thus $N\kappa\mu_\infty(n!) \to 1$ as $n \uparrow \infty$ through the even integers not divisible by 4. By a similar argument $N\kappa\mu_0(n!) \to 1$ as $n\uparrow\infty$ through the integers divisible by 4. It follows that both 0 and ∞ are adjoined as limit points at infinity (but not "the point at infinity") while δ is adjoined as an isolated point. Neither 0 nor ∞ have compact neighborhoods and therefore $\tilde{\underline{\underline{X}}}$ is not locally compact. ///

We consider now the relevant Dirichlet spaces and generators. In fact the situation is very similar to and indeed simpler than Case 4 in Section 12 and would not merit separate consideration except that, as mentioned above, we wish to make comparison with SMP, Section 24. The excursion form is given by

$$N(\varphi,\varphi) = \mu_\infty\kappa(\psi_0+N\kappa\mu_0)\{\varphi(0)-\varphi(\infty)\}^2 + r\kappa(\psi_0+N\kappa\mu_0)\varphi^2(0) + (r\kappa N\kappa\mu_\infty)\varphi^2(\infty)$$

and Q must have the form

$$Q(\varphi,\varphi) = a\varphi^2(0) + b\varphi^2(\infty) + c\{\varphi(0)-\varphi(\infty)\}^2$$

with $a, b, c \geq 0$. The reference Hilbert space for $(\underline{\underline{F}}^e, E^e)$ consists of functions f defined on $[0,\infty]$ with inner product

$$\int_0^\infty dx\, m(x) f^2(x) + f^2(0) + f^2(\infty) .$$

The Dirichlet space $\underline{\underline{F}}^e$ consists precisely of those f whose restrictions to $[0,\infty)$ belong to $\underline{\underline{F}}^{ref}$ and

(13.15)
$$E^e(f,f) = \frac{1}{2}\int_0^\infty dx\, \{f'(x)\}^2$$

$$+ \int_0^\infty dx\, \kappa(x)[\mu_0(x)\{f(x)-f(0)\}^2 + \mu_\infty(x)\{f(x)-f(\infty)\}^2 + r(x)f^2(x)]$$

$$+ af^2(0) + bf^2(\infty) + c\{f(0)-f(\infty)\}^2 .$$

Convergence in (13.15) does not depend on a choice for $f(\infty)$. If $f \in$ domain A^e, then by Theorem 11.1

(13.16)
$$m(x)A^e f(x) = \frac{1}{2}f''(x) - \kappa(x)\{f(x) - \mu_0(x)f(0) - \mu_\infty(x)f(\infty)\} \quad \text{on } (0,\infty)$$

(13.17)
$$A^e f(0) = (\partial f/\partial n)(0) - af(0) - c\{f(0)-f(\infty)\}$$

(13.18)
$$A^e f(\infty) = (\partial f/\partial n)(\infty) - bf(\infty) - c\{f(\infty)-f(0)\} .$$

Exactly as in Section 12

(13.19)
$$(\partial f/\partial n)(0) = \frac{1}{2}f'(0) + \int_0^\infty dx\, \kappa(x)\mu_0(x)\{f(x)-f(0)\}$$

(13.20)
$$(\partial f/\partial n)(\infty) = \int_0^\infty dx\, \kappa(x)\mu_\infty(x)\{f(x)-f(\infty)\}$$

and therefore (13.17), (13.18) are equivalent to

(13.17')
$$A^e f(0) = \frac{1}{2}f'(0) + \int_0^\infty dx\, \kappa(x)\mu_0(x)\{f(x)-f(0)\} - af(0) - c\{f(0)-f(\infty)\}$$

(13.18') $$A^e f(\infty) = \int_0^\infty dx\, \kappa(x)\mu_\infty(x)\{f(x)-f(\infty)\} - bf(\infty) - c\{f(\infty)-f()0\} .$$

Also the same reasoning as for Case 7 in Section 12 establishes

(13.21) $$f'(\infty) = 0.$$

Now (13.21) also follows from the fact that $f'(\infty)$ must exist and $f'(x) \in L^2(dx)$. Reasoning exactly as in Case 4 of Section 12, we see that the generator $A^\sim$ is defined by

(13.22) $$m(x)A^\sim f(x) = \frac{1}{2} f''(x) - \kappa(x)\{f(x) - (b+c+\kappa\, \mu_\infty)^{-1} \int_0^\infty dy\, \kappa(y)\mu_\infty(y)f(y)$$
$$- \mu_0(x)f(0) - (c/[b+c+\kappa\mu_\infty])\mu_\infty(x)f(0)\}$$

supplemented by the boundary condition

(13.23)
$$\frac{1}{2}f'(0) + \int_0^\infty dx\, \kappa(x)\mu_0(x)\{f(x)-f(0)\} + \{c/(b+c+\kappa\mu_\infty)\}\int_0^\infty dx\, \kappa(x)\mu_\infty(x)\{f(x)-f(0)\}$$
$$= af(0) + \{(cb)/(b+c+\kappa\mu_\infty)\}f(0)$$

at least when f and $A^\sim f$ are bounded. In general we cannot strip away this last restriction.

We look now at the Dirichlet space $(\underline{\underline{F}}^\sim, \underline{\underline{E}}^\sim)$. Clearly

$$Mf(\infty) = (b+c+\kappa\mu_\infty)^{-1} \int dx\, \kappa(x)\mu_\infty(x)f(x)$$

$$D_\infty f(\infty) = c/(b+c+\kappa\mu_\infty)f(1) .$$

The perturbed space $\underline{\underline{F}}^0 = \underline{\underline{F}}$ and the perturbed form E^0 is given by

$$E^0(f,f) =$$

$$= \frac{1}{2}\int dx\ \{f'(x)\}^2 + \frac{1}{2}(b+c+\kappa\mu_\infty)^{-1}\int_0^\infty dx\ \kappa(x)\mu_\infty(x)\int_0^\infty dy\ \kappa(y)\mu_\infty(y)\{f(x)-f(y)\}^2$$

$$+ \int_0^\infty dx\ \kappa(x)\{r(x)+\mu_0(x)\}f^2(x)$$

$$+ \{(b+c)/(b+c+\kappa\mu_\infty)\}\int_0^\infty dx\ \kappa(x)\mu_\infty(x)f^2(x) .$$

The full Dirichlet space $\underline{\underline{F}}^\sim = \underline{\underline{F}}^{ref}$ and for $f \in \underline{\underline{F}}^\sim$

(13.24) $$E^\sim(f,f) = \frac{1}{2}\int_0^\infty dx\ \{f'(x)\}^2 + \int_0^\infty dx\ \kappa(x)r(x)f^2(x)$$

$$+ \int_0^\infty dx\ \kappa(x)\mu_0(x)\{f(x)-f(0)\}^2 + af^2(0) + B(f,f)$$

where

$$B(f,f) = \int_0^\infty dx\ \kappa(x)\mu_\infty(x)\{f(x)-Mf(\infty)-D_\infty f(\infty)\}^2$$

$$+ b\{Mf(\infty)+D_\infty f(\infty)\}^2 + c\{f(0)-Mf(\infty)-D_\infty f(\infty)\}^2 .$$

Again we are using the notation and results in Section 5. Now

$$\frac{1}{2}\Theta^r\langle\gamma_r f, \gamma_r f\rangle = 0$$

$$\iint_{\underline{\underline{X}}\times\Delta_r} \kappa\mu 1_{\Delta_s} D_\infty(dx,dy)\{f(x)-\gamma_r f(y)\}^2 = \{c/(b+c+\kappa\mu_\infty)\}\int_0^\infty dx\ \kappa(x)\mu_\infty(x)\{f(x)-f(0)\}^2$$

$$\int \kappa(dx)f^2(x)\mu\{1-M1-D_\infty 1\}(x) = \{b/(b+c+\kappa\mu_\infty)\}\int_0^\infty dx\ \kappa(x)\mu_\infty(x)f^2(x)$$

$$\int_{\Delta_r} \nu(dy)q(y)\{\gamma_r f(y)\}^2 = \{(cb)/(b+c+\kappa\mu_\infty)\}f^2(0)$$

and therefore

$$B(f,f) = \frac{1}{2}(b+c+\kappa\mu_\infty)^{-1}\int_0^\infty dx\ \kappa(x)\mu_\infty(x)\int_0^\infty dy\ \kappa(y)\mu_\infty(y)\{f(x)-f(y)\}^2$$

$$+ \{c/(b+c+\kappa\mu_\infty)\}\int_0^\infty dx\ \kappa(x)\mu_\infty(x)\{f(x)-f(0)\}^2$$

$$+ \{b/(b+c+\kappa\mu_\infty)\}\int_0^\infty dx\ \kappa(x)\mu_\infty(x)f^2(x)$$

$$+ \{(cb)/(b+c+\kappa\mu_\infty)\}f^2(0)$$

and after plugging into (13.25)

$$E^{\sim}(f,f) = \frac{1}{2}\int_0^\infty dx\{f'(x)\}^2 + \int_0^\infty dx\ \kappa(x)r(x)f^2(x) + \int_0^\infty dx\ \kappa(x)\mu_0(x)\{f(x)-f(0)\}^2$$

$$+ \frac{1}{2}(b+c+\kappa\mu_\infty)^{-1}\int_0^\infty dx\ \kappa(x)\mu_\infty(x)\int_0^\infty dy\ \kappa(y)\mu_\infty(y)\{f(x)-f(y)\}^2$$

$$+ \{c/(b+c+\kappa\mu_\infty)\}\int_0^\infty dx\ \kappa(x)\mu_\infty(x)\{f(x)-f(0)\}^2$$

$$+ \{b/(b+c+\kappa\mu_\infty)\}\int_0^\infty dx\ \kappa(x)\mu_\infty(x)f^2(x)$$

$$+ \{a+(cb)/(b+c+\kappa\mu_\infty)\}f^2(0)\ .$$

This agrees with formula (24.13) in SMP. Collapsing the time scale thus annihilates the point ∞ and the net effect is to introduce a "jumping term" and to replace $r(x)$, $\mu_0(x)$, a by $r(x) + \{b/(b+c+\kappa\mu_\infty)\}\mu_\infty(x)$, $\mu_0(x) + \{c/(b+c+\kappa\mu_\infty)\}\mu_\infty(x)$, $\{a+(cb)/(b+c+\kappa\mu_\infty)\}$. Finally, application of Theorems 11.2 and 11.3 yields the same description as before for the generator $A^{\sim}$.

In the remainder of this section we assume that $\kappa(x)$ is not integrable near infinity. On page 24.1 in SMP it was stated that if condition (i) of Theorem 20.2 is satisfied then there is no real loss of generality in assuming that the boundary consists of at most one point other than a point which corresponds to the atom in the terminal algebra. After translating this statement into the terminology of Sections 3 and 4 and taking into account the error discussed in the first remark following Corollary 4.4, we would infer that if the singular boundary $\Delta_s = \phi$ then Δ can consist of at most two points. This is false. Indeed we will now construct an example where $\Delta_s = \phi$ and $\Delta_r = [0,1)$, the half open unit interval.

Let $\kappa(x) = 1$ for all x and let $\{R_n\}_{n=1}^{\infty}$ be a sequence of positive numbers such that the reciprocals converge

(13.25) $$\Sigma_{n=1}^{\infty} R_n^{-1} < +\infty .$$

Define a mapping λ from the half line $\underline{\underline{X}} = (0,\infty)$ into the half open interval $[0,1)$ by the prescription

(13.26) $$\lambda x = (x-(R_1+\ldots+R_n))/R_{n+1} \quad \text{for } R_1+\ldots+R_n \leq x < R_1+\ldots+R_{n+1} .$$

For the auxiliary space $\underline{\underline{M}}$ we take the closed interval

(13.27) $$\underline{\underline{M}} = [0,1] .$$

For each $x \in \underline{\underline{X}}$ let the jumping meausre $\mu(x,\cdot)$ be the point mass concentrated at λx. The excursion form is defined on pairs (Φ,φ) living on $\underline{\underline{T}} \cup \underline{\underline{M}}$ by

(13.28) $$N((\Phi,\varphi),(\Phi,\varphi)) = \frac{1}{2}\int_0^{\infty} dx \int_0^{\infty} dy\, N(x,y)\{\varphi(\lambda x)-\varphi(\lambda y)\}^2 + \int_0^{\infty} dx\, p_0(x)\{\Phi-\varphi(\lambda x)\}^2$$

where $N(x,y) = \sqrt{2}\,\eta_\infty(x\wedge y)p_0(x\vee y)$ with $\eta_\infty(x) = \sinh\sqrt{2}\,x$ and $p_0(x) = e^{-\sqrt{2}\,x}$. Note that the second term always converges for φ bounded. Let $\underline{\underline{W}}_p([0,1])$ be the subspace of functions φ in the Sobolev space of paragraph 12.1 which satisfy the periodic boundary condition

(13.29) $$\varphi(0) = \varphi(1) = 0.$$

Fix $\varphi \in \underline{\underline{W}}_p([0,1])$ and define f on $\underline{\underline{X}} = (0,\infty)$ by $f(x) = \varphi(\lambda x)$. Then

$$\frac{1}{2}\int_0^\infty dx\, \{f'(x)\}^2 + \int_0^\infty dx \int \mu(x, ds)\{f(x) - \varphi(s)\}^2$$

$$= \frac{1}{2}\int_0^\infty dx\, \{f'(x)\}^2 = \frac{1}{2}\Sigma_{n=1}^\infty (1/R_n)\int_0^1 ds\, \{\varphi'(s)\}^2 < +\infty .$$

We conclude from Theorem 3.2 that f has a representation $f = f_0 + f(0)p_0 + N\kappa\mu\varphi$ with $f_0 \in \underline{\underline{F}}_{(e)}$ and that

$$\frac{1}{2}\int_0^\infty dx\, \{f'(x)\}^2 = E(f_0, f_0) + N((\gamma_T f, \varphi), (\gamma_T f, \varphi)) .$$

In particular the excursion space $\underline{\underline{N}}$ contains any pair (Φ, φ) with $\varphi \in \underline{\underline{W}}_p$. Notice that we were able to deduce this without examining the specific formula (13.28). For the Dirichlet space $\underline{\underline{H}}$ we take all pairs (Φ, φ) with $\varphi \in \underline{\underline{W}}_p$ and with $\Phi = \varphi(0)$ on $\underline{\underline{T}}$. Then $\Delta = [0,1]$ with 0 and 1 identified and $\underline{\underline{X}} \cup \Delta$ is topologized in such a way that $0 \in \Delta$ acts also as the endpoint $0 \in [0,\infty)$. Also all of Δ is adjoined as a set of limit points at infinity. It is reassuring that by Appendix A we need not concern ourselves with the details of this adjunction. For harmonic measure we take Lebesgue measure on (0,1) and counting measure on {0}. For φ defined on Δ

$$H\varphi(x) = \varphi(0)p_0(x) + N(\varphi \cdot \lambda)(x)$$

and

$$(13.30) \quad U_{0,\infty}(1,\varphi^2) = \varphi^2(0)\mathrm{Lim}_{v\uparrow\infty} v\int_0^\infty dx\, m(x)(1-vG_v)p_0(x) + \int_0^\infty dy\, \varphi^2(\lambda y)\,.$$

Since p_0 is not in $\underline{\underline{F}}_{(e)}$ the first term is infinite whenever $\varphi(0) \neq 0$. The second term diverges whenever φ is non-null on $(0,1)$. Thus (13.30) diverges for any nontrivial $\varphi \in L^2(\Delta,\nu)$ and therefore for any nontrivial $\varphi \in \underline{\underline{H}}$. This guarantees that the singular boundary $\Delta_s = \phi$. For the Dirichlet form Q we take

$$Q(\varphi,\varphi) = N(\varphi,\varphi) + \frac{1}{2}\int_0^1 dt\, \{\varphi'(t)\}^2 + \int_0^1 dt\, \kappa_\Delta(t)\varphi^2(t) + a\varphi^2(0)$$

with $a \geq 0$ and κ_Δ an integrable function on $(0,1)$. A function f belongs to the Dirichlet space $\underline{\underline{F}}^\sim$ if and only if there exists a function $\gamma f \in \underline{\underline{H}}$, necessarily unique, such that

$$(13.31) \qquad f(0) = \gamma f(0)\ ;\ \int_0^\infty dx\, \{f(x) - \gamma f(\lambda x)\}^2 < +\infty$$

and then

$$(13.32) \qquad E^\sim(f,f) = \frac{1}{2}\int_0^\infty dx\, \{f'(x)\}^2 + \int_0^\infty dx\, \{f(x)-\gamma f(\lambda x)\}^2$$

$$+ af^2(0) + \frac{1}{2}\int_0^1 dt\, \{\gamma f'(t)\}^2 + \int_0^1 dt\, \kappa_\Delta(t)\{\gamma f(t)\}^2\,.$$

By Theorem 10.1 the generator $A^\sim$ is defined by

$$(13.33) \qquad m(x)A^\sim f(x) = \frac{1}{2}f''(x) - f(x) + \gamma f(\lambda x)$$

supplemented by the boundary condition

$$(13.34) \qquad (\partial f/\partial n)(\psi) = \frac{1}{2}\int_0^1 dt\, \gamma f'(t)\psi'(t) + \int_0^1 dt\, \kappa_\Delta(t)\gamma f(t)\psi(t) + a\gamma f(0)\psi(0)\,.$$

for general $\psi \in \underline{\underline{H}}$.

We assume now that f satisfies all the above conditions for being in domain $\tilde{A}$ except possibly (13.34) and consider the problem of simplifying this boundary condition. At present our results on this point are incomplete. It is helpful to separate out the function

$$\eta(x) = f(x) - \gamma f(\lambda x) .$$

Clearly $\eta \in L^2(dx) \cap \underline{\underline{W}}$ and therefore η is continuous and vanishes at infinity. (The point is that η agrees near $+\infty$ with a function whose Fourier transform is integrable.) Thus f itself is bounded and uniformly continuous. For $\psi \in \underline{\underline{H}}$ and with $D_n = (1/n, n)$, we can write

$$(13.35) \qquad \text{Ⓔ}^{ex}[\sigma(D_n)<+\infty;\psi(X^{ex}_{\zeta^*-0});\{f(X_{\sigma(D_n)}) - \gamma f(X^{ex}_{\zeta^*-0})\}]$$

$$= \psi(0)\ell_n\{f(1/n)-f(0)\} + \int_{1/n}^{n} dx\ \{f(x)-\gamma f(\lambda x)\}\psi(\lambda x)$$

$$+ \int_0^{1/n} dx\ \{f(1/n)-\gamma f(\lambda x)\}\psi(\lambda x)H^n 1(x)$$

$$+ \int_n^{\infty} dx\ \{f(n)-\gamma f(\lambda x)\}\psi(\lambda x)H^n 1(x) .$$

The first term on the right in (13.35) is the contribution from the terminal normal derivative. The constant ℓ_n is the mass of the terminal equilibrium distribution ℓ_n^0 which must be concentrated at $1/n$. Just as in Section 12 we have $\ell_n \sim n/2$. Since the right side of (13.33) is integrable, f'' must be integrable near 0. Thus $f'(0)$ exists and so the first term on the right in (13.35) converges to $\frac{1}{2}f'(0)$. The third term goes to zero by trivial estimates. Also the fourth term goes to zero, but this requires some argument. Note first that $H^n 1(x) = e^{-\sqrt{2}(x-n)}$ for $x > n$ and so the fourth term equals

$$\int_n^\infty dx\, e^{-\sqrt{2}(x-n)}\{f(n)-\gamma f(\lambda x)\}$$

which is asymptotically equivalent to

$$\int_n^\infty dx\, e^{-\sqrt{2}(x-n)}\{\gamma f(\lambda n)-\gamma f(\lambda x)\} \tag{13.36}$$

since $\eta(x)$ is continuous and vanishes at infinity. Fix $\varepsilon > 0$ and let $\delta > 0$ be such that $|\gamma f(s)-\gamma f(t)| < \varepsilon$ whenever $(s-t) < \delta$. If $R_1 + \ldots + R_{m-1} \leq n < R_1 + \ldots + R_m$ and if $R^*(n) = \min\{R_m, R_{m+1}, \ldots\}$ then clearly $|\lambda n-\lambda x| < \delta$ whenever $x \leq n + R^*(n)\delta$ and it follows that (13.36) is dominated by

$$\varepsilon + 2\,\|\gamma f\|_\infty \int_{\delta R^*(n)}^\infty dx\, e^{-\sqrt{2}\,x} .$$

Clearly $R^*(n)\uparrow\infty$ as $n\uparrow\infty$ and we conclude that

$$(\partial f/\partial n)(\psi) = \frac{1}{2} f'(0)\psi(0) + \mathrm{Lim}_{n\uparrow\infty}\int_0^n dx\, \{f(n)-\gamma f(\lambda x)\}\psi(\lambda x).$$

and therefore (13.34) is equivalent to

$$\frac{1}{2} f'(0)\psi(0) + \mathrm{Lim}_{n\uparrow\infty}\int_0^n dx\, \{f(x)-\gamma f(\lambda x)\}\psi(\lambda x) \tag{13.34'}$$

$$= \frac{1}{2}\int_0^1 dt\, \gamma f'(t)\psi'(t) + \int_0^1 dt\, \kappa_\Delta(t)\gamma f(t)\psi(t) + a\gamma f(0)\psi(0) .$$

Of course there is no need to restrict n to integers in this argument. Suppose now that $(\gamma f)'$ is in bounded variation on Δ so that $\frac{1}{2}(\gamma f)''$ is in the distribution sense a signed measure in bounded variation on Δ. Then also $\frac{1}{2}(\gamma f\cdot\lambda)''$ is a signed measure in bounded variation on $\underline{\underline{X}}$ and therefore the same must be true for $\frac{1}{2}\eta''(x) - \eta(x)$. Since $\eta(x)$ is bounded, it must have a representation

$$\eta(x) = ce^{-\sqrt{2}\,x} + \int_0^\infty N(x,t)\lambda(dt)$$

where λ is the signed measure $\frac{1}{2}\eta'' - \eta$. This guarantees that $\eta(x)$ is integrable and therefore

(13.37) $$\int_0^\infty dx\ \{f(x) - \gamma f(\lambda x)\}$$

converges absolutely so that (13.34') can now be replaced by

(13.34'') $$\frac{1}{2} f'(0)\psi(0) + \int_0^\infty dx\ \{f(x) - \gamma f(\lambda x)\}\psi(\lambda x)$$

$$= \frac{1}{2}\int_0^1 dt\ \gamma f'(t)\psi'(t) + \int_0^1 dt\ \kappa_\Delta(t)\gamma f(t)\psi(t) + a\gamma f(0)\psi(0)\ .$$

If (13.37) converges then for $t \in \Delta$ the series

(13.38) $$f^{\#}(t) = \Sigma_{n=1}^\infty R_n\{f(R_1 + \ldots + R_{n-1} + tR_n) - \gamma f(t)\}$$

converges absolutely to an integrable function on Δ and (13.34'') is equivalent to

(13.34''') $$\frac{1}{2} f'(0)\psi(0) + \int_0^1 dt\ f^{\#}(t)\psi(t)$$

$$= \frac{1}{2}\int_0^1 dt\ \gamma f'(t)\psi'(t) + \int_0^1 dt\ \kappa_\Delta(t)\gamma f(t)\psi(t) + a\gamma f(0)\psi(0)\ .$$

Taking $\psi \in C^1_{com}((0,1))$ in (13.34''') we conclude that $(\gamma f)'$ is absolutely continuous on $(0,1)$ and that

(13.39) $$\frac{1}{2}(\gamma f)''(t) - \kappa_\Delta(t)\gamma f(t) = -f^{\#}(t) \qquad \text{on } (0,1)\ .$$

Taking $\psi = 1$ in (13.34''') and using (13.39), we get

$$\frac{1}{2} f'(0) - \frac{1}{2}\int_0^1 dt\ (\gamma f)''(t) = a\gamma f(0)\psi(0)$$

which is the same as

(13.40) $$\frac{1}{2}\{f'(0)+(\gamma f)'(0)-(\gamma f)'(1)\} = a\gamma f(0)\ .$$

We conclude that (13.34) can be replaced by (13.39) and (13.40) if it is known that $(\gamma f)'$ is in bounded variation. However we do not have answers to either of the following two questions.

13.1. Does (13.37) converge absolutely for general $f \in \underline{F}^{\sim}$ such that the right side of (13.33) is integrable?

13.2. What are the possibilities for γf when $f \in$ domain $A^{\sim}$?

The above arguments do show that these questions are related. If the answer to 13.1 is yes, then $f \in$ domain $A^{\sim}$ implies that $(\gamma f)'$ is absolutely continuous on Δ.

We finish this section by modifying the above example in order to illustrate the first remark following Corollary 4.4. Let Γ be any subset of $(0,1)$ which has positive Lebesgue measure but is nowhere dense. Let

$$\kappa(x) = 1 \text{ if } x < R_1 \text{ or if } \lambda x \in [0,1)\backslash\Gamma$$
$$0 \text{ if } x \geq R_1 \text{ and } \lambda x \in \Gamma .$$

Let $\underline{H}$ and therefore Δ be the same as before. If φ is the indicator of Γ, then clearly

$$U_{0,\infty}(1,\varphi^2) = \int_0^\infty dx\ \kappa(x) \int_0^{R_1} dy\ N(x,y) I(y \in R_1\Gamma)$$

converges. But it is still true that $U_{0,\infty}(1,\varphi^2) = +\infty$ for all $\varphi \in \underline{\underline{H}}$ and therefore $\Delta_s = \phi$.

14. Brownian Motion on Euclidean Space

In this section $\underline{\underline{X}} = \underline{\underline{R}}^d$, Euclidean d-space and the reference measure is $m(x)\,dx$ with $m(x)$ integrable and everywhere positive. Now $\underline{\underline{W}}$ denotes the collection of locally integrable functions f on $\underline{\underline{R}}^d$ such that the gradient grad f is in $L^2(dx)$. Operators such as grad and the Laplacian Δ will always be understood in the distribution sense. The reflected space $\underline{\underline{F}}^{ref} = \underline{\underline{W}}$ and for $f \in \underline{\underline{F}}^{ref}$

(14.1) $$E(f, f) = \frac{1}{2} \int_{R^d} dx \, \{\text{grad}\, f(x)\}^2 .$$

Of course $\underline{\underline{F}}_{(e)}$ is the E closure of $\underline{\underline{W}} \cap C_{com}(R^d)$ and $\underline{\underline{F}} = \underline{\underline{F}}_{(e)} \cap L^2(m)$. The process associated with $(\underline{\underline{F}}, E)$ is a time changed standard Brownian motion.

We will look successively at the cases $d = 1, 2$ and 3. For $d = 1$ and 2 the Dirichlet space $(\underline{\underline{F}}, E)$ is recurrent and $\underline{\underline{F}}^{ref} = \underline{\underline{F}}_{(e)}$. The semigroup $\{P_t\}$ is strictly Markov and therefore is dominated by no other submarkov semigroup, symmetric or not. In particular $(\underline{\underline{F}}, E)$ has no extensions in the sense of SMP, Section 20. For $d = 3$ the space $(\underline{\underline{F}}, E)$ is transient and $\underline{\underline{F}}_{(e)}$ is a proper subspace of $\underline{\underline{F}}^{ref}$. We will characterize $\underline{\underline{F}}_{(e)}$ and also study extensions of $(\underline{\underline{F}}, E)$. It turns out that the "normal derivative at infinity" is fundamental for the latter. For $d > 3$ the situation is essentially the same as for $d = 3$.

Before beginning we establish once and for all our convention that the Fourier transform Ⓕ is defined for f on R^d by

(14.2) $$Ⓕ f(y) = \int dx \, e^{-ix \cdot y} f(x)$$

so that the inverse Fourier transform is given by

(14.2') $$Ⓕ^{-1} g(x) = (2\pi)^{-d} \int dy \, e^{ix \cdot y} g(y) .$$

Now let d = 1. The argument at the beginning of Section 13 can easily be adapted to show that $\underline{\underline{F}}_{(e)} = \underline{\underline{F}}^{ref}$ and therefore $(\underline{\underline{F}}, E)$ is recurrent by SMP, Theorem 1.9. It then follows by SMP, Theorem 1.6 that P_t is strictly Markov and so $(\underline{\underline{F}}, E)$ has no extensions. Before going on we establish another consequence of recurrence which will turn out to have special significance for us. Suppose that $f \in \underline{\underline{W}}$ and the second derivative f'' is integrable. Integration by parts in (14.1) gives

$$\frac{1}{2}\int_{-\infty}^{+\infty} dx\, f'(x)g'(x) + \frac{1}{2}\int_{-\infty}^{+\infty} dx\, f''(x)g(x) = 0$$

for $g \in C^1_{com}(R)$. But the function 1 can be approximated by such g and it follows that necessarily

$$\int_{-\infty}^{+\infty} dx\, f''(x) = 0 . \tag{14.3}$$

Also this is suggested by, but does not actually follow from Theorem 7.1. We can also prove this directly using Fourier transforms. Any $f \in \underline{\underline{W}}$ is dominated by a linear term near infinity since f' is the sum of a bounded function and an integrable function. In particular f is a tempered distribution and therefore Ⓕf is well defined as a tempered distribution. (This is true for any d.) Also Ⓕ$f''(y) = -y^2$Ⓕ$f(y)$ and since f'' is integrable it follows that $\alpha = \mathrm{Lim}_{y\downarrow 0} y^2$Ⓕ$f(y)$ exists and that $-\alpha = \int dx\, f''(x)$. If $\alpha \neq 0$ then yⒻ$f(y) \sim \alpha/y$ as $y \to 0$. But this is impossible since $f \in \underline{\underline{W}}$ and therefore yⒻ$f(y)$ is square integrable.

Next let d = 2. Let φ be any function on $\underline{\underline{R}}^2$ such that $\varphi \in C^1_{com}(R)$ and $\varphi(0) = 1$. As $N\uparrow\infty$ the functions $\varphi_N(x) = \varphi(x/N)$ belong to $\underline{\underline{F}}$ and converge pointwise to 1 and $E(\varphi_N, \varphi_N)$ is independent of N. Thus $1 \in \underline{\underline{F}}_{(e)}$ and it follows again that $(\underline{\underline{F}}, E)$ is recurrent and no extensions exist. Just as for n = 1 it is true that if $f \in \underline{\underline{W}}$ and if Δf is integrable, then necessarily

(14.4) $$\int_{\underline{\underline{R}}^2} dx\, \Delta f(x) = 0 .$$

From now on in this section we assume $d = 3$. The most direct way to show that $(\underline{\underline{F}}, E)$ is transient is to assume instead that the reference measure is Lebesgue measure dx itself. By SMP, Theorem 8.2 transience does not depend on the choice of a reference measure. It is well-known then that the transition operators have density $p_t(x,y) = (2\pi t)^{-\frac{3}{2}} \exp\{-|x-y|^2/2t\}$ and the change of variables $s = |x-y|^2/2t$ is enough to establish the formula

(14.5) $$\int_0^\infty dt\, p_t(x,y) = 1/2\pi|x-y|$$

from which it follows that Nf is finite almost everywhere for $f \in L^1(dx)$. We will give two additional proofs that $(\underline{\underline{F}}, E)$ is transient.

The first proof was shown to the author by S. Port and C. Stone. Let f_n, $n \geq 1$ be a sequence in $\underline{\underline{W}}$ such that $\sup E(f_n, f_n) < +\infty$. The estimate

$$\int_{|y|\leq\delta} dy\, |ⒻF f_n(y)| \leq \left\{\int dy\, |y|^2 |ⒻF f_n(y)|^2\right\} \int_{|y|\leq\delta} dy\, |y|^{-2}$$

shows that the left side converges to 0 with δ uniformly in n and it follows in particular that the Ⓕf_n cannot converge in the tempered distribution sense to a nonzero measure concentrated at the origin. But then it is impossible that $f_n \to 1$ boundedly and almost everywhere which is enough to guarantee that $(\underline{\underline{F}}, E)$ is transient.

For the second proof let

$$f(x) = Ⓕ^{-1} |y|^{-2} e^{-|y|^2/2}$$

defined in the sense of tempered distributions. Then f is C^∞ and vanishes at infinity, $\operatorname{grad} f \in L^2(dx)$, and $-\Delta f(x) = (2\pi)^{-3/2} e^{-|x|^2/2}$. Thus $f \in \underline{\underline{W}}$, but Δf is integrable and

(14.6) $$\int_{\underline{\underline{R}}^3} dx\, \Delta f(x) \neq 0 .$$

If $(\underline{\underline{F}}, E)$ were recurrent, then the argument for $n = 1$ would show that (14.6) is impossible, so we conclude again that $(\underline{\underline{F}}, E)$ is transient.

Let $D \subset \underline{\underline{R}}^3$ be open and denote the complement $\underline{\underline{R}}^3 \setminus D$ by M. If $f \in \underline{\underline{W}}$ is harmonic on D (that is, if $H^M f = f$), then

$$\frac{1}{2} \int dx \operatorname{grad} f(x) \operatorname{grad} g(x) = 0$$

whenever $g \in \underline{\underline{W}} \cap C_{com}(D)$ and it follows that $\Delta f = 0$ on D in the distribution sense and therefore f is harmonic in the classical sense.

Every function which is bounded and harmonic on $\underline{\underline{R}}^3$ is constant. This can be proved for example by differentiating the Poisson formula for a sphere of radius R and letting $R \uparrow \infty$ to show that the gradient vanishes identically. For a proof in a more general context we refer to the paper of Choquet and Deny [6].

Since $(\underline{\underline{F}}, E)$ is transient the terminal set $\underline{\underline{T}}$ is nonempty. Since every bounded harmonic function in $\underline{\underline{W}}$ is constant, the terminal algebra Ⓣ consists only of a single atom which we denote by ∞. In the notation of Section 3, $\Delta = \{\infty\}$ and since $He_\infty = 1$ the union $\underline{\underline{R}}^3 \cup \{\infty\}$ is topologized so that ∞ acts as the point at infinity. The dead point δ is adjoined as an isolated point and can be ignored.

The calculation (14.5) is enough to insure that the potential operator N has a density

(14.7) $$N(x, y) = 1/2\pi|x-y| .$$

In paragraph 7.1 we let D_n be the open ball of radius n. It is clear that the equillibrium measure ℓ_n^{∞} is a multiple of surface measure $d\sigma_n$ on the sphere ∂D_n. Thus there exists a constant ℓ_n^{∞} such that

$$\int \ell_n^{\infty}(dx)f(x) = \ell_n^{\infty} \int_{\partial D_n} d\sigma_n f(\sigma_n)$$

This constant is determined by

$$1 = \ell_n^{\infty} \int_{\partial D_n} d\sigma_n N(0, \sigma_n)$$

which gives

$$\ell_n^{\infty} = 1/2n\ . \tag{14.8}$$

The formula (7.4') is valid whenever w(x) is integrable and w(x) dx has finite energy. Taking w(x) spherically symmetric and bounded with compact support, we get

$$\gamma_T f(\infty) = \operatorname{Lim}_{n\uparrow\infty} (4\pi n^2)^{-1} \int_{\partial D_n} d\sigma_n f(\sigma_n)\ . \tag{14.9}$$

That is, (14.9) exists for all $f \in \underline{\underline{W}}$ and $f \in \underline{\underline{F}}_{(e)}$ if and only if $\gamma_T f(\infty) = 0$. Of course there is no need to restrict to integers n in taking the limit. For this case the alternative formula (7.7') also reduces to (14.9).

Our prescription (7.24) for the terminal normal derivative becomes

$$(\partial f/\partial n)(\infty) = \operatorname{Lim}_{n\uparrow\infty} (1/2n) \int_{\partial D_n} d\sigma_n \{f(\sigma_n) - \gamma f(\infty)\}\ . \tag{14.10}$$

and an easy special case of Theorem 7.2 guarantees that if $f \in \underline{\underline{W}}$ and if Δf is integrable, then the limit (17.10) exists and

$$(\partial f/\partial n)(\infty) + \frac{1}{2}\int dx\, \Delta f(x) = 0. \tag{14.11}$$

We do not know if such results have been obtained previously.

The generator A is defined by

$$m(x)Af(x) = \frac{1}{2}f''(x)\ , \tag{14.12}$$

supplemented by the boundary condition

$$\gamma f(\infty) = 0 \tag{14.13}$$

at least when f is bounded. Since there may exist harmonic functions in $L^2(m)$, the restriction to f bounded cannot be stripped away without replacing it by a requirement such as "$f \in \underline{\underline{W}}$."

The only possibilities for $(\underline{\underline{F}}^{\sim}, E^{\sim})$ are

$$\underline{\underline{F}}^{\sim}_{(e)} = \underline{\underline{W}}$$

$$E^{\sim}(f,f) = \frac{1}{2}\int dx\,\{\text{grad } f(x)\}^2 + a\{\gamma f(\infty)\}^2$$

with $a \geq 0$. The generator $A^{\sim}$ is defined by

$$m(x)A^{\sim}f(x) = \frac{1}{2}f''(x) \tag{14.14}$$

supplemented by

$$(\partial f/\partial n)(\infty) = a\gamma f(\infty) \tag{14.15}$$

at least for f bounded and again the restriction cannot be stripped away in general.

We have noted above that the existence of $f \in \underline{\underline{W}}$ for which Δf is integrable and (14.6) is valid can be used to establish the transience of $(\underline{\underline{F}}, E)$. Conversely we can now show that the transience of $(\underline{\underline{F}}, E)$ can be used to give a noncomputational proof for the existence of such f. The point is that if $f \in$ domain $A^{\sim}$ and if $a > 0$, then the identity (14.15) implies $f \in \underline{\underline{F}}_{(e)}$ unless (14.6) is true. But not every $f \in$ domain $A^{\sim}$ can belong to $\underline{\underline{F}}_{(e)}$.

15. Brownian Motion in a Half Plane.

In this section $\underline{\underline{X}} = \underline{\underline{R}}^{2,+}$, the open half plane. A typical point $x \in \underline{\underline{R}}^{2,+}$ is a pair (a, b) with $b > 0$. In previous sections we chose a bounded measure for reference measure to avoid certain technical problems. It turns out that it is more convenient here to choose Lebesgue measure dx for reference measure.

The Sobolev space $\underline{\underline{W}}$ and the Dirichlet form E are defined exactly as in Section 14 with $\underline{\underline{R}}^{2,+}$ playing the role of $\underline{\underline{R}}^d$. Again $f \in \underline{\underline{W}}$ is harmonic if and only if it is harmonic in the classical sense. We take for granted the well-known classical result that every bounded harmonic function h(x) has a unique representation

$$h(a,b) = \int_{-\infty}^{+\infty} dt\ c_b(a-t)\gamma h(t) \tag{15.1}$$

where γh is a bounded function on $\underline{\underline{R}}$ which is unique up to dt equivalence and where

$$c_b(t) = b\{\pi(b^2+t^2)\}^{-1} \tag{15.2}$$

Thus we can identify the terminal set $\underline{\underline{T}}$ with $\underline{\underline{R}}$ and the terminal algebra Ⓣ with the Borel algebra on $\underline{\underline{R}}$, two sets being equivalent if their symmetric difference is null for Lebesgue measure dt. At the same time in this section we establish

15.1. Convention. The one sided limits $X_{\zeta-0}$, X_{ζ^*} are understood to be the appropriate limits on the real axis $\underline{\underline{R}}$, rather than a dead point ∂.

It is clear that the 1-balayage of dx onto the boundary $\underline{\underline{R}}$ is translation invariant and therefore either it is not a Radon measure or else it is a multiple of Lebesgue measure on $\underline{\underline{R}}$. The latter is the case but seems to require a nontrivial proof. Rather than estimate the kernel for representing 1- harmonic functions, we will use the following identity involving the Feller kernels of Section 2

(15.3) $$U_{0,1}(1,\varphi^2) = \frac{1}{2}U_{0,1}\langle\varphi,\varphi\rangle + U_{0,1}(\varphi,\varphi).$$

The identity (15.3) itself depends only on symmetry. The real point is that H1 = 1 and therefore

(15.4) $$U_{0,1}(1,\varphi^2) = \int_{\underline{\underline{R}}^{2,+}} dx\, H_1\varphi^2(x)\ .$$

Thus to show that the 1-balayage of dx is Radon and therefore a multiple of dt, it suffices to show that both terms on the right side of (15.3) converge for at least one φ which is bounded away from 0 on an interval.

We use the partial Fourier transform

$$\textcircled{F}_I f(\lambda,b) = \int_{-\infty}^{+\infty} da\, e^{-i\lambda a} f(a,b)\ .$$

From the identity

$$\int_{-\infty}^{+\infty} dt\, e^{-i\lambda t}(1+t^2)^{-1} = \pi e^{-|\lambda|}$$

which can be established using the calculus of residues, follows after change of variables

(15.5) $$\textcircled{F}_1\, H\varphi(\lambda,b) = e^{-b|\lambda|}\,\textcircled{F}\varphi(\lambda)$$

at least for $\varphi \in L^2(dt)$. Therefore by the Plancherel formula

(15.6) $$\int_{\underline{\underline{R}}^{2,+}} dx\,\{H\varphi(x)\}^2 = (2\pi)^{-1}\int_0^\infty db \int_{-\infty}^{+\infty} d\lambda e^{-b|\lambda|}\,|\textcircled{F}\varphi(\lambda)|^2$$

(15.6 continued) $$= (2\pi)^{-1} \int_{-\infty}^{+\infty} d\lambda |\lambda|^{-1} |Ⓕ\varphi(\lambda)|^2$$

which converges if $\varphi \in$ Ⓢ, the Schwartz space of rapidly decreasing functions, and if in addition

(15.7) $$\int_{-\infty}^{+\infty} dt\, \varphi(t) = 0.$$

Before continuing we note that $H\varphi$ is never integrable for $\varphi \geq 0$ and nontrivial and therefore the 0-balagage of dx is not Radon. This follows since by (15.6)

$$\int_{\underline{\underline{R}}^{2,+}} dx\, H\varphi(x) = \text{Lim}_{a\uparrow\infty} \int_{\underline{\underline{R}}^{2,+}} dx\, H\varphi(x) He^{-(\cdot)^2/2a}(x)$$

$$= \text{Lim}_{a\uparrow\infty} (2\pi)^{-\frac{1}{2}} a^{\frac{1}{2}} \int_{-\infty}^{+\infty} d\lambda |\lambda|^{-1} Ⓕ\varphi(\lambda) e^{-a|\lambda|^2/2}$$

$$= \text{Lim}_{\lambda\to 0} |\lambda|^{-1} Ⓕ\varphi(\lambda) = +\infty .$$

Also for $\varphi \in$ Ⓢ we compute

(15.8) $$\frac{1}{2}\int_{\underline{\underline{R}}^{2,+}} dx\, \{\text{grad}\, H\varphi(x)\}^2 = (4\pi)^{-1} \int_0^{\infty} db \int_{-\infty}^{+\infty} d\lambda\, |\lambda|^2 e^{-b|\lambda|} |Ⓕ\varphi(\lambda)|^2$$

$$= (4\pi)^{-1} \int_{-\infty}^{+\infty} |\lambda| |Ⓕ\varphi(\lambda)|^2$$

which is finite.

Now consider $\varphi \in$ Ⓢ satisfying the mean condition (15.7).* Then $H\varphi \in L^2(\underline{\underline{R}}^{2,+}, dx)$ and therefore also $H_1\varphi = H\varphi - G_1 H\varphi$ is. Thus

$$U_{0,1}(\varphi, \varphi) = \int_{\underline{\underline{R}}^{2,+}} dx\, H\varphi(x) H_1\varphi(x)$$

converges. Also by SMP, (20.53) and (20.24)

$$U_{0,1}\langle\varphi,\varphi\rangle \le U_{0,\infty}\langle\varphi,\varphi\rangle = \int_{\underline{\underline{R}}^{2,+}} dx\,\{\text{grad } H\varphi(x)\}^2 .$$

This is enough to guarantee that the 1-balagage of dx is indeed a finite multiple of dt. Therefore we choose dt itself for harmonic measure. Then the excursion space $\underline{\underline{N}}$ is the subspace of $\varphi \in L^2(dt)$ such that Ⓕ$\varphi \in L^2(|\lambda|d\lambda)$ and for such φ

(15.9)
$$N(\varphi,\varphi) = (4\pi)^{-1}\int_{-\infty}^{+\infty} d\lambda\,|\lambda|\,|\text{Ⓕ}\varphi(\lambda)|^2 .$$

We also have

(15.9')
$$N(\varphi,\varphi) = (4\pi)^{-1}\int_{-\infty}^{+\infty} ds \int_{-\infty}^{+\infty} dt\,(s-t)^{-2}\{\varphi(s)-\varphi(t)\}^2 .$$

To prove this we need the identity

(15.10)
$$\text{Lim}_{\varepsilon\downarrow 0}\int_{|t|>\varepsilon} dt\, t^{-2}(1-e^{-i\lambda t}) = \pi|\lambda|$$

which can be established as follows. The supremum of the left side is dominated by a multiple of $1+\lambda^2$ and (15.10) is certainly valid with π replaced by some constant α on the right. To determine α, multiply both sides by $e^{-\lambda^2/2}$ and integrate from 0 to ∞ to get

$$\alpha = \int_0^\infty dt\, t^{-2}\sqrt{2\pi}\,(1-e^{-t^2/2}) = \sqrt{2\pi}\int_0^\infty du\,(2u)^{-3/2}(1-e^{-u})$$

$$= \frac{1}{2}\sqrt{\pi}\,2\int_0^\infty du\, u^{-\frac{1}{2}} e^{-u} = \pi .$$

Now for $\varphi \in$ Ⓢ the right side of (15.9') is the limit as $\varepsilon\downarrow 0$ of

$$(2\pi)^{-1} \int_{-\infty}^{+\infty} dt \int_{-\infty}^{+\infty} ds\, I(|s-t| \geq \varepsilon)(s-t)^{-2}\{\varphi(t)^2 - \varphi(s)\varphi(t)\}$$

which by the Plancherel theorem

$$= (2\pi)^{-1} \int_{-\infty}^{+\infty} du\, I((u) \geq \varepsilon) u^{-2} \int_{-\infty}^{+\infty} d\lambda |\textcircled{F}\varphi(\lambda)|^2 \{1 - e^{-i\lambda \cdot u}\}$$

and we get the right side of (15.9) after applying (15.10). It is clear from (15.9) that Ⓢ is dense in $\underline{\underline{N}}$ relative to the N norm in the sense of SMP, paragraph 1.6.1 and from this it follows directly that (15.9') is valid for general $\varphi \in \underline{\underline{N}}$. Indeed (15.9') is valid for general $\varphi \in \underline{\underline{N}}_{(e)}$, the extended Dirichlet space of SMP, Section 1. A little bit more is true. Since killing measure is null and since any Brownian trajectory starting in $\underline{\underline{R}}^{2,+}$ must eventually hit the real line $\underline{\underline{R}}$, it follows again by SMP, (20.53) that if φ is such that $H\varphi \in \underline{\underline{W}}$ then

(15.11) $$N(\varphi, \varphi) = \frac{1}{2} \int_{-\infty}^{+\infty} dt \int_{-\infty}^{+\infty} ds\; u_{0,\infty}(t, s)\{\varphi(t) - \varphi(s)\}^2$$

with $u_{0,\infty}$ the Feller density. But it is easy to check that Ⓢ is "rich enough" for (15.11) and (15.9') together to compel

(15.12) $$u_{0,\infty}(t, s) = (2\pi)^{-1}(t-s)^{-2} .$$

Thus (15.9') is valid for all φ. It is also true that if (15.9') converges, then $\varphi \in \underline{\underline{N}}_{(e)}$, but this requires an additional argument which we postpone until Section 18 when we consider the one dimensional symmetric Cauchy process. We are not sure to what extent (15.9) is valid. For $\varphi = 1$ it does not even make sense since Ⓕ$1(\lambda)$ is not a function.

There exists a potential density

(15.13) $$N(x, y) = (1/\pi) \log \{|x^* - y| / |x - y|\}$$

where x^* is the conjugate to x defined by reflection across the real axis. That is,

$$(a, b)^* = (a, -b) .$$

To verify this note first that by an elementary symmetry argument the transition density is given by

$$p_t(x, y) = (2\pi t)^{-1}(\exp\{-|x-y|^2/2t\} - \exp\{-|x^* -y|^2/2t\}) .$$

From the calculation

$$\int_0^T dt\,(2\pi t)^{-1} e^{-|x|^2/2t} = (2\pi)^{-1} \int_{|x|/\sqrt{T}}^{\infty} du\,(2/u) e^{-u^2/2}$$

$$= \pi^{-1} \log |x|^{-1} - \pi^{-1} \log T^{-\frac{1}{2}} + \pi^{-1} \int_{|x|/\sqrt{T}}^{\infty} du\, u \log u e^{-u^2/2}$$

follows

$$\int_0^T dt\, p_t(x, y) = (1/\pi) \log \{|x^*-y|/ |x-y|\} + \pi^{-1} \int_{|x-y|/\sqrt{T}}^{|x^*-y|/\sqrt{T}} du\, u \log u e^{-u^2/2}$$

and then (15.13) follows after passage to the limit $T \uparrow \infty$.

We modify somewhat the general approach used in Section 7 to study boundary functions and normal derivatives. For $\varepsilon > 0$ let D_ε be the upper half plane

$$D_\varepsilon = \{a, b) : b > \varepsilon\} .$$

This does not have compact closure and, it does have infinite capacity. Nevertheless we will use the D_ε, $\varepsilon > 0$ in place of the sequence $\{D_n\}_{n \geq 1}$ in paragraph 7.1.

For treating boundary functions we use a strengthened form of (7.7') with $i = (0,1)$ as reference point. If $f \in \underline{\underline{F}}^{ref} \cap C_b$ and if $\varepsilon > 0$ then $H^{\sim\varepsilon} f(x)$ is the unique bounded harmonic function in the closure of the half plane D_ε which agrees with f on the line $b = \varepsilon$. Thus for $b > \varepsilon$

$$H^{\sim\varepsilon} f(a, b) = \int_{-\infty}^{+\infty} dt\, c_{b-\varepsilon}(t-a) f(a) \tag{15.14}$$

and it is routine to extend (15.14) to general $f \in \underline{\underline{F}}^{ref} = \underline{\underline{W}}$. For the special case $(a, b) = (0, 1)$ we get

$$H^{\sim\varepsilon} f(0, 1) = \int_{-\infty}^{+\infty} dt\, c_1(t)\{c_{1-\varepsilon}(t)/c_1(t)\} f(\varepsilon, t) . \tag{15.15}$$

It is easy to check that the ratios

$$c_{1-\varepsilon}(t)/c_1(t) \to 1 \tag{15.16}$$

uniformly in t as $\varepsilon \downarrow 0$. If $f = f_0 \in \underline{\underline{F}}_{(e)}$ then by SMP, Theorem 7.3 the functions $H^{\sim\varepsilon} f_0(x) \to 0$ quasi-everywhere as $\varepsilon \downarrow 0$. By Harnack's inequality this is true everywhere in $\underline{\underline{R}}^{2,+}$ and in particular at $x = (0, 1)$. We conclude that if $f_0 \in \underline{\underline{F}}_{(e)}$, then

$$f_0(\varepsilon, \cdot) \to 0 \quad \text{in} \quad L^1(c_1)$$

as $\varepsilon \downarrow 0$. If $f = h$ is harmonic, then actually

$$h(\varepsilon, \cdot) \to \gamma h \quad \text{in} \quad L^2(c_1) \tag{15.17}$$

as $\varepsilon \downarrow 0$. To prove this we note first that $\gamma h \in L^2(c_1)$. This is because

$$E_x\{\gamma h(X_{\zeta-0})\}^2 \tag{15.18}$$

$$= h^2(x) + \int_{\underline{\underline{R}}^{2,+}} dy\, N(x, y)\{\text{grad } h(y)\}^2$$

is finite for quasi-every x and therefore by Harnack's inequality for all x and in particular for $x = (0,1)$ since $\{\text{grad } h(x)\}^2$ is integrable. (See Section 1 in SMP.) Since

$$h(\varepsilon, t) = C_\varepsilon \gamma h(t)$$

where C_ε is the Cauchy operator,

(15.19) $$C_\varepsilon \varphi(t) = \int_{-\infty}^{+\infty} ds\, c_\varepsilon(t-s)\varphi(s) \quad .$$

the convergence result (15.17) will follow if we can show that the operators C_ε converge to the identity in the strong operator topology on $L^2(c_1)$. But this follows since the functions $c_\varepsilon(t)$ form an approximation to the identity in the usual sense and since

(15.20) $$\int_{-\infty}^{+\infty} dt\, c_1(t) C_\varepsilon \varphi^2(t) = \int_{-\infty}^{+\infty} dt\, c_1(t) \int_{-\infty}^{+\infty} ds\, c_\varepsilon(s-t)\varphi^2(s)$$

$$= \int_{-\infty}^{+\infty} ds\, c_{1+\varepsilon}(s)\varphi^2(s)$$

$$\leq \int_{-\infty}^{+\infty} ds\, c_1(s)\varphi^2(s)(1+\varepsilon) \; .$$

We summarize in

Theorem 15.1. *If* $f \in \underline{\underline{W}}$, *then* $\gamma f \in L^2(c_1)$ *and as* $\varepsilon \downarrow 0$

$$f(\varepsilon, \cdot) \to \gamma f \quad \text{in} \quad L^1(c_1) \; .$$

///

Remark. We have already seen that if f is harmonic, then L^1 convergence can be replaced by L^2 convergence in Theorem 15.1. We do not know if this is true for general $f \in \underline{\underline{W}}$. Also if f is harmonic then the limit exists in a nontangential sense almost everywhere. (See [49] or [50] for a precise definition.) It is clear that this cannot be true for general $f \in \underline{\underline{W}}$ since such f can only be specified up to sets of capacity zero and any countable dense subset of $\underline{\underline{R}}^{2,+}$ has capacity zero. It might be interesting to find an appropriate weakening of nontangential convergence which does work for general $f \in \underline{\underline{W}}$. ///

In order to treat the normal derivative by the technique of section 7, we must find the joint distribution of the hitting position $X_{\sigma(D_\varepsilon)}$ and the initial position X_{ζ^*} relative to the measure Ⓟ of SMP, Section 5. The equillibrium distribution $\ell^1_\varepsilon(dx)$ is a constant ℓ^1_ε times Lebesgue measure on the line $b = \varepsilon$. This is clear from continuity of trajectories and from translation invariance in the direction of the real axis. The constant ℓ^1_ε is determined by the identity

$$\ell^1_\varepsilon \int_{-\infty}^{+\infty} da\, N(\{a,\varepsilon\},\{0,\varepsilon\}) = 1$$

or equivalently

$$\ell^1_\varepsilon (1/\pi)\int_{-\infty}^{+\infty} da\, \log\{\sqrt{a^2+4\varepsilon^2}/a\} = 1 .$$

The elementary calculation

$$(1/\pi)\int_{-\infty}^{+\infty} da\, \frac{1}{2}\log\{1+(4\varepsilon^2/a^2)\} = (2\pi)^{-1}\int_{-\infty}^{+\infty} da\, a(4\varepsilon^2)(2a^{-3})/\{1+(4\varepsilon^2/a^2)\}$$

$$= (4\varepsilon^2/\pi)\int_{-\infty}^{+\infty} da\, a^{-2}/\{1+(4\varepsilon^2/a^2)\} = 2\varepsilon$$

then establishes

$$\ell^1_\varepsilon = 1/2\varepsilon .$$

Comparing this with the corresponding result in Section 13, we see that the approximate Markov process for the upper half plane can be constructed by letting the a-coordinate be the stationary Brownian motion with Lebesgue measure as distribution at each constant time and by letting the b-coordinate be independent and governed by the corresponding approximate Markov process for the half line.

Now we derive an explicit formula for the integral

(15.21) $$Ⓔ[\varphi(X_{\zeta*}) \; ; \; \sigma(D_\varepsilon) < +\infty \; ; \; f(X_{\sigma(D_\varepsilon)})]$$

for φ bounded and continuous on the terminal boundary $\underline{\underline{R}}$ and for $f \in C_{com}(\underline{\underline{X}})$. The restrictions on φ and f guarantee that (15.21) converges and can be computed from

(15.22) $$Ⓔ[\varphi(X_{\zeta*});\sigma(D_\varepsilon) < +\infty \; ; \; f(X_{\sigma(D_\varepsilon)})]$$

$$= \mathrm{Lim}_{\delta\downarrow 0} Ⓔ[\varphi(\mathrm{abs}X_{\sigma(D_\delta)});\sigma(D_\varepsilon) < +\infty \; ; \; f(X_{\sigma(D_\varepsilon)})]$$

$$= \mathrm{Lim}_{\delta\downarrow 0}(2\delta)^{-1}\int_{-\infty}^{+\infty} dt\, \varphi(t)\, Ⓔ_{(t,\delta)}[\sigma(D_\varepsilon) < +\infty \; ; \; f(X_{\sigma(D_\varepsilon)})]$$

Here $\mathrm{abs}X_{\sigma(D_\delta)}$ is the abscissa, or a-coordinate of the initial position $X_{\sigma(D_\delta)}$. The main step is to establish for $0 < \delta < \varepsilon$ and for t real the identity

(15.23) $$Ⓔ_{(t,\delta)}[\sigma(D_\varepsilon) < +\infty \; ; \; f(X_{\sigma(D_\varepsilon)})] = \int_{-\infty}^{+\infty} ds\, q_{\delta,\varepsilon}(s-t)f(s)$$

with $q_{\delta,\varepsilon}(u)$ given by

(15.24) $$q_{\delta,\varepsilon}(u) = \varepsilon^{-1}\{e^{\pi u/\varepsilon} \sin(\pi\delta/\varepsilon)\}/\{1 + e^{2\pi u/\varepsilon} + 2e^{\pi u/\varepsilon} \cos(\pi\delta/\varepsilon)\}$$

It is well-known, and easy to prove directly using the strong Markov property for Brownian motion, that the left side of (15.23), as a function of $z = t + i\delta$, is the bounded harmonic finction on the closed strip $0 \leq \mathrm{Im}\, z \leq \varepsilon$ which vanishes on the real line Im $z = 0$ and agrees with f on the line Im $z = \varepsilon$. Thus (15.23) will follow if we can prove that

(15.25) $$F(t+i\delta) = \int_{-\infty}^{+\infty} ds\, q_{\delta,\varepsilon}(s-t)F(s+i\varepsilon)$$

whenever F is a bounded harmonic function on the above strip which vanishes on the real line. The transformation

(15.26) $$w = e^{\pi z/\varepsilon} = e^{\pi t/\varepsilon}\{\cos(\pi\delta/\varepsilon) + i\sin(\pi\delta/\varepsilon)\}$$

maps the strip onto the upper half plane in such a way that the upper boundary corresponds to the left half line $\tau < 0$ and the lower boundary to the right half line $\tau > 0$. Applying the classical formula for bounded harmonic functions in the upper half plane, we get

$$F(z) = \int_{-\infty}^{0} d\tau\, c_{\mathrm{Im}\, w}(\tau - \mathrm{Re}\, w)F(-(\varepsilon/\pi)\log\tau + i\varepsilon)$$

with w given by (15.26). Finally (15.25) and therefore (15.23) follow after substituting $\tau = -e^{\pi s/\varepsilon}$ and observing that

$$c_{\mathrm{Im}\, w}(\tau - \mathrm{Re}\, w) = \pi^{-1}\{e^{\pi u/\varepsilon}\sin(\pi\delta/\varepsilon)\}/\{[e^{\pi u/\varepsilon}\sin(\pi\delta/\varepsilon)]^2$$
$$+ [e^{\pi u/\varepsilon}\cos(\pi\delta/\varepsilon) + e^{\pi s/\varepsilon}]^2\}.$$

Before continuing we prove that $q_{\delta,\varepsilon}(u)$ has a simple Fourier transform

(15.27) $$Ⓕ q_{\delta,\varepsilon}(\lambda) = (\sinh\delta\lambda)/(\sinh\varepsilon\lambda).$$

To check this make the substitution $x = e^{\pi u/\varepsilon}$ in the integral

$$\int_{-\infty}^{+\infty} du\, e^{-i\lambda u} q_{\delta,\varepsilon}(u)$$

to get instead

(15.28) $$\pi^{-1}\int_0^\infty dx\, e^{i\lambda\epsilon \log x/\pi}\, \sin(\pi\delta/\varepsilon)/\{1+x^2+2x\cos(\pi\delta/\varepsilon)\}\,.$$

In general for $0<\theta<\pi$ and for a real we have

(15.29) $$\int_0^\infty dx\, e^{ia\log x}\{1+u^2+2u\cos\theta\}^{-1} = \{\pi\sinh\theta a\}/\{\sin\theta\sinh\pi a\}$$

To prove this note first that the left side is an even function of a and therefore we need only consider $a>0$. Denote the left side by I and let C_R be the integral for the analytic continuation of the integrand over the contour consisting of the circle $|z| = R$ traversed counter - clockwise from arg. $z = 0$ to arg $z = 2\pi$ with a cut from 0 to R. The integrand has simple poles at $z = e^{i(\pi+\theta)}$ and $z = e^{i(\pi-\theta)}$ and so by the calculus of residue

$$C_R = 2\pi i\{e^{iai(\pi-\theta)}/(e^{i(\pi-\theta)}+e^{i\theta}) + e^{iai(\pi+\theta)}/(e^{i(\pi+\theta)}+e^{-i\theta})\}$$

$$= \pi e^{-\pi a}\{e^{a\theta}/(\sin\theta) + e^{-a\theta}/(-\sin\theta)\} = 2\pi e^{-\pi a}(\sinh a\theta)/(\sin\theta)$$

On the other hand consideration of the change in the argument along the cut establishes

$$\mathrm{Lim}_{R\uparrow\infty}\, C_R = (1-e^{-2\pi a})I$$

and (15.29) follows. A passage to the limit $\theta\downarrow 0$ is easily justified to establish

(15.29') $$\int_0^\infty dx\, e^{ia\log x}(1+u)^{-2} = (\pi a)/\sinh\pi a\,.$$

Now (15.27) follows after applying (15.29) to (15.28).

After taking the limit inside the integral sign in (15.23) we get

(15.30)
$$Ⓔ[\varphi(X_{\zeta^*})\ ;\ \sigma(D_\varepsilon) < +\infty\ ;\ f(X_{\sigma(D_\varepsilon)})]$$

$$= \text{Lim}_{\delta\downarrow 0}(2\delta)^{-1}\int_{-\infty}^{+\infty} dt\ \varphi(t)\int_{-\infty}^{+\infty} ds\ q_{\delta,\varepsilon}(s-t)f(s)$$

$$= \int_{-\infty}^{+\infty} dt\ \varphi(t)\int_{-\infty}^{+\infty} ds\ q_\varepsilon(s-t)f(s)$$

where

(15.31)
$$q_\varepsilon(u) = \text{Lim}_{\delta\downarrow 0}(2\delta)^{-1} q_{\delta,\varepsilon}(u) = (\pi/2\varepsilon^2)e^{\pi u/\varepsilon}\{1+e^{\pi u/\varepsilon}\}^{-2}$$

The last passage to the limit is easy to justify. For example the result (15.29') can be applied to compute the Fourier transforms

(15.32)
$$Ⓕ q_\varepsilon(\lambda) = \lambda\{2\ \sinh \lambda\varepsilon\}^{-1}$$

Thus by the continuity theorem for Fourier transforms $(2\delta)^{-1}q_{\delta,\varepsilon}(u) \to q_\varepsilon(u)$ in the weak * topology generated by the pairing with bounded continuous functions, and this suffices. Notice that (15.32) and (15.27) together give

(15.33)
$$Ⓕ q_\varepsilon(\lambda) = Ⓕ q_\delta(\lambda)\, Ⓕ q_{\delta,t}(\lambda)\ .$$

which is consistent with the probabilistically obvious identity

(15.33')
$$q_\varepsilon(u) = \int_{-\infty}^{+\infty} dt\ q_\delta(t)q_{\delta,\varepsilon}(u-t)\ .$$

It turns out that the distribution (15.31) does appear in the literature and its distribution function is referred to as the logistic distribution function. See No. 54 in [25]. This was pointed out to the author by T. Harris.

Now we are ready to study the normal derivative. For our "test functions" we take $\psi \in$ Ⓢ satisfying the mean condition (15.7), the latter being required to guarantee that $H\psi(x) \in L^2(dx)$. The main point is to establish the two identities

$$(15.34)\ \textcircled{E}[\sigma(D_\varepsilon) < +\infty\ ; \psi(X_{\zeta*})\{\gamma f(X_{\zeta-0}) - f(X_{\sigma(D_\varepsilon)})\}] = \frac{1}{2}\int_{\underline{\underline{R}}^{2,+}} dx\, \Delta f(x) H^\varepsilon H\psi(x)$$

$$(15.35)\ \mathrm{Lim}_{\varepsilon\downarrow 0}\textcircled{E}[\sigma(D_\varepsilon); \psi(X_{\zeta*})\{\gamma f(X_{\zeta*}) - \gamma f(X_{\zeta-0})\}] = \frac{1}{2}\int_{\underline{\underline{R}}^{2,+}} dx\, \mathrm{grad} f(x)\, \mathrm{grad}\, H\psi(x)$$

for $f \in \underline{\underline{W}}$ such that $\Delta f \in L^2(dx)$. We note first that for fixed $\varepsilon > 0$ the following integrals converge.

$$(15.36)\qquad \textcircled{E}[\sigma(D_\varepsilon) < +\infty\ ;\ \psi(X_{\zeta*})\gamma f(X_{\zeta*})] = (2\varepsilon)^{-1}\int_{-\infty}^{+\infty} dt\ \psi(t)\gamma f(t)\ .$$

$$(15.37)\qquad \textcircled{E}[\sigma(D_\varepsilon) < +\infty\ ;\ \psi(X_{\zeta*})f(X_{\sigma(D_\varepsilon)})] = \int_{-\infty}^{+\infty} dt \int_{-\infty}^{+\infty} ds\ \psi(t) f(s,\varepsilon) q_\varepsilon(t-s)$$

$$= \int_{-\infty}^{+\infty} dt\ \psi(t) Q_\varepsilon f(\cdot,\varepsilon)(t)$$

$$(15.38)\ \textcircled{E}[\sigma(D_\varepsilon) < +\infty\ ;\ \psi(X_{\zeta*})\gamma f(X_{\zeta-0})] = \int_{-\infty}^{+\infty} dt \int_{-\infty}^{+\infty} du \int_{-\infty}^{+\infty} ds\ \psi(t) q_\varepsilon(t-u) c_\varepsilon(u-s)\gamma f(s)$$

$$= \int_{-\infty}^{+\infty} dt\ \psi(t) Q_\varepsilon C_\varepsilon \gamma f(t)\ .$$

For (15.36) convergence follows because $\gamma f \in L^2(c_1)$ by Theorem 15.1 and because $\psi \in S$. For (15.37) and (15.38) this will follow once we show that Q_ε is a bounded operator both on $L^1(c_1)$ and $L^2(c_1)$. (For C_ε this follows from (15.20).) Clearly it suffices to show that

$$(15.39)\qquad \int_{-\infty}^{+\infty} ds\ c_1(s) q_\varepsilon(t-s)$$

is dominated by a multiple of $c_1(t)$ and for this it suffices to show that

$$(15.40) \qquad \int_{-\infty}^{+\infty} ds\, (1+s^2)^{-1} e^{\pi(t-s)/\varepsilon}\{1+e^{\pi(t-s)/\varepsilon}\}^{-2} = O(t^{-2})$$

as $t\uparrow\infty$. But this is clear since the left side of (15.40) is dominated by

$$e^{-\pi t/2\varepsilon}\int_{-\infty}^{t/2} ds\,(1+s^2)^{-1} + (1+t^2/4)^{-1}\int_{t/2}^{t} ds\, e^{-\pi(t-s)/\varepsilon}$$
$$+ (1+t^2)^{-1}\int_t^{\infty} ds\, e^{\pi(t-s)/\varepsilon} .$$

The identity (3.10) implies in the present context

$$(15.41) \quad \frac{1}{2}\int_{\underline{\underline{R}}^{2,+}} dx \operatorname{grad} f(x) \operatorname{grad} H\psi(x) = \frac{1}{2}\, Ⓔ\{\psi(X_{\zeta*})-\psi(X_{\zeta-0})\}\{\gamma f(X_{\zeta*})-\gamma f(X_{\zeta-0})\} .$$

Now that the convergence problems have been taken care of, the identity (15.35) follows directly from (15.41) and symmetry. Clearly there exists a sequence $f_n \in \underline{\underline{W}}$ such that $f_n \to f$ pointwise and relative to the W-norm, such that $\Delta f_n \in L^2(dx) \cap L^1(dx)$ for each n, and such that $\Delta f_n \to \Delta f$ in $L^2(dx)$. At the same time $\gamma f_n \to \gamma f$ in $L^2(c_1)$ and therefore it suffices in (15.34) to consider the case when $\Delta f(x) \in L^1(dx)$. But then we can argue exactly as in the proof of Theorem 7.1. Finally we combine (15.34) and (15.35) to get

$$(15.42) \qquad \operatorname{Lim}_{\varepsilon\downarrow 0} Ⓔ[\sigma(D_\varepsilon) < +\infty\, ;\, \psi(X_{\zeta*})\{\gamma_T f(X_{\zeta*}) - f(X_{\sigma(D_\varepsilon)})\}$$

$$= \frac{1}{2}\int_{\underline{\underline{R}}^{2,+}} dx\, \Delta f(x) H\psi(x) + \frac{1}{2}\int_{\underline{\underline{R}}^{2,+}} dx \operatorname{grad} f(x) \operatorname{grad} H\psi(x).$$

Adapting the definition in Section 6 we _define_ the _normal derivative_ by

(15.43) $$(\partial f/\partial n)(\psi) = \mathrm{Lim}_{\varepsilon \downarrow 0} \text{Ⓔ}[\alpha(D_\varepsilon) < +\infty\, ;\, \psi(X_{\zeta *})\{f(X_{\sigma(D_\varepsilon)}) - \gamma_T f(X_{\zeta *})\}]$$

$$= \mathrm{Lim}_{\varepsilon \downarrow 0} \int_{-\infty}^{+\infty} dt \int_{-\infty}^{+\infty} ds\; q_\varepsilon(t-s)\psi(t)\{f(\varepsilon, s) - \gamma_T f(t)\}$$

and we are ready for

Theorem 15.2. *Let* $f \in \underline{\underline{W}}$ *be such that* $\Delta f \in L^2(dx)$ *and let* $\psi \in S$. *Then the normal derivative* (15.43) *exists and*

(15.44) $$-(\partial f/\partial n)(\psi) = \frac{1}{2}\int_{\underline{\underline{R}}^{2,+}} dx\, \Delta f(x) H_1\psi(x) + \frac{1}{2}\int_{\underline{\underline{R}}^{2,+}} dx\, \text{grad}\, f(x)\, \text{grad}\, H_1\psi(x)\, .$$

If in addition ψ *satisfies the mean condition* (15.7) *then* (15.44) *can be replaced by*

(15.44') $$-(\partial f/\partial n)(\psi) = \frac{1}{2}\int_{\underline{\underline{R}}^{2,+}} dx\, \Delta f(x) H\psi(x) + \frac{1}{2}\int_{\underline{\underline{R}}^{2,+}} dx\, \text{grad}\, f(x)\, \text{grad}\, H\psi(x).$$

///

Proof. We have already treated the case when ψ satisfies the mean condition (15.7). For the general case we proved indirectly. The previous argument can be modified to establish (15.44) but with the left side defined in terms of Ⓔ$_1$, the approximate Markov process associated with the process $(\underline{\underline{F}}, E_1)$. Therefore we will be done if we can show that the left side of (15.44) can also be computed by (15.43). The point is that even if ψ does not satisfy (15.7) it is still true that ψ is square integrable on $\underline{\underline{R}}^{2,+}\setminus D_1$ and therefore (15.44') and (15.43) are valid whenever $\Delta f(x)$ is integrable on D_1. Also in this case (15.44) and (15.44') agree. Finally it suffices to observe that in general there exists f_1 satisfying this additional condition such that $f_1 = f$ on $\underline{\underline{R}}^{2,+}\setminus D_1$ and therefore (15.43) is identical for f and f_1. ///

It is routine to check that (15.44) and (15.44') reduce to the classical formula when f is smooth.

The generator A is defined by

(15.46) $$Af = \frac{1}{2}\Delta f$$

supplemented by the requirement

(15.47) $$f \in \underline{\underline{W}}$$

and the boundary condition

(15.48) $$\gamma f = 0 .$$

With the help of Theorem 6.1 it is easy to see that we can suppress (15.47) for f bounded. But certainly (15.47) cannot be omitted in the general case.

In the remainder of this section we study extensions $(\underline{\underline{F}}^{\sim}, E^{\sim})$ of $(\underline{\underline{F}}, E)$. Let $(\underline{\underline{H}}, Q)$ be a pair satisfying conditions 3.4.1 and 3.4.2. This means

15.2.1. $\underline{\underline{H}}$ is a linear subset of $\underline{\underline{N}}$ and Q - N is contractive on $\underline{\underline{H}}$.

15.2.2. $(\underline{\underline{H}}, Q)$ is a Dirichlet space on $L^2(\underline{\underline{R}}, dt)$. ///

In practice 15.2.1 is relatively easy to check, but 15.2.2 depends on resolving an interesting technical problem. If $\underline{\underline{H}}$ = N, then the usual compactification of $\underline{\underline{R}}^{2,+}$ as a subset of $\underline{\underline{R}}^2$ is suitable for the construction in Section 3. In the general case some other compactification must be used. We will not consider this question here.

In order to be able to apply the formula (15.43) for the normal derivative, we impose the relatively innocuous condition

15.2.3. $\underline{\underline{H}} \cap (S)$ is dense in $\underline{\underline{H}}$. ///

The extended Dirichlet space $\underline{\underline{F}}^{\sim}_{(e)}$ is precisely the subspace of $f \in \underline{\underline{W}}$ such that the boundary function $\gamma_T f \in \underline{\underline{H}}$ and for such f

(15.49) $$E^{\sim}(f,f) = \frac{1}{2} \int_{\underline{\underline{R}}^{2,+}} dx \, \{\text{grad } f(x)\}^2 + (Q-N)(\gamma_T f, \gamma_T f) .$$

This is a simple special case of Theorem 5.1. By Theorem 9.1 the generator $A^{\sim}$ is defined by

(15.50) $$A^{\sim} f(x) = \frac{1}{2} \Delta f(x)$$

supplemented by (15.47) and the conditions

(15.51) $$\gamma f \in \underline{\underline{H}}$$

(15.52) $$(\partial f/\partial n)(\psi) = \{Q-N\}(\gamma f, \psi) \quad \text{for} \quad \psi \in \underline{\underline{H}} \cap (S) .$$

Again (15.47) can be omitted when f is bounded.

The choice $(\underline{\underline{H}}, Q) = (\underline{\underline{N}}, N)$ corresponds to the classical reflected Brownian motion. This can be verified indirectly as follows. Let $P^0_t, t > 0$ be the transition semigroup for standard Brownian motion on the full plane $\underline{\underline{R}}^2$. Then the operators $P^{\sim}_t, t > 0$ for the reflected Brownian motion is given by

$$P^{\sim}_t f(x) = P^0_t (f \circ \pi)(x) \quad \text{for} \quad x \in \underline{\underline{R}}^{2,+}$$

where π is the mapping defined by

$$\pi(a, b) = (a, |b|) .$$

It is easily checked that the $P^{\sim}_t, t > 0$ form a symmetric Markovian semigroup on

$L^2(\underline{\underline{R}}^{2,+}, dx)$. Also by Lemma 1.1 in SMP a given function $f \in L^2(\underline{\underline{R}}^{2,+}, dx)$ belongs to the associated Dirichlet space $\underline{\underline{F}}^r$ if and only if the limit as $t \downarrow 0$ of

$$(15.53)\quad (1/t) \int_{\underline{\underline{R}}^{2,+}} dx \, \{f(x) - P_t^{\sim} f(x)\} f(x) = \frac{1}{2}(1/t) \int_{\underline{\underline{R}}^2} dx \, \{f \circ \pi(x) - P_t^0 f \circ \pi(x)\} f \circ \pi(x)$$

is finite, in which case the Dirichlet norm $E^r(f,f)$ is given by this limit. It follows that $f \in \underline{\underline{F}}^r$ if and only if $f \circ \pi$ belongs to $\underline{\underline{W}}(\underline{\underline{R}}^2) \cap L^2(\underline{\underline{R}}^2, dx)$ and then

$$E^r(f,f) = \frac{1}{4} \int_{\underline{\underline{R}}^2} dx \, \{\operatorname{grad} f \circ \pi(x)\}^2 = \frac{1}{2} \int_{\underline{\underline{R}}^{2,+}} dx \, \{\operatorname{grad} f(x)\}^2 .$$

It follows directly that $\underline{\underline{F}}^r$ is contained in $\underline{\underline{F}}^{\sim}$ described above and that $E^r = E^{\sim}$ on $\underline{\underline{F}}^r$. Therefore our assertion will follow if we can show that for any $f \in \underline{\underline{F}}^{\sim}$ the extension $f \circ \pi \in \underline{\underline{W}}(\underline{\underline{R}}^2)$. For this it suffices to show that

$$(15.54)\qquad \int_{\underline{\underline{R}}^2} dx \, f(\pi x)(\partial\varphi/\partial a)(x) = - \int_{\underline{\underline{R}}^2} dx \, (\partial f/\partial a)(\pi x)\varphi(x)$$

$$(15.55)\qquad \int_{\underline{\underline{R}}^2} dx \, f(\pi x)(\partial\varphi/\partial b)(x) = - \int_{\underline{\underline{R}}^{2,+}} dx \, (\partial f/\partial b)(x)\{\varphi(x) - \varphi(x^*)\}$$

for $\varphi \in C^{\infty}_{com}(\underline{\underline{R}}^2)$. For (15.54) it suffices to replace φ by $\varphi\psi_\varepsilon$ where

$$\begin{aligned} \psi_\varepsilon(a,b) &= 0 && \text{for } |b| \le \varepsilon \\ \psi_\varepsilon(a,b) &= 1 && \text{for } |b| \ge 2\varepsilon ,\ \varphi(a,b) \ne 0 \\ (\partial\psi_\varepsilon/\partial a)(a,b) &= 0 && \text{whenever } \varphi(a,b) \ne 0 . \end{aligned}$$

in which case (15.54) is certainly valid and then pass to the limit $\varepsilon \downarrow 0$. This argument would have to be refined before we could apply it to (15.55) since the contribution from $(\partial\psi_\varepsilon/\partial b)\varphi$ cannot be ignored. Instead we will establish (15.55) with the help of Theorem 14.1. We show first that for $\varepsilon > 0$

$$(15.56) \qquad \int_{D_\varepsilon} dx\, f(x)(\partial\varphi/\partial b)(x) + \int_{D_\varepsilon} dx\, (\partial f/\partial b)(x)\varphi(x) = -\int_{-\infty}^{+\infty} dt\, f(t,\varepsilon)\varphi(t,\varepsilon)\ .$$

This is elementary when f is continuously differentiable. But using convolutions it is easy to approximate f by such functions f_n both in the Sobolev space $\underline{\underline{W}}(D_{\varepsilon/2})$ and in $L^2(D_\varepsilon, dx)$. It is immediately clear that the left sides for f_n converge to the left side for f. Also at least for a subsequence $H^{\sim\varepsilon}|f-f_n| \to 0$ quasi-everywhere on D_ε and therefore everywhere and in particular at $i = (0,1)$ which guarantees that $f_n(\cdot,\varepsilon) \to f(\cdot,\varepsilon)$ in $L^1(c_{1-\varepsilon})$ and establishes convergence for the right sides of (15.56). Now (15.56) together with its analogue in the lower half plane gives

$$(15.57) \qquad \int_{\underline{\underline{R}}^2} dx\, I(|b|>\varepsilon)f(\pi x)(\partial\varphi/\partial b)(x) = -\int_{\underline{\underline{R}}^{2,+}} dx\, I(b>\varepsilon)(\partial f/\partial b)(x)\{\varphi(x)-\varphi(x^*)\}$$

$$-\int_{-\infty}^{+\infty} dt\, f(t,\varepsilon)\{\varphi(t,\varepsilon)-\varphi(t,-\varepsilon)\}\ .$$

Finally it suffices to observe that $\sup_{\varepsilon>0}|\varphi(t,\varepsilon) - \varphi(t,-\varepsilon)|$ is dominated by a multiple of $c_1(t)$ and that $f(t,\varepsilon) \to \gamma f(t)$ in $L^1(c_1)$ as $\varepsilon\downarrow 0$ which is enough to guarantee that the second term on the right in (15.57) converges to 0 as $\varepsilon\downarrow 0$, and (15.55) follows. This completely justifies our assertion that $(\underline{\underline{F}}^{\sim}, E^{\sim})$ corresponds to the classical reflecting Brownian motion. The boundary condition (15.52) now reduces to

$$(15.58) \qquad \operatorname{Lim}_{\varepsilon\downarrow 0}\int_{-\infty}^{+\infty} ds\, q_\varepsilon(t-s)\{f(\varepsilon,s)-\gamma f(t)\} = 0$$

in the sense of tempered distributions.

We consider now the general case when the corresponding transition semigroup $\tilde{P}_t$, $t > 0$ commutes with translations in the a-direction. The Dirichlet space $(\underline{\underline{H}}, Q)$ is translation invariant on $\underline{\underline{R}}$ and therefore is associated with a symmetric infinitely divisible process on $\underline{\underline{R}}$. Thus by the well-known Levy-Khintchine classification there are three possibilities.

15.3.1. Pure Diffusion. $\underline{\underline{H}} = \underline{\underline{W}}(\underline{\underline{R}}) \cap L^2(\underline{\underline{R}}, dt)$ and for $\varphi \in \underline{\underline{H}}$

$$Q(\varphi, \varphi) = N(\varphi, \varphi) + \frac{1}{2}\sigma^2 \int_{-\infty}^{+\infty} dt \{\varphi'(t)\}^2$$

where $\sigma > 0$.

15.3.2. Pure Jump. There is a Radon measure $\ell(dt)$ on $\underline{\underline{R}} - \{0\}$ which is symmetric about the origin and satisfies the integrability condition

$$\int_0^1 \ell(dt)t^2 + \int_1^\infty \ell(dt) < +\infty .$$

The Dirichlet space $\underline{\underline{H}}$ is the subspace of $\varphi \in \underline{\underline{N}}$ such that

$$\int_{-\infty}^{+\infty} dt \int_{-\infty}^{+\infty} \ell(du)\{\varphi(t) - \varphi(t+u)\}^2 < +\infty$$

and for $\varphi \in \underline{\underline{H}}$

$$Q(\varphi, \varphi) = N(\varphi, \varphi) + \frac{1}{2}\int_{-\infty}^{\infty} dt \int_{-\infty}^{+\infty} \ell(du)\{\varphi(u) - \varphi(u+t)\}^2 .$$

15.3.3. Mixture. $\underline{\underline{H}}$ is as in 15.3.1, the measure ℓ is as in 15.3.2 and $\sigma > 0$. For $\varphi \in \underline{\underline{H}}$

$$Q(\varphi, \varphi) = N(\varphi, \varphi) + \frac{1}{2}\sigma^2 \int_{-\infty}^{+\infty} dt \{\varphi'(t)\}^2 + \frac{1}{2}\int_{-\infty}^{+\infty} dt \int_{-\infty}^{+\infty} \ell(du)\{\varphi(t) - \varphi(t+u)\}^2 .$$

///

It would be a relatively easy exercise to derive this classification of translation invariant Dirichlet spaces on $L^2(\underline{\underline{R}}, dt)$ from our results in SMP, Chapter II without appealing to the classical Lévy-Khintchine result. We do not bother with this here but we will give the analogous derivation in the setting of Chapter 23 below. Just as in Section 14 there are additional possibilities if we replace dx by a bounded reference measure and ask only that the pair $(\underline{\underline{F}}^{\sim}_{(e)}, E^{\sim})$ be translation invariant. This is because the pair $(\underline{\underline{H}}, Q)$ described above may be transient.

At least for the pure diffusion case 15.3.1 it is possible to formulate the boundary condition (15.52) quite differently. With the help of Fourier transforms it is easy to see that for each $\varphi \in \underline{\underline{H}}$ there exists a unique function $\Omega\varphi \in L^2(\underline{\underline{R}})$ such that

$$N(\varphi,\psi) = -\int_{-\infty}^{+\infty} dt\, \Omega\varphi(t)\psi(t)$$

for all $\psi \in \underline{\underline{N}}$ and therefore in $\underline{\underline{H}}$. (Indeed Ω is the generator for the symmetric Cauchy process on $\underline{\underline{R}}$, but this need not concern as here.) If $f \in$ domain $A^{\sim}$, then for $\psi \in$ Ⓢ satisfying the mean condition (15.7) the boundary condition (15.52) can be written

$$-\frac{1}{2}\int dx \text{ grad } f(x) \text{ grad } H\psi(x) - \frac{1}{2}\int dx\, \Delta f(x) H\psi(x) = \frac{1}{2}\sigma^2 \int_{-\infty}^{+\infty} dt\, \gamma f'(t)\psi'(t)$$

and therefore

(15.59) $$-\frac{1}{2}\sigma^2\int_{-\infty}^{+\infty} dt\, \gamma f'(t)\psi'(t) = N(\gamma f, \psi) + \frac{1}{2}\int dx\, \Delta f(x) H\psi(x)$$

$$= -\int_{-\infty}^{+\infty} dt\, \Omega\gamma f(t)\psi(t) + \frac{1}{2}\int dx\, \Delta f(x) H\psi(x).$$

There exists a locally integrable function $\pi\Delta f(t)$ on $\underline{\underline{R}}$ such that

$$(15.60) \qquad \int dx\, \Delta f(x) H\psi(x) = \int_{-\infty}^{+\infty} dt\, \pi\Delta f(t)\psi(t)$$

for all such ψ. We cannot conclude this directly from Fubini's theorem since $H|\psi|(x)$ is not square integrable, but we can argue indirectly. The point is that

$$\frac{1}{2}\int dx\, \Delta f(x)H\psi(x) = \frac{1}{2}\int dx\, \Delta f(x)H_1\psi(x) + \frac{1}{2}\int dx\, \Delta f(x)G_1H\psi(x)$$

$$= \frac{1}{2}\int dx\, \Delta f(x)H_1\psi(x) + \int dx\, G_1\{\tfrac{1}{2}\Delta f(x)-f(x)\}H\psi(x)$$

$$+ \int dx\, f(x)G_1H\psi(x)$$

$$= \frac{1}{2}\int dx\, \Delta f(x)H_1\psi(x) + \int dx\, \{H_1\gamma f(x)-f(x)\}H\psi(x)$$

$$+ \int dx\, f(x)G_1H\psi(x)$$

$$= \int dx\, \{\tfrac{1}{2}\Delta f(x)-f(x)\}H_1\psi(x) + \int dx\, H_1\gamma f(x)H\psi(x)$$

and both terms in the last line converge absolutely for general $\psi \in \underline{\underline{N}}$ and in particular for general $\psi \in$ Ⓢ. For the first term this follows since both factors are square integrable and for the second term this follows from (15.3). This establishes the existence of locally integrable $\pi\Delta f(t)$ satisfying (15.60) for the special ψ described above, and so (15.59) can be replaced by

$$(15.61) \qquad -\frac{1}{2}\sigma^2 \int_{-\infty}^{+\infty} dt\, \gamma f'(t)\psi'(t) = \int_{-\infty}^{+\infty} dt\, \{-\Omega\gamma f(t)+\tfrac{1}{2}\pi\Delta f(t)\}\psi(t)\,.$$

It follows that $\gamma f'(t)$ is absolutely continuous and that

$$(15.62) \qquad \frac{1}{2}\sigma^2\gamma f''(t) + \Omega\gamma f(t) = \frac{1}{2}\pi\Delta f(t)$$

Thus if $f \in$ domain $A^{\sim}$ then

(15.63) $$\gamma f'(t) \in L^2(dt)$$

(15.64) $\gamma f'(t)$ is absolutely continuous; $\gamma f''$ is tempered

and whenever $\psi \in$ Ⓢ satisfies the mean condition (15.7)

(15.65) $$\frac{1}{2}\int dx\, \Delta f(x) H\psi(x) = \int_{-\infty}^{+\infty} dt\, \{\frac{1}{2}\sigma^2 \gamma f''(t) + \Omega\gamma f(t)\}\, \psi(t)\ .$$

Thus we have replaced (15.52) by (15.64) and (15.65).

Finally we continue to assume that Q is a local form but we drop the requirement of translation invariance. It is clear from our construction in Section 3 and from Theorem 11.10 in SMP that Q is local if and only if the process $\{x_t^{\sim}\}$ has continuous trajections. We will restrict attention to the case where Q can be represented

(15.66) $$Q(\varphi,\varphi) = N(\varphi,\varphi) + \int_{-\infty}^{+\infty} \nu(dt)\{\varphi'(t)\}^2$$

with ν a Radon measure and where $\underline{\underline{H}}$ is an appropriate closure of $C^1_{com}(\underline{\underline{R}})$. Condition 15.2.1 is automatic but it is not always possible to choose $\underline{\underline{H}}$ so that 15.2.2 is satisfied. Following M. Fukushima [17] we say that Q is <u>closable</u> if such a choice is possible. It is easy to see that Q is <u>closable</u> if and only if the following condition is satisfied.

<u>15.4. Conditions for Closability</u>. If $\{\varphi_n\}_{n=1}^{\infty}$ is a sequence in $C^1_{com}(\underline{\underline{R}})$ such that

(15.67) $$\operatorname{Lim}_{m,n\uparrow\infty} Q(\varphi_m-\varphi_n, \varphi_m-\varphi_n) = 0$$

(15.68) $$\operatorname{Lim}_{n\uparrow\infty} \int_{-\infty}^{+\infty} dt \, \{\varphi_n(t)\}^2 = 0$$

then in addition

(15.69) $$\operatorname{Lim}_{n\uparrow\infty} Q(\varphi_n, \varphi_n) = 0.$$

///

We finish with a relatively weak partial result which does at least indicate what is involved in checking 15.4. First some terminology. Let $\nu_0(t)$ be the density for the absolutely continuous part of ν. The regular set Ⓡ(ν) is the open set of points in a neighborhood of which the reciprocal $1/\nu$ is integrable. The singular set is the complement Ⓢ$(\nu) = \underline{\underline{R}}$ - Ⓡ(ν).

Theorem 15.3. (i) If ν is absolutely continuous and if

(15.70) $$\nu_0(t) = 0 \quad \text{a.e.} \quad \text{on} \quad S(\nu),$$

then the formal Dirichlet form (15.66) is closable.

(ii) If (15.66) is closable, then ν is absolutely continuous. ///

Proof. It is an easy consequence of Feller's classification for one dimensional diffusions [15] that the stated conditions on ν in (i) are necessary and sufficient for the difference Q-N to be closable and this would be enough to prove (i). For a direct proof of (i) let $\{\varphi_n\}_{n=1}^{\infty}$ satisfy (15.67) and (15.68) and define $\psi_n = \psi_n'$ so that $\psi_n \to \psi$ in $L^2(\nu)$. We need only show that $\psi = 0$. But if $a < b$ is such that the closed interval $[a, b]$ belongs to the regular set Ⓡ(ν), then the estimate

$$\left| \int_a^b dt \, \psi_n(t) \right| \leq \int_a^b dt \, \nu^{-1}(t) \int_a^b dt \, \nu(t) \psi_n^2(t)$$

guarantees that $\int_a^b dt\, \psi(t) = 0$ whenever $\varphi_n(a), \varphi_n(b) \to 0$ and therefore $\psi = 0$ in $L^2(\nu)$. To prove (ii) assume that ν has a nontrivial singular piece and let ψ_n, $n \geq 1$ be a sequence of continuous functions with fixed compact support in $\underline{\underline{R}}$ such that $\psi = \operatorname{Lim}_{n\uparrow\infty} \psi_n$ exists and is nontrivial in $L^2(\underline{\underline{R}}, \nu)$ but $\int_{-\infty}^{+\infty} dt\, \{\psi_n(t)\}^2 \to 0$ and also $\int_{-\infty}^{+\infty} dt\, \psi_n(t) = 0$ for each n. Then the integrated sequence $\varphi_n(t) = \int_{-\infty}^{t} ds\, \psi_n(s)$ violates 15.4. The only nontrivial point is that the excursion norms $N(\varphi_n, \varphi_n)$ are dominated by the norms $\int_{-\infty}^{+\infty} dt\, \{\psi_n(t)\}^2$ and therefore $Q(\varphi_n - \varphi_m, \varphi_n - \varphi_m) \to 0$ but $Q(\varphi_n, \varphi_n) \to \int \nu(dt)\{\psi(t)\}^2$. ///

16. A Theorem of Beurling and Deny

In this section dx is Lebesgue measure on d-dimensional Euclidean space $\underline{\underline{R}}^d$. The following theorem is stated informally on pages 212, 213 in the fundamental paper [2] of A. Beurling and J. Deny.

Theorem 16.1. Let $(\underline{\underline{F}}, E)$ be a Dirichlet space on $L^2(\underline{\underline{R}}^d, dx)$ such that $C^{\infty}_{com}(\underline{\underline{R}}^d)$ is a dense subset of $\underline{\underline{F}}$. Then for $g \in C^{\infty}_{com}(\underline{\underline{R}}^d)$

$$E(g,g) = \Sigma^d_{i,j=1} \int \nu_{i_j}(dx)\partial_i g(x)\partial_j g(x) \tag{16.1}$$

$$+ \frac{1}{2}\int\int J(dx,dy)\{g(x)-g(y)\}^2 + \int \kappa(dx)g^2(x)$$

where ν_{i_j} are signed Radon measures on $\underline{\underline{R}}^d$ with $\nu_{ij} = \nu_{ji}$, where κ is a Radon measure on $\underline{\underline{R}}^d$ and where $J(dx, dy)$ is a symmetric Radon measure on the complement of the diagonal in $\underline{\underline{R}}^d \times \underline{\underline{R}}^d$. Moreover the representation is unique. ///

Of course $\partial_i g(x)$ denotes the partial derivative with respect to the ith coordinate. It is clear that κ must be the killing measure and J the Lévy kernel of SMP. Also the ν_{ij} must form a nonnegative definite matrix in the sense that

$$\Sigma_{i,j} \int \nu_{ij}(dx)g_i(x)g_j(x) \geq 0$$

whenever $g_1, \ldots, g_d \in C_{com}(\underline{\underline{R}}^d)$. Proofs of Theorem 16.1, at least for the special case when E is assumed to be "local", have appeared in Japanese, for example in [55], but to our knowledge not in any other language. The proof which follows is a slight refinement of one which was communicated privately to the author by S. Watanabe and was intended to handle the special case when E is local.

Proof. Fix $g \in C_{com}(\underline{R}^d)$ and let $u_1, \ldots, u_d \in C^{\infty}_{com}(\underline{R}^d)$ agree with the coordinate functions $x_1, \ldots, x_d$ in a neighborhood of the support of g. By SMP, Lemma 1.1

$$(16.2) \qquad E(g,g) = \operatorname{Lim}_{t\downarrow 0}[(1/t)\int dx\,\{1-P_t 1(x)\}g^2(x)$$
$$+ (1/2t)\int dx \int P_t(x,dy)\{g(x)-g(y)\}^2]$$

and by SMP, Lemma 11.1 the first term on the right in (16.2) converges to the last term in (16.1). Therefore we need only concern ourselves with the second term on the right in (16.2). It follows from SMP, Theorem 11.3 that for each pair i, j there exists a unique signed measure ν^0_{ij} in bounded variation on $\underline{R}^d$ such that for general $f \in C_{com}(\underline{R}^d)$

$$(16.3)\ \operatorname{Lim}_{t\downarrow 0}(1/t)\int dx\, f(x)\int P_t(x,dy)\{u_i(x)-u_i(y)\}\{u_j(x)-u_j(y)\} = 2\int \nu^0_{ij}(dx)f(x)\ .$$

Let $J(dx, dy)$ be the Lévy kernel of SMP, Section 10. For $f, f' \in C^{\infty}_{com}(\underline{R}^d)$ with disjoint supports

$$\operatorname{Lim}_{t\downarrow 0}(1/t)\int dx\int P_t(x,dy)f(x)f'(y)$$
$$= -\operatorname{Lim}_{t\downarrow 0}(1/2t)\int dx\int P_t(x,dy)\{f(x)-f(y)\}\{f'(x)-f'(y)\}$$
$$= -E(f,f') = \int\int J(dx,dy)f(x)f'(y)$$

and from this it follows that

$$\operatorname{Lim}_{t\downarrow 0}(1/t)\int dx\int P_t(x,dy)F(x,y) = \int\int J(dx,dy)F(x,y)$$

whenever $F(x, y)$ is continuous with compact support contained in the complement of the diagonal in $\underline{R}^d \times \underline{R}^d$. It follows from SMP, (11.32) that the signed measures

ν_{ij} defined by

(16.5) $$\int \nu_{ij}(dx)f(x) = \int \nu^0_{ij}(dx)f(x) - \frac{1}{2}\int\int J(dx, dy)f(x)\{u_i(x)-u_i(y)\}\{u_j(x)-u_j(y)\}$$

depend locally on u_i, u_j and therefore are well defined independent of the choice of u_i, u_j on a neighborhood of the support of g. Let $u(x) \in C_{com}(\underline{R}^d)$ be identically 1 in a neighborhood of 0 and such that $0 \leq u(x) \leq 1$. Then

$$\mathrm{Lim}_{t\downarrow 0}(1/t)\int dx \int P_t(x, dy)(1-\theta(x-y))\{g(x)-g(y)\}^2$$

$$= \int\int J(dx, dy)(1-\theta(x-y))\{g(x)-g(y)\}^2$$

$$\mathrm{Lim}_{t\downarrow 0}(1/t)\int dx \int P_t(x, dy)(1-\theta(x-y))\partial_i g(x)\partial_j g(x)\{u_i(x)-u_i(y)\}\{u_j(x)-u_j(y)\}$$

$$= \int\int J(dx, dy)(1-\theta(x-y))\partial_i g(x)\partial_j g(x)\{u_i(x)-u_i(y)\}\{u_j(x)-u_j(y)\}$$

and therefore

$$\mathrm{Lim}_{t\downarrow 0}(1/2t)\int dx \int P_t(x, dy)\{g(x)-g(y)\}^2$$

$$= \Sigma_{i,j}\int \nu^0_{i_j}(dx)\partial_i g(x)\partial_j g(x) + \frac{1}{2}\int\int J(dx, dy)(1-\theta(x-y))\{g(x)-g(y)\}^2$$

$$- \frac{1}{2}\int\int J(dx, dy)(1-\theta(x-y))\Sigma_{i,j}\partial_i g(x)\partial_j g(x)\{u_i(x)-u_i(y)\}\{u_j(x)-u_j(y)\}$$

$$+ \mathrm{Lim}_{t\ 0}(1/2t)\int dx \int P_t(x, dy)\theta(x-y)[\{g(x)-g(y)\}^2$$

$$- \Sigma_{i,j}\partial_i g(x)\partial_j g(x)\{u_i(x)-u_i(y)\}\{u_j(x)-u_j(y)\}] .$$

Since by SMP, Lemma 1.1

$$\sup_{t>0}(1/2t)\int dx \int P_t(x, dy)\{u_i(x)-u_i(y)\}^2 \leq E(u_i, u_i)$$

it follows with the help of Taylor' theorem that the last term on the right converges to zero as support $\theta \downarrow \{0\}$ and (16.1) follows. Uniqueness in (16.1) is clear from our derivation. ///

It is natural to ask what the possibilities are in (16.1). That is, when is the form (16.1), defined initially on $C^{\infty}_{com}(\underline{\underline{R}}^d)$, closable in the sense of Section 15. Apparently little is known about this at present, even when $J = 0$. In the next section we get some information in a special case.

17. Brownian Motion on $\underline{\underline{R}}^2\backslash\underline{\underline{R}}$.

In this section $\underline{\underline{X}} = \underline{\underline{R}}^2\backslash\underline{\underline{R}}$, the plane with the real line deleted. That is,

$$\underline{\underline{R}}^2\backslash\underline{\underline{R}} = \{(a,b) : b \neq 0\} .$$

The situation is not very different from Section 15 and would not warrant separate consideration except that we take a different point of view here. The reference measure is Lebesgue measure dx viewed as living either on $\underline{\underline{R}}^2$ or $\underline{\underline{X}}$, depending on context.

We begin by considering a Dirichlet space $(\underline{\underline{F}}^\sim, E^\sim)$ on $(\underline{\underline{R}}^2, dx)$ which satisfies the hypotheses of Theorem 16.1 and which is local. This means that $E(f,g) = 0$ whenever $f, g \in C^\infty_{com}(\underline{\underline{R}}^d)$ have disjoint supports. By SMP, Theorem 11.10 this is equivalent to assuming $J = 0$ in (15.1) and also equivalent to assuming that the associated process has continuous trajectories. (The equivalence with the latter was first established by M. Fukushima [56].) Let $(\underline{\underline{F}}, E)$ be the absorbed Dirichlet space for $\underline{\underline{X}}$. We assume that $(\underline{\underline{F}}, E)$ is the Dirichlet space associated with standard Brownian motion on $\underline{\underline{X}}$. This is equivalent to requiring that

$$E^\sim(f,f) = \frac{1}{2} \int dx \, \{\operatorname{grad} f(x)\}^2 \tag{17.1}$$

whenever $f \in C^\infty_{com}(\underline{\underline{X}})$. Let $\underline{\underline{H}}$ be the restriction $\gamma\underline{\underline{F}}^\sim$ of $\underline{\underline{F}}^\sim$ to $\underline{\underline{R}}$ in the sense of SMP, Section 8 and for $\varphi \in \underline{\underline{H}}$ let

$$Q(\varphi, \varphi) = E^\sim(H\varphi, H\varphi)$$

with H the hitting operator for $\underline{\underline{R}}$. By the results in SMP, Section 8

$$E^\sim(f,f) = E(f - H\gamma f, f - H\gamma f) + Q(\gamma f, \gamma f) \tag{17.2}$$

for general $f \in \underline{\underline{F}}^\sim$. The arguments at the beginning of Section 17 show that for harmonic measure we can choose Lebesgue measure dt and it follows again by SMP, Section 8 that $(\underline{\underline{H}}, Q)$ is a Dirichlet space on $L^2(\underline{\underline{R}}, dt)$. Let $(\underline{\underline{N}}, N)$ be as in Section 15. It then follows from the First Structure Theorem in SMP (since $\kappa = 0$ on $\underline{\underline{X}}$) that $\underline{\underline{H}}$ is contained in $\underline{\underline{N}}$ and that for $\varphi \in \underline{\underline{H}}$

$$Q(\varphi, \varphi) = 2N(\varphi, \varphi) + D(\varphi, \varphi) \tag{17.3}$$

where D is a contractive bilinear form on $\underline{\underline{H}}$. The factor 2 results from the fact that $\underline{\underline{X}}$ consists of both half planes. It is clear that $C^{\infty}_{com}(\underline{\underline{R}})$ is dense in $\underline{\underline{H}}$ and therefore $(\underline{\underline{F}}^{\sim}, E^{\sim})$ is completely determined by the restriction of D to $C^{\infty}_{com}(\underline{\underline{R}})$. Our first result is

<u>Lemma 17.1.</u> D <u>is local. That is,</u> $D(\varphi, \psi) = 0$ <u>whenever</u> $\varphi, \psi \in C^{\infty}_{com}(\underline{\underline{R}})$ <u>have disjoint supports.</u> ///

<u>Proof.</u> It suffices to show that

$$Q(\gamma f, \gamma g) = 2N(\gamma f, \gamma g) \tag{17.4}$$

whenever $f, g \in C^{\infty}_{com}(\underline{\underline{R}}^2)$ have disjoint supports. The point is that $E^{\sim}(f, g) = 0$ and therefore

$$Q(\gamma f, \gamma g) = E^{\sim}(f, g) - E(f - Hf, g - Hg)$$

is unchanged if we replace $E^{\sim}$ by the Dirichlet form for standard Brownian in $\underline{\underline{R}}^2$. But then (17.4) follows from the calculations at the beginning of Section 15.

///

Combining Lemma 17.1 with Theorem 16.1, we conclude that for $\varphi \in C^{\infty}_{com}(\underline{\underline{R}})$

$$Q(\varphi, \varphi) = 2N(\varphi, \varphi) + \int \nu(dt) \, \{\varphi'(t)\}^2 + \int \kappa\,(dt)\, \varphi^2(t)$$

where κ, are Radon measures on $\underline{\underline{R}}$. But Q must be closable on $C^{\infty}_{com}(\underline{\underline{R}})$ and therefore on $C^{1}_{com}(\underline{\underline{R}})$ and the results at the end of Section 15 are applicable. (The presence of κ is irrelevant for this.) In particular $\nu(dt)$ has a density $\nu(t)$ and (15.70) is a sufficient condition for a given function $\nu(t) \geq 0$ to occur. Thus for general $g \in C^{\infty}_{com}(\underline{\underline{R}}^2)$

$$(17.5) \qquad E^{\sim}(g, g) = \frac{1}{2} \int dx \, \{\operatorname{grad} g(x)\}^2 + \int_{-\infty}^{+\infty} dt \, \nu(t) \{\partial_1 g(t, 0)\}^2 + \int_{-\infty}^{+\infty} \kappa(dt) \{g(t, 0)\}^2 .$$

We summarize in

Theorem 17.2. Let $(\underline{\underline{F}}^{\sim}, E^{\sim})$ be a Dirichlet space on $(\underline{\underline{R}}^2, dx)$ satisfying the hypotheses of Theorem 16.1 and such that (17.1) is valid for $f \in C^{\infty}_{com}(\underline{\underline{X}})$. Then there exists a unique Radon measure $\kappa(dt)$ on $\underline{\underline{R}}$ and a unique nonnegative function $\nu(t)$ on $\underline{\underline{R}}$ such that (17.5) is valid for general $g \in C^{\infty}_{com}(\underline{\underline{R}}^2)$. Conversely if $\kappa(dt)$ is any Radon measure on $\underline{\underline{R}}$ and if $\gamma(t) \geq 0$ or $\underline{\underline{R}}$ satisfies (15.70), then there exists $(\underline{\underline{F}}^{\sim}, E^{\sim})$ as above satisfying (17.5). ///

Remark. We do not know if (15.70) is also a necessary condition for $\nu(t)$. ///

The special case when $\kappa = 0$, $\nu = 1$ is closely related to the example based on 15.3.1. Also it is identical with example 2.1 in [30] where the "skew product" is used to construct the process directly and where interesting properties of the sample paths are derived.

Chapter IV. Stable Processes

Symmetric stable processes in one dimension are considered in Sections 18, 19 and 20. The multi-dimensional isotropic case is considered in Section 21. As for Brownian motion in the upper half plane, Section 15, our attitude is that these processes have intrinsic interest, and we go into considerable detail. Our basic computational tool is a factorization of the generator, first introduced by J. Elliott [11] for the Cauchy process in one dimension and then extended by S. Watanabe [52] to the case $\alpha \neq 1$ and again by Elliott [12] to the multi-dimensional case. In each case we first study the process on the full line or Euclidean space and then apply our general theory to a particular class of processes which dominate the absorbed process for the open unit interval or ball.

The Cauchy process is singled out for special attention in Section 18. The reason for this is that the nontrivial factor is the restricted Hilbert transform which can be explicitly inverted, and this permits us to get more precise information than in the other cases. In particular, we can give a complete description of the generator, as long as we use the "correct" reference measure.

The case $0 < \alpha < 1$ is treated in Section 19. The main feature here is that the killing density is integrable, and therefore the regular boundary Δ_r must be empty. Thus the dominating Dirichlet spaces are quite different from their analogues for $\alpha \geq 1$. This is also true for the generators if we use the formulation in Theorems 11.2 and 11.3. However, Theorem 11.4 together with the strong Feller property lead naturally to a formulation which is essentially the same as for $\alpha \geq 1$. This is in contrast to Cases 4 and 5 in Section 12 where the corresponding formulation seemed quite unnatural.

The case $1 < \alpha < 2$ is treated rather briefly in Section 20 since the main features are essentially the same as for the Cauchy process. (However, see the remark at the end of this section.)

Our treatment of the multi-dimensional case in Section 21 parallels the one-dimensional case, but our results are less precise. However, we do consider in some detail the problem of establishing the strong Feller property. Also we apply the technique of Theorem 11.4 to formulate boundary conditions for the generator in the singular range, $0 < \alpha < 1$. The approach of Theorem 11.2 and 11.3 does not seem to yield an explicit description for the generator.

18. One Dimensional Cauchy Process.

In our treatment of this example we were greatly influenced by the paper of Elliott [11], although our general approach is quite different. The results in [11] concern certain sub-Markovian semigroups which dominate the absorbing semigroup for the interval $\underline{I} = (-1, +1)$ and the basic technique focuses on an explicit construction of the resolvent. We restrict attention to the symmetric case, and we rely on Dirichlet space techniques.

We begin by considering the symmetric Cauchy process on the full line $\underline{R}$. For $t > 0$ let $c_t(x)$ and C_t be defined by (15.2) and (15.19). It is clear that the C_t form a continuous semigroup of symmetric contractions on $L^2(dx)$. Since

$$c_t > 0 \; ; \; \int_{-\infty}^{+\infty} dx\, c_t(x) = 1,$$

it follows that the C_t are Markovian and also irreducible in the sense of SMP, Section 1. Let $(\underline{C}, C)$ be the associated Dirichlet space on $L^2(dx)$. From (15.5) it follows that

$$\text{Ⓕ} C_t f(\lambda) = e^{-t|\lambda|} \text{Ⓕ} f(\lambda)$$

and we conclude from the Plancherel theorem that

$$C(f, f) = (2\pi)^{-1} \int_{-\infty}^{+\infty} d\lambda\, |\lambda| \{\text{Ⓕ} f(\lambda)\}^2 \tag{18.1}$$

and therefore by (15.9) and (15.9')

$$C(f, f) = (2\pi)^{-1} \int_{-\infty}^{+\infty} dx \int_{-\infty}^{+\infty} dy (x-y)^{-2} \{f(x) - f(y)\}^2 . \tag{18.1'}$$

Regularity of $(\underline{C}, C)$ follows from (18.1). Also (18.1') implies that $C(f, f)$ is always well defined, although possibly infinite.

Lemma 18.1. The extended Dirichlet space $\underline{C}_{(e)}$ is the set of functions f for which the right side of (18.1') converges, and then (18.1') is valid. The reflected space $\underline{\underline{C}}_{ref} = \underline{\underline{C}}_{(e)}$, and the Dirichlet space $(\underline{\underline{C}}, C)$ is recurrent. ///

Proof. We begin by showing that $1 \in \underline{\underline{C}}_{(e)}$, using an argument shown to the author in private conversation by S. Port and C. Stone. Since $1/|\lambda|$ is not integrable in a neighborhood of 0, there exists a sequence of functions $\varphi_n(\lambda) \geq 0$ belonging to $L^2(d\lambda)$ and satisfying

$$(2\pi)^{-1}\int_{-\infty}^{+\infty} d\lambda\, \varphi_n(\lambda) = 1;\ (2\pi)^{-1}\int_{-\infty}^{+\infty} du|\lambda|\ |\varphi_n(\lambda)|^2 \leq 1;\ \operatorname{supp}\varphi_n \subset [-\tfrac{1}{n}, +\tfrac{1}{n}].$$

Indeed, we need only let $\varphi_n(\lambda) = (1/|\lambda|)I(a_n \leq |\lambda| \leq 1/n)$ with $a_n < 0$ determined by the condition

$$(2\pi)^{-1}\int_{a_n}^{1/n} d\lambda/\lambda = \tfrac{1}{2} \quad \text{or} \quad -\log(na_n) = \pi.$$

Then $f_n = Ⓕ^{-1}\varphi_n$ belongs to $\underline{\underline{C}}$ and $C(f_n, f_n) \leq 1$ for all n while the Plancherel theorem guarantees that

$$\text{(18.2)} \qquad \operatorname{Lim}_{n\uparrow\infty} \int dx f_n(x) g(x) = \operatorname{Lim}_{n\uparrow\infty} (2\pi)^{-1}\int_{-\infty}^{+\infty} d\lambda\, \varphi_n(\lambda) Ⓕg(\lambda)^*$$

$$= Ⓕg(0) = \int_{-\infty}^{+\infty} dx\, g(x)$$

for all $g \in Ⓢ$. Since (18.2) guarantees that $f_n \to 1$ weakly in $L^2(\underline{\underline{R}}, m)$ whenever $m(x) > 0$ is integrable, it follows by the classical theorem of Banach and Saks that the Cesàro sums of a subsequence converge almost everywhere, and therefore $1 \in \underline{\underline{C}}_{(e)}$. (The theorem of Banach and Saks is stated and proved by Riesze and Nagy [45, p. 80]. The proof is reproduced in SMP, p. 1.12.) Since $(\underline{\underline{C}}_{(e)}, C)$ is an "honest Hilbert space" in the transient case, this is enough to guarantee that $(\underline{\underline{C}}, C)$ is recurrent in the sense of SMP, Section 1. It follows directly from Fatou's lemma that (18.1') is valid for general f in $\underline{\underline{C}}_{(e)}$. To complete the proof choose

$f_n \in \underline{\underline{C}}$ uniformly bounded and each nonnegative with compact support such that $C(f_n, f_n)$ stays bounded and $f_n \to L$ almost everywhere. (The existence of such f_n follows directly from regularity of $(\underline{\underline{C}}, C)$ once it is shown that $1 \in \underline{\underline{C}}_{(e)}$.) Consider g satisfying $0 \le g \le 1$ such that the right side of (18.1') converges. Clearly $g \wedge f_n$ belongs to $\underline{\underline{C}}$, the norms $C(g \wedge f_n, g \wedge f_n)$ are bounded and $g \wedge f_n \to g$ almost everywhere. This together with an obvious splitting and approximation by truncations is enough to guarantee that any g for which the right side converges belongs to $\underline{\underline{C}}_{(e)}$, and the theorem is completely proved. ///

Remark 1. The last argument can be used to identify $\underline{\underline{C}}_{(e)}$ and $\underline{\underline{C}}_{ref}$ in general when $(\underline{\underline{C}}, C)$ is recurrent, and therefore $1 \in \underline{\underline{C}}_{(e)}$. This is much simpler than the argument given in SMP, Section 16. ///

Remark 2. It is completely elementary that if $(\underline{\underline{F}}, E)$ is a translation invariant Dirichlet space on $L^2(\underline{\underline{R}}, dx)$ then there exists a nonnegative continuous function $\psi(\lambda)$ such that

$$E(f, f) = (2\pi)^{-1} \int_{-\infty}^{+\infty} d\lambda \psi(\lambda) |\textcircled{F} f(\lambda)|^2,$$

and $\underline{\underline{F}}$ is the set of $f \in \underline{\underline{F}}$ for which this converges. Of course this follows from the general Levy-Khintchin representation for infinitely divisible processes, but it is much more elementary. The proof of Lemma 18.1 combines with the argument in Section 13 for $d = 3$ to show that $(\underline{\underline{F}}, E)$ is transient if and only if the reciprocal $\psi^{-1}(\lambda)$ is integrable in a neighborhood of the origin. Of course this follows, directly from more general (and also much deeper) results of F. Spitzer, and also S. Port and C. Stone. It turns out that also integrability near infinity is an important condition. Singletons are nonpolar for $(\underline{\underline{F}}, E)$ if and only if the reciprocal $\psi^{-1}(\lambda)$ is integrable near infinity. This follows from the general criterion of H. Kesten. (See [3] or [33].) We give here a proof which depends on symmetry. If $\psi^{-1}(\lambda)$ is integrable near infinity, then $\{1+\psi(\lambda)\}^{-1}$ is integrable and the estimate

$$\begin{aligned}\{f(0)\}^2 &= \{(2\pi)^{-1}\int_{-\infty}^{+\infty}d\lambda \text{Ⓕ}f(\lambda)\}^2 \\ &\leq (2\pi)^{-2}\int_{-\infty}^{+\infty}d\lambda\{1+\psi(\lambda)\}|\text{Ⓕ}f(\lambda)|^2\int_{-\infty}^{+\infty}d\lambda\{1+\psi(\lambda)\}^{-1} \\ &= E_1(f,f)(2\pi)^{-1}\int_{-\infty}^{+\infty}d\lambda\{1+\psi(\lambda)\}^{-1}\end{aligned}$$

valid for $f \in L^2(\underline{\underline{R}}, dx)$ such that Ⓕf has compact support, shows that ε_0 has finite energy and therefore $\{0\}$ is nonpolar. If ψ is not integrable near infinity let $\alpha_n > n$ be determined by the condition $\int_n^{n+\alpha_n} d\lambda\{1+\psi(\lambda)\}^{-1} = 1$, and let $f_n \in \underline{\underline{F}} \cap C(\underline{\underline{R}})$ be such that except for a $d\lambda$ null set

$$\text{Ⓕ}f_n(\lambda) = \begin{cases} 1 & \text{if } n \leq \lambda \leq n+\alpha_n \\ 0 & \text{otherwise.}\end{cases}$$

Then $f_n(0) = 1$ and $E_1(f_n, f_n) = 1$ and $f_n \to 0$ weakly relative to E_1. After replacing by a subsequence and taking Cesàro means as in the proof of Lemma 18.1, we get a sequence $g_n \in \underline{\underline{F}} \cap C(\underline{\underline{R}})$ such that $E_1(g_n, g_n) \to 0$ and still $g_n(0) = 1$ for all n. This implies that the singleton $\{0\}$ is polar. For the Cauchy process, $\psi(\lambda) = (2\pi)^{-1}|\lambda|$ and therefore singletons are polar. ///

Certainly the local generator Ⓐ^R can be defined by

$$\text{Ⓐ}^R f(x) = \text{P.V.}\ \pi^{-1}\int_{-\infty}^{+\infty}dy\{f(y)-f(x)\}/(x-y)^2$$

where P.V. denotes the principal value integral, defined by inserting the indicator $I(|x-y| > \varepsilon)$ and then passing to the limit $\varepsilon \downarrow 0$. Much more interesting for us is the connection with the Hilbert transform

$$\text{Ⓗ}f(x) = \text{P.V.}\ \pi^{-1}\int dy(x-y)^{-1}f(y). \tag{18.3}$$

We take for granted the classical result that $\textcircled{H}$ is bounded on $L^p(dx)$, $1 < p < +\infty$, and that the principal value integral converges in $L^p(dx)$ for $1 < p < \infty$ and almost everywhere for $f \in L^p(dx)$, $1 \leq p < +\infty$. Also, at least for $f \in L^2(dx)$,

$$\textcircled{F}\textcircled{H}f(\lambda) = -i\,\mathrm{sgn}\,\lambda\,\textcircled{F}f(\lambda) \tag{18.4}$$

and therefore if also $D_x\textcircled{H}f \in L^2(dx)$

$$\textcircled{F}\textcircled{D}_x\,Hf(\lambda) = |\lambda|\textcircled{F}f(\lambda). \tag{18.5}$$

(See for example Chapter VI in the book of E. Stein and G. Weiss [50].) If $f \in L^2(dx)$, then even when $D_x\textcircled{H}f$ is not in $L^2(dx)$, it is a tempered distribution and (18.5) remains valid if the Fourier transform $\textcircled{F}$ is extended to tempered distributions u by the convention

$$\int dx\,\textcircled{F}u(x)\varphi(x) = \int dx\,u(x)\textcircled{F}\varphi(x) \tag{18.6}$$

for $\varphi \in \textcircled{S}$, $u \in \textcircled{S}'$. (See Chapter I in [50].) Thus for $f \in L^2(dx)$

$$\textcircled{A}^R f(x) = -D_x\textcircled{H}f(x) \tag{18.7}$$

at least if the local generator (that is, (6.1)) is interpreted in the distribution sense. It is also true that

$$\textcircled{A}^R f(x) = -\textcircled{H}D_x f(x) \tag{18.7'}$$

whenever the right side makes sense as a tempered distribution. Thus by Theorem 8.1 the generator A^R for $(\underline{\underline{C}}, C)$ is defined by

$$A^R f(x) = -\textcircled{H}D_x f(x) = -D_x\textcircled{H}f(x) \tag{18.8}$$

supplemented by the condition that $f \in \underline{\underline{C}}_{(e)}$. In fact it is clear from the Plancherel theorem that in this case the last condition can be suppressed.

We turn our attention now to the open interval $\underline{\underline{I}} = (-1,+1)$, beginning with some preliminary results. We are particularly interested in the Hilbert transform on $\underline{\underline{I}}$

(18.9) $$\textcircled{H}^I f(x) = \text{P.V.}\ \pi^{-1}\int_{-1}^{+1} dy\,(x-y)^{-1} f(y),$$

which we consider only for $x \in \underline{\underline{I}}$. This operator has been studied in detail by several authors [1], [26], [48] and [51]. We will follow Söhngen [48] and study $\textcircled{H}^I$ by exploiting its connection with the classical Hilbert transform $\textcircled{H}^T$ for the torus $T = (-\pi, +\pi]$. This is defined by

(18.10) $$\textcircled{H}^T F(\theta) = \text{P.V.}\ (2\pi)^{-1} \cdot \int_{-\pi}^{+\pi} dt\, F(t) \cot \tfrac{1}{2}(\theta - t).$$

While we are at it we will prove a little bit more about $\textcircled{H}^I$ than we actually need. We begin by summarizing some properties of $\textcircled{H}^T$ that are relevant for us.

Theorem 18.2. (i) *If* $F \in L^1(\underline{\underline{T}}, d\theta)$, *then the principal value integral in* (18.10) *converges almost everywhere. If* $F \in L^2(\underline{\underline{T}}, d\theta)$, *then the convergence is also in the* $L^2(\underline{\underline{T}}, d\theta)$ *sense.*

(ii) $\textcircled{H}^T 1 = 0$ *and for* $n \geq 1$

$$\textcircled{H}^T \cos n\theta = \sin n\theta; \textcircled{H}_T \sin n\theta = -\cos n\theta.$$

(iii) *If* F *belongs to the classical Sobolev space* $\underline{\underline{W}}(\underline{\underline{T}})$ *and if* $F(-\pi+0) = F(\pi)$, *then also* $\textcircled{H}^T F$ *belongs to* $\underline{\underline{W}}(\underline{\underline{T}})$ *and* $\textcircled{H}^T F(-\pi+0) = \textcircled{H}^T F(\pi)$.

(iv) *Let* G *be given in* $L^2(\underline{\underline{T}}, d\theta)$. *Then the equation*

(18.11) $$G = \textcircled{H}^T F$$

has a solution $F \in L^2(\underline{\underline{T}}, d\theta)$ *if and only if*

(18.12) $$\int_{-\pi}^{+\pi} d\theta\, G(\theta) = 0.$$

In this case the most general solution is given by

$$F(\theta) = \alpha - \textcircled{H}^{T} G(\theta)$$

where α is an arbitrary constant. ///

For (i) we refer to A. Zygmund's book [54, I, Chapt. 7]. The remainder of the theorem follows from

(18.13)
$$\Sigma_{n=1}^{\infty} r^n \{\sin n\theta \cos nt - \cos n\theta \sin nt\}$$
$$= \Sigma_{n=1}^{\infty} r^n \sin n(\theta - t)$$
$$= \text{Im. } re^{i(\theta-t)}/\{1-re^{i(\theta-t)}\}$$
$$= r\sin(\theta-t)/\{1+r^2-2r\cos(\theta-t)\}$$

once it is shown that at least for a dense set of $F \in L^2(\underline{\underline{T}}, d\theta)$

(18.14)
$$H^T F(\theta) = \text{Lim}_{r\uparrow 1} \pi^{-1} \int_{-\pi}^{+\pi} dt F(t) r \sin(\theta - t)\{1+r^2-2r\cos(\theta-t)\}^{-1}.$$

A proof of (18.14) for F continuous can also be found in Zygmund's book [54, I, p. 103].

Next we establish a connection between $\textcircled{H}^I$ and $\textcircled{H}^T$ in the pointwise sense.

Theorem 18.3. (i) If $f \in L^1(\underline{\underline{I}}, dx)$, then $f(\cos\theta)|\sin\theta| \in L^1(\underline{\underline{T}}, d\theta)$ and the following two relations are equivalent modulo null sets

(18.15)
$$g(x) = \textcircled{H}^I f(x)$$

(18.16)
$$g(\cos\theta)\sin\theta = -\textcircled{H}^T\{f(\cos t)|\sin t|\}(\theta).$$

(ii) If $f \in L^1(\underline{\underline{I}}, (1-x^2)^{-\frac{1}{2}})$, then also $f(\cos\theta) \in L^1(\underline{\underline{T}}, d\theta)$ and (18.15) is equivalent modulo null sets with

(18.17) $$g(\cos\theta) = -\text{Ⓗ}^T\{f(\cos t)\operatorname{sgn} t\}(\theta).$$ ///

Proof. For $\varepsilon > 0$ consider

(18.18) $$I_\varepsilon(x) = \pi^{-1}\int_{-1}^{+1} dy I(|x-y| > \varepsilon)(x-y)^{-1} f(y)$$

and make the change of variables $y = \cos t$, $x = \cos\theta$ to get

$$I_\varepsilon(\cos\theta) = \pi^{-1}\int_0^\pi dt \sin t I(|\cos\theta - \cos t| > \varepsilon) f(\cos t)(\cos\theta - \cos t)^{-1}.$$

Next apply the trigonometric identities

(18.19) $$\sin\theta/\{\cos\theta - \cos t\}$$
$$= -\{\sin\tfrac{1}{2}(\theta+t)\cos\tfrac{1}{2}(\theta-t) + \sin\tfrac{1}{2}(\theta-t)\cos\tfrac{1}{2}(\theta+t)\}/\{2\sin\tfrac{1}{2}(\theta-t)\sin\tfrac{1}{2}(\theta+t)\}$$
$$= -\tfrac{1}{2}\cot\tfrac{1}{2}(\theta-t) - \tfrac{1}{2}\cot\tfrac{1}{2}(\theta+t)$$

(18.20) $$\sin t/\{\cos\theta - \cos t\}$$
$$= -\{\sin\tfrac{1}{2}(\theta+t)\cos\tfrac{1}{2}(\theta-t) - \sin\tfrac{1}{2}(\theta-t)\cos\tfrac{1}{2}(\theta+t)\}/\{2\sin\tfrac{1}{2}(\theta-t)\sin\tfrac{1}{2}(\theta+t)\}$$
$$= -\tfrac{1}{2}\cot\tfrac{1}{2}(\theta-t) + \tfrac{1}{2}\cot\tfrac{1}{2}(\theta+t)$$

to get on the one hand

$$I_\varepsilon(\cos\theta)\sin\theta$$
$$= -\pi^{-1}\int_0^\pi dt\, I(|\cos\theta - \cos t| > \varepsilon) f(\cos t)\sin t\{\tfrac{1}{2}\cot\tfrac{1}{2}(\theta-t) + \tfrac{1}{2}\cot\tfrac{1}{2}(\theta+t)\}$$
$$= -(2\pi)^{-1}\int_{-\pi}^{+\pi} dt\, I(|\cos\theta - \cos t| > \varepsilon) f(\cos t)|\sin t|\cot\tfrac{1}{2}(\theta-t)$$

and on the other hand

$$I_\varepsilon(\cos\theta)$$

$$= -\pi^{-1}\int_0^\pi dt\, I(|\cos\theta - \cos t| > \varepsilon) f(\cos t)\{\tfrac{1}{2}\cot\tfrac{1}{2}(\theta-t) - \tfrac{1}{2}\cot\tfrac{1}{2}(\theta+t)\}$$

$$= -(2\pi)^{-1}\int_{-\pi}^{\pi} dt\, I(|\cos\theta - \cos t| > \varepsilon) f(\cos t)\,\mathrm{sgn}\, t\,\cot\tfrac{1}{2}(\theta - t).$$

Except for special values of θ, which of course can be ignored, the difference between the integrals appearing on the right and the corresponding integrals with $I(|\cos t - \cos\theta| > \varepsilon)$ replaced by $I(|t-\theta| > \alpha)$ with α determined by $|\cos(\theta+\alpha) - \cos\theta| = \varepsilon$ is dominated by an expression of the form

$$(1/\alpha)\int_{\theta+\alpha-M\alpha^2}^{\theta+\alpha+M\alpha^2} dt\,|F(t)| \quad \text{or} \quad (1/\alpha)\int_{\theta-\alpha-M\alpha^2}^{\theta-\alpha+M\alpha^2} dt\,|F(t)|$$

with $F(t)$ integrable. It follows from the classical theorem on differentiating an indefinite integral [50, p. 61] that for almost every θ the difference converges to zero. The theorem follows directly since the calculation can be reversed. ///

Before continuing we use Theorem 18.3 to establish some special formulae. For $n \geq 0$ define the polynomials $Q_n(x)$ and $P_n(x)$ by

$$\cos n\theta = Q_n(\cos\theta) \tag{18.21}$$

$$\sin(n+1)\theta = \sin\theta P_n(\cos\theta). \tag{18.22}$$

By the change of variables $x = \cos\theta$ it is easy to see that the Q_n are orthogonal in $L^2((1-x^2)^{-\frac{1}{2}}, \underline{\underline{I}})$ and the P_n are orthogonal in $L^2((1-x^2)^{\frac{1}{2}}, \underline{\underline{I}})$. Taking $g(\cos\theta)\sin\theta = \sin(n+1)\theta$ and using $\sin(n+1)\theta = Ⓗ^T \cos(n+1)\theta$, we get $g(x) = -Ⓗ^I f(x)$ with $f(\cos\theta)|\sin\theta| = \cos(n+1)\theta$ and therefore for $n \geq 0$

$$P_n(x) = -Ⓗ^I\{Q_{n+1}(t)\,(1-t^2)^{-\frac{1}{2}}\}(x). \tag{18.23}$$

Taking $g(\cos\theta) = \cos n\theta$ and using $\cos n\theta = -Ⓗ^T \sin n\theta$, we get $g(x) = Ⓗ^I f(x)$ with

$f(\cos t)\operatorname{sgn} t = \sin nt$ and therefore for $n \geq 1$

(18.24) $$Q_n(x) = Ⓗ^I\{(1-t^2)^{\frac{1}{2}} P_{n-1}(t)\}(x).$$

Finally, from $Ⓗ^T 1 = 0$ follows $Ⓗ^I f = 0$ when $f(\cos\theta)|\sin\theta| = 1$ and therefore

(18.25) $$Ⓗ^I\{(1-t^2)^{-\frac{1}{2}}\} = 0.$$

The identity (18.24) appears in [11, Lemma 1.2] where the reader is referred to [1] for a proof.

Our main goal is to establish inversion formulae for $Ⓗ^I$. If $g(x) \in L^2(\underline{\underline{I}}, (1-x^2)^{\frac{1}{2}})$, then $g(\cos\theta)\sin\theta \in L^2(\underline{\underline{T}}, d\theta)$. By Theorem 18.3(i)

(18.26) $$g(x) = -Ⓗ^I f(x)$$

with $f(x) \in L^2(\underline{\underline{I}}, (1-x^2)^{\frac{1}{2}})$ if and only if

(18.27) $$g(\cos\theta)\sin\theta = Ⓗ^T\{f(\cos t)|\sin t|\}(\theta).$$

Since $g(\cos\theta)\sin\theta$ is an odd function, such f exists and the most general f can be represented

(18.28) $$f(\cos\theta)|\sin\theta| = \alpha - Ⓗ^T\{g(\cos t)\sin t\}(\theta).$$

Moreover α can be determined from f by

(18.29) $$\alpha = (2\pi)^{-1}\int_{-\pi}^{+\pi} d\theta |\sin\theta| f(\cos\theta)$$

$$= \pi^{-1}\int_{-1}^{+1} dt f(t).$$

The estimate

$$\{\int_{-1}^{+1} dx\, |g(x)|\}^2 \leq \int_{-1}^{+1} dx\, g^2(x)(1-x^2)^{\frac{1}{2}} \int_{-1}^{+1} dy(1-y^2)^{-\frac{1}{2}}$$

shows that $g(x)(1-x^2)^{\frac{1}{2}} \in L^1(\underline{I}, (1-x^2)^{-\frac{1}{2}})$. Therefore Theorem 18.3(ii) can be applied to replace (18.28) by

(18.30) $$f(x)(1-x^2)^{\frac{1}{2}} = \alpha + \textcircled{H}^I\{g(y)(1-y^2)^{\frac{1}{2}}\}(x).$$

This is a general inversion formula for (18.26). Suppose in addition that $g(x) \in L^2(\underline{I}, (1-x^2)^{-\frac{1}{2}})$. Then $g(\cos\theta) \in L^2(\underline{T}, d\theta)$ and by Theorem 18.3(ii) the identity (18.26) is valid with $f(x) \in L^2(\underline{I}, (1-x^2)^{-\frac{1}{2}})$ if and only if

(18.31) $$g(\cos\theta) = \textcircled{H}^T\{f(\cos t)\operatorname{sgn} t\}(\theta).$$

From Theorem 18.2 it is clear that such f exists if and only if

(18.32) $$\int_{-1}^{+1} dt(1-t^2)^{-\frac{1}{2}} g(t) = 0$$

and then f is unique and given by

(18.33) $$f(\cos\theta)\operatorname{sgn}\theta = -\textcircled{H}^T\{g(\cos t)\}(\theta).$$

or equivalently

(18.33') $$\{f(\cos\theta)/|\sin\theta|\}\sin\theta = -\textcircled{H}^T\{(g(\cos t)/|\sin t|)|\sin t|\}(\theta).$$

But $g(x)(1-x^2)^{-\frac{1}{2}} \in L^1(\underline{I}, dx)$ and therefore Theorem 18.3(i) can be applied to replace (18.33') with

(18.34) $$f(x)(1-x^2)^{-\frac{1}{2}} = \textcircled{H}^I\{g(y)(1-y^2)^{-\frac{1}{2}}\}(x).$$

We summarize in

<u>Theorem 18.4.</u> <u>Let</u> $g(x) \in L^2(\underline{I}, (1-x^2)^{\frac{1}{2}})$. <u>Then</u> (18.26) <u>has solutions</u> $f(x) \in L^2(\underline{I}, (1-x^2)^{\frac{1}{2}})$ <u>and the most general such solution is given by</u> (18.30). <u>If also</u> $g(x) \in L^2(\underline{I}, (1-x^2)^{-\frac{1}{2}})$, <u>then</u> (18.26) <u>has a solution</u> $f(x) \in L^2(\underline{I}, (1-x^2)^{-\frac{1}{2}})$ <u>if and only if</u>

g(x) <u>satisfies the auxiliary condition</u> (18.32). <u>In this case such</u> f(x) <u>is unique and is given by</u> (18.34). ///

<u>Remark.</u> Theorem 18.4 implies in particular that if $g(x) \in L^2(\underline{I}, (1-x^2)^{-\frac{1}{2}})$ satisfies (18.32), then

(18.35) $$(1-x^2)^{\frac{1}{2}} Ⓗ^I\{g(y)(1-y^2)^{-\frac{1}{2}}\}(x)$$
$$= \alpha(1-x^2)^{-\frac{1}{2}} + (1-x^2)^{-\frac{1}{2}} Ⓗ^I\{g(y)(1-y^2)^{\frac{1}{2}}\}(x)$$

for some constant α. In general α need not vanish. For example, let $g(x) = x$. Since $x = Q_1(x) = \frac{1}{2}P_1(x)$ we have by (18.23)

$$(1-x^2)^{\frac{1}{2}} Ⓗ^I\{y(1-y^2)^{-\frac{1}{2}}\} = -(1-x^2)^{\frac{1}{2}}$$

and by (18.24)

$$(1-x^2)^{-\frac{1}{2}} Ⓗ^I\{y(1-y^2)^{\frac{1}{2}}\} = \frac{1}{2}Q_2(x)(1-x^2)^{-\frac{1}{2}}$$
$$= \frac{1}{2}(2x^2-1)(1-x^2)^{-\frac{1}{2}}$$
$$= -(1-x^2)^{\frac{1}{2}} + \frac{1}{2}(1-x^2)^{-\frac{1}{2}}$$

and therefore $\alpha = \frac{1}{2}$. Also (18.32) is crucial for the truth of (18.35). This follows from our general results but also can be deduced from consideration of the special case $g(x) \equiv 1$. By (18.25) the left side vanishes while the second term on the right $= x(1-x^2)^{-\frac{1}{2}}$ which violates (18.35). ///

Our final preliminary result relates the operators

(18.36) $$Ⓐf(x) = -D_x Ⓗ^I f(x)$$

(18.37) $$Ⓐ^{\sim} f(x) = -Ⓗ^I D_x f(x).$$

It is understood that every $f \in$ domain Ⓐ$^\sim$ is continuous on the closed interval $[-1,+1]$.

Lemma 18.5. Let f be absolutely continuous on the closed interval $[-1,+1]$. Then in the distribution sense on the open interval $\underline{\underline{I}}$

(18.38) $$Ⓐ^\sim f(x) = Ⓐf(x) + f(-1)\{\pi(1+x)\}^{-1} + f(1)\{\pi(1-x)\}^{-1}. \qquad ///$$

Proof. We note first that if μ is a bounded measure on $\underline{\underline{R}}$, then the Hilbert transform Ⓗμ is well defined as a tempered distribution by the formula

(18.39) $$Ⓗ\mu(\varphi) = (2\pi)^{-1}\int_{-\infty}^{+\infty} d\lambda(-i\,\mathrm{sgn}\,\lambda)Ⓕ\mu(\lambda)Ⓕ\varphi(\lambda)^*.$$

(It is understood here that φ is real valued.) In particular if $\mu = \varepsilon_x$, the point mass concentrated at x, then Ⓕ$\varepsilon_x(\lambda) = e^{-ix\lambda}$ and therefore

$$Ⓗ\varepsilon_x(\varphi) = (2\pi)^{-1}\int_{-\infty}^{+\infty} d\lambda(-i\,\mathrm{sgn}\,\lambda)e^{-ix\lambda}Ⓕ\varphi(\lambda)^*$$

$$= (2\pi)^{-1}\int_{-\infty}^{+\infty} d\lambda(-i\,\mathrm{sgn}\,\lambda)e^{-ix\lambda}Ⓕ\varphi(-\lambda)$$

$$= (2\pi)^{-1}\int_{-\infty}^{+\infty} d\lambda(i\,\mathrm{sgn}\,\lambda)e^{ix\lambda}Ⓕ\varphi(\lambda)$$

which is equivalent to

(18.40) $$Ⓗ\varepsilon_x(\varphi) = -Ⓗ\varphi(x).$$

A special case gives

(18.41) $$Ⓗ\varepsilon_{-1}(x) = \{\pi(1+x)\}^{-1};Ⓗ\varepsilon_1(x) = -\{\pi(1-x)\}^{-1}$$

in the distribution sense on $\underline{\underline{I}}$. Now we need only observe that in the distribution sense on $\underline{\underline{I}}$,

$$\text{(H)}^I D_x f(x) = \text{(H)}\{D_x f - f(-1)\varepsilon_{-1} + f(1)\varepsilon_1\}(x)$$

$$= D_x \text{(H)} f(x) - f(-1)\text{(H)}\varepsilon_{-1}(x) + f(1)\text{(H)}\varepsilon_1(x)$$

$$= D_x \text{(H)}^I f(x) - f(-1)\{\pi(1+x)\}^{-1} - f(1)\{\pi(1-x)\}^{-1}. \qquad ///$$

Now we are ready to study the absorbed process for the open interval $\underline{\underline{I}} = (-1, +1)$. In developing the L^2 theory we will use $(1-x^2)^{-\frac{1}{2}}dx$ as a reference measure rather than dx in order to fully exploit our information about the operator (H)^I. In the language of SMP we are dealing with the process obtained from the symmetric Cauchy process on the line by absorption combined with an appropriate random time change. By the results in Sections 7 and 8 of SMP the process under consideration is symmetric with associated Dirichlet space $(\underline{\underline{F}}, E)$ on $L^2(\underline{\underline{I}}, (1-x^2)^{-\frac{1}{2}})$ determined as follows. The Dirichlet norm is

$$E(f, f) = \tfrac{1}{2}J\langle f, f\rangle + \int_{-1}^{+1} dx\,\kappa(x) f^2(x) \tag{18.42}$$

where

$$\kappa(x) = \int dy\, I(|y| > 1)\pi^{-1}(x-y)^{-2} = \{\pi(1+x)\}^{-1} + \{\pi(1-x)\}^{-1}$$

$$J\langle f, f\rangle = \pi^{-1}\int_{-1}^{+1} dx \int_{-1}^{+1} dy\,(x-y)^{-2}\{f(x)-f(y)\}^{-2}.$$

The function space $\underline{\underline{F}}$ is the set of functions f for which (18.42) converges. The additional requirement $f(x) \in L^2(\underline{\underline{I}}, (1-x^2)^{-\frac{1}{2}})$ can be omitted since $\kappa(x)$ dominates $(1-x^2)^{-\frac{1}{2}}$. The latter condition guarantees also that the $L^2(\underline{\underline{I}}, (1-x^2)^{-\frac{1}{2}})$ infinitesimal generator A is surjective, and the Green's operator $(-A)^{-1}$ is bounded on $L^2(\underline{\underline{I}}, (1-x^2)^{-\frac{1}{2}})$. Our first result is

Theorem 18.6. The $L^2(\underline{\underline{I}}, (1-x^2)^{-\frac{1}{2}})$ generator A is defined by

$$Af(x) = (1-x^2)^{\frac{1}{2}}\text{(A)}f(x). \tag{18.43}$$

///

Proof. It suffices to show that the operator Ⓐ has no annihilators belonging to $L^2(\underline{\underline{I}}, (1-x^2)^{-\frac{1}{2}})$. But if D_x Ⓗ$^I h = 0$, then Ⓗ$^I h$ = const. and by (18.23) the only possibility is $h = \{ax+b\}(1-x^2)^{-\frac{1}{2}}$ which belongs to $L^2(\underline{\underline{I}}, (1-x^2)^{-\frac{1}{2}})$ only when $h = 0$. ///

No boundary conditions occur in Theorem 18.6. This is what we expect from our general theory since the terminal boundary is empty. The easiest way to see the latter is to note that singletons are polar for the Cauchy process and therefore exits from $\underline{\underline{I}}$ must be by means of "jumps from the interior". In fact P. W. Millar [39] has proved a much more general result along these lines. For a symmetric translation invariant process on $\underline{\underline{R}}$, bounded harmonic functions exist (equivalently, "continuous passage" is possible) if and only if the Gaussian component is present.

It turns out that an alternative formulation is possible in terms of the operator Ⓐ$^\sim$ which does involve boundary conditions. The basic result is

Theorem 18.7. The $L^2(\underline{\underline{I}}, (1-x^2)^{-\frac{1}{2}})$ generator A is defined by

(18.44) $$Af(x) = (1-x^2)^{\frac{1}{2}} Ⓐ^\sim f(x)$$

supplemented by the boundary conditions

(18.45) $$f(-1) = f(+1) = 0.$$ ///

Proof. If $f \in$ domain A, then by Theorem 18.6, the function $g(x) = -$Ⓗ$^I f(x)$ is absolutely continuous on $[-1, +1]$. Indeed, $D_x g(x) \in L^2(\underline{\underline{I}}, (1-x^2)^{\frac{1}{2}})$ which is slightly stronger. By the second part of Theorem 18.4, the function $f(x)$ satisfies (18.34). From the relation

(18.46) $$\int_0^{\pi} d\theta \{D_\theta g(\cos\theta)\}^2 = \int_{-1}^{+1} dx(1-x^2)^{-\frac{1}{2}}\{(1-x^2)^{\frac{1}{2}} D_x g(x)\}^2$$
$$= \int_{-1}^{+1} dx(1-x^2)^{\frac{1}{2}}\{D_x g(x)\}^2$$

it follows that $g(\cos\theta)$ satisfies the hypothesis of Theorem 18.2(iii). By Theorem 18.3(i) the equation (18.34) is equivalent to

$$f(\cos\theta)\operatorname{sgn}\theta = -Ⓗ^{I}\{g(\cos t)\}(\theta),$$

and by Theorem 18.2(iii) the function $f(\cos\theta)\operatorname{sgn}\theta \in \underline{\underline{W}}(\underline{\underline{T}})$, and also it is periodic. This guarantees on the one hand (by (18.46)) that $D_x f(x) \in L^2(I, (1-x^2)^{\frac{1}{2}})$ and in particular that $f \in$ domain $\tilde{Ⓐ}$ and on the other hand that f satisfies the boundary conditions (18.45). The theorem now follows directly from Theorem 18.6 and Lemma 18.5. ///

In the process of proving Theorem 18.7 we showed that $(1-x^2)^{\frac{1}{2}}Ⓐf(x) \in L^2(\underline{\underline{I}}, (1-x^2)^{-\frac{1}{2}})$ if and only if

(18.47) $$D_x f(x) \in L^2(\underline{\underline{I}}, (1-x^2)^{\frac{1}{2}}).$$

Thus f belongs to domain A if and only if it satisfies (18.47) and (18.45). This implies immediately that the conclusion of Theorem 8.4 is valid. (It is easy to show directly that $C^{\#} = C_0(\underline{\underline{I}})$.) It rarely happens that one can give such a precise description for an L^2 generator. The boundary conditions (18.45) are not superfluous since the functions 1 and arcsin x are annihilators for Ⓐ.

There exists a potential density $N^I(x,y)$, and it is given by

(18.48) $$N^I(x,y) = (2\pi)^{-1}\log\{1-xy+(1-x^2)^{\frac{1}{2}}(1-y^2)^{\frac{1}{2}}\}$$
$$-(2\pi)^{-1}\log\{1-xy-(1-x^2)^{\frac{1}{2}}(1-y^2)^{\frac{1}{2}}\}.$$

We refer to [11, Section 3] for a proof. It is clear that at least for bounded $v(x)$

(18.49) $$f(x) = \int_{-1}^{+1} dy N^I(x,y)v(y)$$

if and only if $-(1-x^2)^{-\frac{1}{2}}Af(x) = v(x)$. We will use this equivalence to establish two identities which are special cases of ones established in [52]. Our only purpose is to illustrate a computational technique. The results themselves will not be used below. The first identity is

(18.50) $$N^I 1(x) = (1-x^2)^{\frac{1}{2}} \quad \text{for } x \in \underline{\underline{I}},$$

a special case of (2.8) in [52]. Note that the left side is the expectation of the first exit time from $\underline{\underline{I}}$ for the original Cauchy process on $\underline{\underline{R}}$. For a proof denote the left side by $f(x)$ and apply Theorem 18.7 to conclude that f satisfies the boundary condition (18.45) and also $Ⓗ^I D_x f(x) = 1$. By the first part of Theorem 18.4 we have

(18.51) $$D_x f(x) = \alpha(1+x^2)^{-\frac{1}{2}} - (1-x^2)^{-\frac{1}{2}} Ⓗ^I\{(1-y^2)^{\frac{1}{2}}\}(x)$$

$$= \alpha(1+x^2)^{-\frac{1}{2}} - x(1-x^2)^{-\frac{1}{2}}$$

using (18.24) in the last step. Finally (18.51) and (18.45) directly imply (18.50).

The second identity gives an explicit expression for the hitting operator for the complement $\underline{\underline{R}}\setminus I$, for the full Cauchy process on $\underline{\underline{R}}$. We show that for $x \in \underline{\underline{I}}$ and for g defined on $\underline{\underline{R}}\setminus \underline{\underline{I}}$

(18.52) $$H^{\sim I} g(x) = \int_{|u|\geq 1} du\, h(x,u) g(u)$$

where

(18.53) $$h(x;u) = \pi^{-1}(1-x^2)^{\frac{1}{2}}(u^2-1)^{-\frac{1}{2}}|u-x|^{-1}.$$

This is a special case of (3.4) in [52]. We have already noted that continuous passage is not possible for the Cauchy process. Thus by the results in SMP, Section 10, the identity (18.52) is valid with $h(x,u)$ given by

$$h(x,u) = \int_{-1}^{+1} dy N^I(x,y)\pi^{-1}(u-y)^{-2}$$

and (18.53) will be established if we can show that

(18.54) $$D_x \textcircled{H}^I \{(1-y^2)^{\frac{1}{2}}(u^2-1)^{-\frac{1}{2}}|u-y|^{-1}\}(x) = (u-x)^{-2}.$$

It suffices to show that for $u > 1$

(18.55) $$\textcircled{H}^I \{(1-y^2)^{\frac{1}{2}}(u-y)^{-1}\}(x) = (u^2-1)^{\frac{1}{2}}(u-x)^{-1}-1$$

In verifying (18.55) we use the technique of Theorem 3.2 in [52]. For fixed $u > 1$ and $x \in \underset{\approx}{I}$ consider

(18.56) $$f(z) = (z^2-1)^{\frac{1}{2}}(z-u)^{-1}(z-x)^{-1}$$

as a function of the complex variable z. This function is well defined and analytic outside the "cut" from -1 and $+1$ and is fixed by the choice $f(z) > 0$ for $z > u$. Moreover,

(18.57) $$zf(z) \to 1 \quad \text{as} \quad |z| \to +\infty$$

$$\operatorname{Lim}_{\varepsilon \downarrow 0} f(y+i\varepsilon) = i(1-y^2)^{\frac{1}{2}}(u-y)^{-1}(x-y)^{-1}$$

$$\operatorname{Lim}_{\varepsilon \downarrow 0} f(y-i\varepsilon) = -i(1-y^2)^{\frac{1}{2}}(u-y)^{-1}(x-y)^{-1}.$$

The integral of $f(z)$ over a small circle centered at x cancels out because of the cut and it follows that the integral of $f(z)$ about a contour which winds about the cut once and excludes u, has the same value as

(18.58) $$-2i \operatorname{Lim}_{\varepsilon \downarrow 0} \int_{-1}^{+1} dy I(|x-y| \geq \varepsilon)(1-y^2)^{\frac{1}{2}}(u-y)^{-1}(x-y)^{-1}.$$

The integral of $f(z)$ around a small circle about u gives $2\pi i(u^2-1)^{\frac{1}{2}}(u-x)^{-1}$ and it follows from the first line in (18.57) that the integral of $f(z)$ around a large circle centered at the origin gives $2\pi i$. Thus (18.58) is the same as

(18.58')
$$2\pi i - 2\pi i(u^2-1)^{\frac{1}{2}}(u-x)^{-1}$$

and (18.55) follows.

We return now to our study of L^2 generators, and we continue to use $(1-x^2)^{-\frac{1}{2}}dx$ as a reference measure. As our notation suggests $(1-x^2)^{\frac{1}{2}}Ⓐ$ is the local generator for $(\underline{\underline{F}}, E)$. This will follow once we show that

(18.59)
$$\tfrac{1}{2}J<f,\varphi> + \int_{-1}^{+1}dx\kappa(x)f(x)\varphi(x) = -\int_{-1}^{+1}dxⒶf(x)\varphi(x)$$

at least for $f \in \underline{\underline{F}}^{env}$ and $\varphi \in C^{\infty}_{com}(\underline{\underline{I}})$. To prove (18.59) note first that any $f \in \underline{\underline{F}}^{env}$ can be extended to a function in $\underline{\underline{C}}$ by locally reflecting across the boundaries -1 and $+1$. This follows from the monotone nature of the kernel $(x-y)^{-2}$. Now fix $\varphi \in C^{\infty}_{com}(\underline{\underline{I}})$ and let $f^{\sim}$ be an extension of f which belongs to $\underline{\underline{C}}$. By (18.7)

$$\tfrac{1}{2}J<f,\varphi> + (2\pi)^{-1}\int_{-1}^{+1}dx\varphi(x)\int dy I(|y|>1)(x-y)^{-2}\{f(x)-f^{\sim}(y)\}$$
$$= \int_{-1}^{+1}dx\varphi(x)D_xⒽf^{\sim}(x)$$

and (18.59) follows after a routine passage to the limit in $f^{\sim}$. Notice that this argument fails if we change the order of $Ⓗ^I$ and D_x, which is consistent with Lemma 18.5. It now follows directly from Theorem 18.6 that there are no other Dirichlet spaces $(\underline{\underline{F}}^{\sim}, E^{\sim})$ on $L^2(\underline{\underline{I}}, (1-x^2)^{-\frac{1}{2}})$ with the same local generator as $(\underline{\underline{F}}, E)$. (This could also be deduced from the fact observed above that the terminal boundary is empty.)

__Remark.__ Our notation is inconsistent with Section 6 since the local generator is $(1-x^2)^{\frac{1}{2}}Ⓐ$ and not Ⓐ. ///

Since $\kappa(x)$ is not integrable, there are many nontrivial extensions $(\underline{\underline{F}}^{\sim}, E^{\sim})$. For example let $\underline{\underline{M}}$ and μ be chosen at the beginning of Section 3 in such a way that

$$\int_{-1}^{+1} dx\kappa(x)\int_{-1}^{+1} dy N(x,y)\kappa(y)\int_{\underline{\underline{M}}}\mu(y,dz)\varphi^2(z) = +\infty$$

for every nontrivial φ on $\underline{\underline{M}}$. Then every pair $(\underline{\underline{H}}, Q)$ satisfying 3.4.1 and 3.4.2 determines an extension of $(\underline{\underline{F}}, E)$. Since $\underline{\underline{F}}^{env}$ contains functions which are not continuous the set $\underline{\underline{M}}$ can be quite complicated, and one has little "a priori" control over the possibilities. Nevertheless, our general theory can be applied to give a precise description of the associated Dirichlet space and boundary conditions. We refer to Section 13 where details are given in a similar but technically less complicated context. It is shown in [11] that at least for one particular choice of M and μ one can obtain sharper results. In particular, the strong Feller property for the resolvents can be established. In the remainder of this section we give details for that choice.

The starting point in [11] is the operator $\text{Ⓐ}^{\sim}$. The idea there is to construct all resolvents (not necessarily symmetric) with generator contained in $\text{Ⓐ}^{\sim}$. Our results for Ⓐ together with Lemma 8.5 imply that for absolutely continuous $f \in \underline{\underline{F}}^{env}$ and for $\varphi \in C^{\infty}_{com}(\underline{\underline{I}})$

$$-\int dx\,\text{Ⓐ}^{\sim} f(x)\varphi(x) = -\int dx\,\text{Ⓐ} f(x)\varphi(x) - \int_{-1}^{+1} dx\,\varphi(x)\{f(-1)(\pi(1+x))^{-1} + f(1)(\pi(1-x))^{-1}\}$$

$$= \tfrac{1}{2} J\langle f,\varphi\rangle + \int_{-1}^{+1} dx\{\pi(1+x)\}^{-1}\varphi(x)\{f(x)-f(-1)\}$$

$$+ \int_{-1}^{+1} dx\{\pi(1-x)\}^{-1}\varphi(x)\{f(x)-f(1)\}.$$

Thus the symmetric resolvents constructed in [11] fit into the general scheme of Chapter II if we make the choice

(18.60) $$\Delta = \{-1, +1\}$$

$$\kappa\mu_{-1}(x) = \{\pi(1+x)\}^{-1};\ \kappa\mu_1(x) = \{\pi(1-x)\}^{-1}.$$

It follows easily from Theorem 4.3 that the singular boundary Δ_s is empty and therefore $(\underline{\underline{F}}^{\sim}, E^{\sim})$ is always an extension in the sense of SMP, Section 20. (See Remark 4 on page 20.33 in SMP.) In order to save space and at the same time facilitate comparison with [11], we assume always that $\underline{\underline{H}}$ is the one-dimensional subspace of functions on Δ which vanish at -1. Thus from a formal point of view we are replacing (18.60) by

(18.60') $$\Delta = \{+1\}$$

$$\kappa r(x) = \{\pi(1+x)\}^{-1};\ \kappa\mu_1(x) = \{\pi(1-x)\}^{-1}.$$

The function space $\underline{\underline{F}}^{\sim}$ is always the set of $f \in L^2(\underline{\underline{I}}, (1-x^2)^{\frac{1}{2}})$ for which there exists a constant $\gamma f(1)$, necessarily unique, such that

(18.61) $$\tfrac{1}{2}J\langle f, f\rangle + \int_{-1}^{+1} dx\{\pi(1+x)\}^{-1} f^2(x) + \int_{-1}^{+1} dx\{\pi(1-x)\}^{-1}\{f(x) - \gamma f(1)\}^2$$

converges. The hitting operator is given by

(18.62) $$H\varphi(x) = \int_{-1}^{+1} dy N^I(x, y)\{\pi(1-y)\}^{-1}\varphi(1).$$

The most general Q in paragraph 3.4 has the form

(18.63) $$Q(\varphi, \varphi) = \{\int_{-1}^{+1} dx \int_{-1}^{+1} dy N^I(x, y)\pi^{-2}(1+x)^{-1}(1-y)^{-1} + a\}\varphi^2(1)$$

with $a \geq 0$ and then

(18.64) $$E^{\sim}(f, f) = \tfrac{1}{2}J\langle f, f\rangle + \int_{-1}^{+1} dx\{\pi(1+x)\}^{-1} f^2(x)$$

$$+ \int_{-1}^{+1} dx\{\pi(1-x)\}^{-1}\{f(x) - \gamma f(1)\}^2 + a\{\gamma f(1)\}^2.$$

This corresponds to paragraph 9.1 in [11] for the choices

$$\tau p(dx) = \{\pi(1-x)\}^{-1}dx \ ; \ p_2 - \tau = a$$

$$\sigma = \gamma = 0.$$

(Strictly speaking our situation is not covered by [11] since it is assumed there that $\tau p(dx)$ is a bounded measure. But this is a minor point.) Our general Theorem 10.1 gives immediately

Theorem 18.8. *The* $L^2(\underline{I}, (1-x^2)^{-\frac{1}{2}})$ *generator* $A^{\sim}$ *is defined by*

$$A^{\sim}f(x) = (1-x^2)^{\frac{1}{2}}(\text{Ⓐ}f(x) + \gamma f(1)\{\pi(1-x)\}^{-1}) \tag{18.65}$$

supplemented by the requirement that $f \in \underline{F}^{\sim}$ *and that* f *satisfies the boundary condition*

$$\int_{-1}^{+1} dx\{\pi(1-x)\}^{-1}\{f(x) - \gamma f(1)\} = a\gamma f(1). \tag{18.66}$$

In the next theorem we exploit the special features of (18.60) to sharpen this considerably.

Theorem 18.9. *The* $L^2(\underline{I}, (1-x^2)^{-\frac{1}{2}})$ *generator* $A^{\sim}$ *is defined by*

$$A^{\sim}f(x) = (1-x^2)^{\frac{1}{2}}\text{Ⓐ}^{\sim}f(x) \tag{18.67}$$

supplemented by the two boundary conditions

$$f(-1) = 0 \ ; \ \int_{-1}^{+1} dx\{\pi(1-x)\}^{-1}\{f(x) - f(1)\} = af(1). \tag{18.68}$$

In particular, every $f \in$ *domain* $A^{\sim}$ *is absolutely continuous.* ///

Proof. We note first that by (18.25) the function

$$\xi_1(x) = \tfrac{1}{2} + \pi^{-1}\arcsin x$$

belongs to domain $Ⓐ^\sim$ and $Ⓐ^\sim \xi_1(x) = 0$. If $f \in$ domain $A^\sim$, then by Theorem 18.8 the difference $f(x) - \gamma f(0)\xi_1(x) \in$ domain A and the direct part of the theorem follows after applying Theorem 18.7.

The proof of the converse depends on showing that $\xi_1(x) \in \underline{\underline{F}}^\sim$. But this is clear since $\varphi(x) = \tfrac{1}{2}(1+x)$ belongs to $\underline{\underline{F}}^\sim$ and the difference $\xi_1(x) - \varphi(x)$ belongs to domain A by Theorem 18.7 and therefore to $\underline{\underline{F}}$. Now let absolutely continuous f satisfy (18.68) and also $(1-x^2)^{\frac{1}{2}} Ⓐ^\sim f(x) \in L^2(\underline{\underline{I}}, (1-x^2)^{-\frac{1}{2}})$. Then $f(x) - f(1)\xi_1(x)$ belongs to $\underline{\underline{F}}$ by the previous argument. Thus $f \in \underline{\underline{F}}^\sim$ and $\gamma f(1) = f(1)$ which is enough to guarantee that $f \in$ domain $A^\sim$. ///

Theorem 18.9 is more or less equivalent to a special case of Theorem 9.1 in [11]. It implies in particular that the associated resolvent operators and transition operators have the strong Feller property.

It is instructive to look more closely at our derivation of (18.66). Clearly $f \in \underline{\underline{F}}^\sim$ for which the right side of (18.67) belongs to $L^2(\underline{\underline{I}}, (1-x^2)^{-\frac{1}{2}})$ belongs to domain $A^\sim$ if and only if

$$E^\sim(f, H1) = -\int_{-1}^{+1} dx\, Ⓐ^\sim f(x) H1(x).$$

This can be written

$$\begin{aligned} \tfrac{1}{2} J\langle f, H1\rangle &+ \int_{-1}^{+1} dx \{\pi(1+x)\}^{-1} f(x) H1(x) \\ &+ \int_{-1}^{+1} dx \{\pi(1-x)\}^{-1} \{f(x) - \gamma f(1)\} H1(x) + \int_{-1}^{+1} dx\, Ⓐ^\sim f(x) H1(x) \\ &= \int_{-1}^{+1} dx \{\pi(1-x)\}^{-1} \{f(x) - \gamma f(1)\} - a\gamma f(1), \end{aligned}$$

and after comparing with (18.66) we see that our results in Section 7 imply the identity

(18.69) $$\tfrac{1}{2}J\langle f, Hl\rangle + \int_{-1}^{+1} dx\{Ⓐf(x)+\kappa(x)f(x)\}Hl(x) = 0.$$

In fact it is possible to prove (18.69) directly without appealing to Section 7. Clearly (18.69) reduces to (18.59) if we replace Hl by $\varphi \in C^{\infty}_{com}(\underline{\underline{I}})$. Since singletons are polar for $(\underline{\underline{C}}, C)$, the absorbed space $\underline{\underline{F}}$ is dense in the enveloping space $\underline{\underline{F}}^{env}$ relative to the norm $J\langle f, f\rangle$ and it follows that Hl can be approximated by functions in $C^{\infty}_{com}(\underline{\underline{I}})$ to establish (18.69). However, the appropriate analogue to (18.69) is still valid in Section 20 when singletons are nonpolar, and this approximation argument is invalid. Thus the right condition for (18.69) is not that singletons be polar for the original process, but rather the weaker condition that the terminal boundary for the absorbed process be empty. Of course it is the latter condition that is used in the argument of Section 7.

19. One Dimensional Stable Process, $0 < \alpha < 1$.

In this and the next section we study one-dimensional symmetric stable processes with index α, $\alpha \neq 1$ or 2. The excepted cases correspond to the Cauchy process and to Brownian motion which we have already studied. Our treatment is strongly influenced by the paper of S. Watanabe [52], but just as in Section 18 our general approach is different. In this section we restrict attention to the case $0 < \alpha < 1$.

Again we consider first the process on the full line with Lebesgue measure dx as reference measure. The transition operators C_t are defined by

$$C_t f(x) + \int_{-\infty}^{+\infty} dy c_t(x-y) f(y)$$

where the densities c_t are defined by inverse Fourier inversion

$$c_t(x) = (2\pi)^{-1} \int_{-\infty}^{+\infty} d\lambda e^{-i\lambda x} e^{-t|\lambda|^\alpha} \tag{19.1}$$

$$= \pi^{-1} \int_0^\infty d\lambda \cos \lambda x e^{-t\lambda^\alpha} .$$

Once it is shown that the c_t are nonnegative, then it is quite elementary that the operators C_t, $t > 0$, form a symmetric Markovian semigroup on $L^2(\underline{R}, dx)$. We give now a simple proof based on a criterion of Pólya and taken from the book of K. L. Chung [7, p.183]. The point is that we can integrate by parts to get

$$c_t(x) = (\pi x)^{-1} \int_0^\infty d\lambda \sin \lambda x \{\alpha \lambda^{\alpha-1} e^{-t\lambda^\alpha}\},$$

and since the expression inside the brackets is nonnegative and decreasing (this is false for $\alpha > 1$), the expression on the right can be represented as an alternating series whose first term is positive. This shows at the same time that the $c_t(x)$ are strictly positive, and therefore the associated Dirichlet space is irreducible.

It is clear that the Dirichlet norm is given by

(19.2) $$C(f,f) = (2\pi)^{-1}\int_{-\infty}^{+\infty} d\lambda\,|\lambda|^{\alpha}\,|Ⓕf(\lambda)|^2$$

and that the Dirichlet space $\underline{\underline{C}}$ is the subspace of $f \in L^2(\underline{\underline{R}}, dx)$ for which (19.2) converges. The second argument in Section 14 for $d = 3$ shows that $(\underline{\underline{C}}, C)$ is transient. (See the second remark following Lemma 18.1.) It turns out that a potential density $N(x)$ exists and can be computed explicitly

(19.3) $$\begin{aligned} N(x) &= \pi^{-1}\Gamma(1-\alpha)\sin(\pi\alpha/2)|x|^{\alpha-1} \\ &= \{\cos(\pi\alpha/2)/2\Gamma(\alpha)\}|x|^{\alpha-1}. \end{aligned}$$

This follows since for $x > 0$

$$\begin{aligned} N(x) &= \int_0^{\infty} dt\, c_t(x) \\ &= (2\pi)^{-1}\int_0^{\infty} dt \int_{-\infty}^{+\infty} d\lambda\, e^{i\lambda x} e^{-t|\lambda|^{\alpha}} \\ &= (2\pi)^{-1}\int_{-\infty}^{+\infty} d\lambda\, e^{i\lambda x}|\lambda|^{-\alpha} \\ &= \pi^{-1}\mathrm{Re.}\int_0^{\infty} d\lambda\, e^{i\lambda x}\lambda^{-\alpha}, \end{aligned}$$

and this integral can be shifted to the positive imaginary axis to give

$$\begin{aligned} N(x) &= \pi^{-1}\mathrm{Re.}\int_0^{\infty} i\,d\tau\, e^{-\tau x} e^{-(i\pi\alpha)/2}\tau^{-\alpha} \\ &= \pi^{-1}\Gamma(1-\alpha)x^{\alpha-1}\mathrm{Re.}\, e^{i\pi(1-\alpha)/2} \\ &= \pi^{-1}\Gamma(1-\alpha)\sin(\pi\alpha/2)x^{\alpha-1}. \end{aligned}$$

Just as in Section 18 we factor the generator into the product of two operators. At the same time we derive an alternative formula for the Dirichlet norm.

The role of the Hilbert transform is played here by the operator

(19.4) $$Ⓙf(x) = (c\ \ /\alpha)\int_{-\infty}^{+\infty} dy\,|x-y|^{-\alpha}\mathrm{sgn}(x-y)f(y)$$

where for typographic convenience we have introduced the constant

$$c = \pi^{-1}\Gamma(\alpha+1)\sin\tfrac{1}{2}\pi\alpha.$$

If $f \in L^1(\underline{\underline{R}}, dx)$, then the right side converges almost everywhere to the sum of a bounded function and an integrable function, and so Ⓙf is well defined as a tempered distribution. However, (19.4) does not make sense as it stands for general f bounded or in $L^2(\underline{\underline{R}}, dx)$. To get around this technical problem we introduce the mollified operator

$$\text{(19.4')} \qquad Ⓙ^m f(x) = (c/\alpha)\int_{-1}^{+1} dy\,|x-y|^{-\alpha}\operatorname{sgn}(x-y)f(y)$$
$$+ (c/\alpha)\int dy\,I(|y|\geq 1)\{|x-y|^{-\alpha}\operatorname{sgn}(x-y)+|y|^{-\alpha}\operatorname{sgn} y\}f(y).$$

The function $Ⓙ^m f(x)$ is well defined as a tempered distribution for $f \in L^p(\underline{\underline{R}}, dx)$, any $p \geq 1$. Moreover, if $f(x)$ is integrable, then

$$\text{(19.5)} \qquad D_x Ⓙ^m f(x) = D_x Ⓙf(x),$$

and this is why we can use $Ⓙ^m$ in place of Ⓙ when f is not integrable.

Lemma 19.1. For $f \in L^1(\underline{\underline{R}}, dx)$

$$ⒻⒿf(\lambda) = -i(\operatorname{sgn}\lambda)|\lambda|^{\alpha-1}Ⓕf(\lambda).$$

Proof. The main point is to show that

$$\text{(19.6)} \qquad (c/\alpha)\operatorname{Lim}_{R\uparrow\infty}\int_{-R}^{+R} dx\,e^{-i\lambda x}|x|^{-\alpha}\operatorname{sgn} x = -i|\lambda|^{\alpha-1}\operatorname{sgn}\lambda$$

with the left side suitably controlled. We will only give here a verification of (19.6). Since the left side of (19.6) is certainly an odd function of λ, it suffices to consider $\lambda > 0$. But then

$$\mathrm{Lim}_{R\uparrow\infty}\int_{-R}^{+R}dxe^{-i\lambda x}|x|^{-\alpha}\mathrm{sgn}\,x = -2i\int_0^\infty dx\,\sin\lambda x\,x^{-\alpha}$$

$$= -2i\mathrm{Im.}\int_0^\infty dxe^{i\lambda x}x^{-\alpha}$$

with the integral defined in the Cauchy-Riemann sense, and after again shifting the contour to the positive imaginary axis, we see that this

$$= -2i\mathrm{Im}\int_0^\infty id\tau e^{-\tau\lambda}\tau^{-\alpha}e^{-i\pi\alpha/2}$$

$$= -2i\Gamma(1-\alpha)\lambda^{\alpha-1}\cos(\pi\alpha/2) = -i\lambda^{\alpha-1}(c/\alpha)^{-1} \qquad ///$$

Just as in Section 18 we have

(19.7) $$A^R f = -D_x Ⓙ^m f$$

and if f is integrable

(19.7') $$A^R f = -D_x Ⓙf.$$

Also if $f \in$ Ⓢ, then

$$C(f,f) = (2\pi)^{-1}\int_{-\infty}^{+\infty}d\lambda(i\lambda)(-i|\lambda|^{\alpha-1}\mathrm{sgn}\,\lambda)|Ⓕf(\lambda)|^2$$

$$= \int dxf(x)D_x Ⓙf(x)$$

$$= -(c/\alpha)\int_{-\infty}^{+\infty}dxD_x f(x)\int dyf(y)|x-y|^{-\alpha}\mathrm{sgn}(x-y)$$

$$= -(c/\alpha)\int dyf(y)\int dx|x-y|^{-\alpha}\mathrm{sgn}(x-y)D_x\{f(x)-f(y)\}$$

$$= -c\int dyf(y)\int dx|x-y|^{-\alpha-1}\{f(x)-f(y)\}$$

and we conclude that

(19.8) $$C(f,f) = \tfrac{1}{2}c\int_{-\infty}^{+\infty}dx\int_{-\infty}^{+\infty}dy|x-y|^{-\alpha-1}\{f(x)-f(y)\}^2$$

and this extends by passage to the limit to general f in the extended space $\underline{\underline{C}}_{(e)}$. Just as in Section 18 we are not sure to what extent (19.2) in valid for general $f \in \underline{\underline{C}}_{(e)}$. Of course (19.8) could have been derived directly without introducing the operator Ⓙ, but we will be interested in the latter for its own sake.

By definition the reflected space $\underline{\underline{C}}^{ref}$ is the totality of functions f for which (19.8) converges (with the usual convention for equivalence of functions). We have already verified that $(\underline{\underline{C}}, C)$ is transient, and it follows from the general result of Choquet and Deny [6] that every bounded harmonic function for $(\underline{\underline{C}}, C)$ is constant. Thus $\underline{\underline{C}}_{(e)}$ has codimension one in $\underline{\underline{C}}^{ref}$, and the constants form a complementary subspace. As in Section 14 for $d \geq 3$ the terminal set $\underline{\underline{T}}$ is a singleton $\{\infty\}$, and we can identify ∞ with the usual "point at infinity". We postpone our discussion of the boundary function $\gamma f(\infty)$ and also the normal derivative $(\partial f/\partial n)(\infty)$ until after we have obtained some information about the generator for the absorbed process.

Again we introduce the restricted operator

(19.9) $$Ⓙ^{I} f(x) = (c/\alpha)\int_{-1}^{+1} dy\,|x-y|^{-\alpha}\,\mathrm{sgn}(x-y) f(y)$$

for the open interval $\underline{\underline{I}} = (-1,+1)$. Clearly (19.9) is a bounded operator on each $L^p(\underline{\underline{I}}, dx)$, $p \geq 1$. Also we denote the absorbed Dirichlet space for $\underline{\underline{I}}$ by $(\underline{\underline{F}}, E)$, but now we continue to use dx as reference measure. By the general results in SMP, Section 7, the space $(\underline{\underline{F}}, E)$ is precisely the restriction of $(\underline{\underline{C}}, C)$ to $\underline{\underline{I}}$, and it follows that

(19.10) $$E(f,f) = \tfrac{1}{2} J\langle f,f\rangle + \int_{-1}^{+1} dx\,\kappa(x) f^2(x)$$

and $\underline{\underline{F}}$ is the subspace of $f \in L^2(\underline{\underline{I}}, dx)$ for which (19.10) converges. Of course

(19.11) $$J\langle f,f\rangle = c\int_{-1}^{+1} dx \int_{-1}^{+1} dy\,|x-y|^{-1-\alpha}\{f(x)-f(y)\}^2$$

$$\kappa(x) = (c/\alpha)\{(1-x)^{-\alpha}+(1+x)^{-\alpha}\}.$$

Since $\kappa(x)$ is bounded from below the requirement $f \in L^2(\underline{\underline{I}}, dx)$ can be suppressed. The existence of a potential kernel $N(x)$ for $(\underline{\underline{C}}, C)$ certainly guarantees the existence of one for $(\underline{\underline{F}}, E)$ which we denote by $N^I(x,y)$. Apparently $N^I(x,y)$ can be computed explicitly only for $\alpha = 1$. (See, however, Theorem 3.3 in [52].) It is also true that for $f \in \underline{\underline{F}}$ and $\varphi \in C^\infty_{com}(\underline{\underline{I}})$ we have

$$E(f,\varphi) = \int_{-1}^{+1} dx\, \varphi(x) D_x \textcircled{J}^I f(x).$$

which suggests that the local generator is given by

$$\textcircled{A} f(x) = -D_x \textcircled{J}^I f(x). \tag{19.12}$$

In fact this can be proved exactly as in Section 18. Now we are ready for

<u>Theorem 19.2</u>. <u>The generator</u> A <u>is defined by</u>

$$Af(x) = \textcircled{A} f(x). \tag{19.13}$$

///

The only thing to show is that (19.13) need not be supplemented by any side conditions. This depends on the uniqueness result, Lemma 2.5, in [52], and this in turn depends on the following result which is stated in [52] as Lemma 2.1. If $0 < \alpha < 2$, $\alpha \neq 1$, and if $x \in \underline{\underline{I}}$, then for $m = 0, 1, \ldots$

$$\int_{-1}^{+1} dy\, |x-y|^{1-\alpha} P_m(y)(1-y^2)^{\frac{1}{2}\alpha-1} = \lambda_m P_m(x) \tag{19.14}$$

where

$$\lambda_m = \{\pi\Gamma(m+\alpha-1)\}/\{\sin(\tfrac{1}{2}\pi\alpha)\Gamma(\alpha-1)\Gamma(m+1)\} \tag{19.15}$$

and where $\{P_m(x)\}_{m=0}^{\infty}$ are orthogonal polynomials for the weight $(1-x^2)^{\frac{1}{2}\alpha-1}$ on

(-1,+1). In fact we can let $P_m(x)$ be the Gegenbauer polynomial

$$P_m(x) = P_m^{(\frac{1}{2}\alpha-\frac{1}{2})}(x)$$

determined by the identity

$$(1-2xw+w^2)^{-2} = \Sigma_{m=0}^{\infty} P_m^{\nu}(x)w^m,$$

but there is no need for us to bother with this. Apparently this result was first established in a paper of Pólya and Szegö [43]. However, the technique of Theorem 3.2 in [52], which we have already applied in Section 18, can be modified to give a rather simple proof of (19.14). We note first that

(19.16) $$Bf(x) = \int_{-1}^{+1} dy(1-y^2)^{\frac{1}{2}\alpha-1}|x-y|^{1-\alpha}f(y)$$

is a bounded symmetric operator on $L^2(\underline{\underline{I}}, (1-x^2)^{\frac{1}{2}\alpha-1})$, and therefore (19.14) will follow if we can establish

$$Bx^m = \lambda_m x^m + \text{lower order}$$

or equivalently

(19.17) $$D_x^m Bx^m = \Gamma(m+1)\lambda_m.$$

Just as in Section 18 the function

$$f(z) = (z^2-1)^{\frac{1}{2}\alpha-1}(z-x)^{1-\alpha}z^m$$

is well defined and analytic on the complement of the cut from -1 to +1. Thus $\int_\Gamma dzf(z)$ is the same for every contour which winds once about the cut. Letting Γ shrink in the obvious way, we can reason essentially as in Section 18 and get

$$\int_\Gamma dzf(z) = 2i\sin(\tfrac{1}{2}\pi\alpha)Bx^m.$$

On the other hand,

$$D_x^m f(z) = \{\Gamma(\alpha+m-1)/\Gamma(\alpha-1)\}(z^2-1)^{\frac{1}{2}\alpha-1}(z-x)^{1-\alpha-m}z^m$$

behaves asymptotically like z^{-1} Thus after considering Γ a large circle centered at $z = 0$ and then passing to the limit, we get

$$D_x^m \int_\Gamma dz f(z) = 2\pi i\{\Gamma(\alpha+m-1)/\Gamma(\alpha-1)\}$$

which proves (19.17) and therefore (19.14). The uniqueness result is

Lemma 19.3. *If* $f \in L^2(\underline{I}, dx)$ *satisfies*

(19.18) $$D_x \textcircled{J}^I f(x) = 0 \quad \text{on } \underline{I},$$

then $f = 0$. ///

Proof. Clearly (19.18) implies

(19.19) $$\int_{-1}^{+1} dy(1-y^2)^{\frac{1}{2}\alpha-1}|x-y|^{-\alpha}\operatorname{sgn}(x-y)g(y) = \text{constant}$$

where $g(x) = (1-x^2)^{1-\frac{1}{2}\alpha}f(x)$. Since

$$\int_{-1}^{+1} dx \int_{-1}^{+1} dy\,|x-y|^{-\alpha}(1-y^2)^{\frac{1}{2}\alpha-1}|g(y)|$$

converges, we can integrate (19.19) to get

(19.19') $$Bg(x) = ax+b$$

for some constants a, b. Since $g(x) \in L^2(\underline{I}, (1-x^2)^{\frac{1}{2}\alpha-1})$ we can apply (19.14) and conclude that $g(x) = a'x+b'$ which is impossible for $f(x) \in L^2(\underline{I}, dx)$ unless $f = 0$.

///

Now it is easy to complete the proof of Theorem 19.2. If $f \in L^2(\underline{I}, dx)$

is such that $D_x \textcircled{J}^I f(x) \in L^2(\underline{\underline{I}}, dx)$, then by surjectivity of A there exists $f^\sim \in$ domain A such that $f^\sim - f$ satisfies the hypothesis of Lemma 19.3, and therefore $f = f^\sim$.

Remark. The truth of Theorem 19.2 depends crucially on our choice of a reference measure. This is clear from the proof of Lemma 19.3. For example if we replace dx by $(1-x^2)^{1-\frac{1}{2}\alpha}dx$, then $f(x) = (1-x^2)^{\frac{1}{2}\alpha-1}$ satisfies $D_x \textcircled{J}^I f(x) = 0$ but does not belong to domain A. ///

Theorem 18.7 does not have an exact analogue here. For example if $0 < \alpha < \frac{1}{2}$, then $\kappa(x) \in L^2(\underline{\underline{I}}, dx)$, and therefore $1 \in$ domain A, but fails to satisfy the boundary conditions (18.45). It may be that there is an analogue if the reference measure is chosen correctly. In any case an analogue to Theorem 18.7 is valid if we restrict attention to $f \in$ domain A with Af bounded. The main point is to show that if $f = N^I g$ with g bounded, then $f \in C_0(\underline{\underline{I}})$. But then f can be represented $f = (1-H^{\sim 1})Ng(x)$ where $H^{\sim 1}$ is the hitting operator for the complement of $\underline{\underline{I}}$, and it suffices either to use the formula for $H^{\sim 1}$ derived below or to argue as in Section 21 below.

Next we use Theorem 19.2 to establish the formula

$$h(x,u) = \pi^{-1}\sin(\tfrac{1}{2}\pi\alpha)(1-x^2)^{\frac{1}{2}\alpha}(u^2-1)^{-\frac{1}{2}\alpha}|x-u|^{-1} \tag{19.21}$$

which is the exact analogue of (18.53). Just as for the Cauchy process

$$h(x,u) = c\int_{-1}^{+1} dy N^I(x,y)|y-u|^{-\alpha-1},$$

and so (19.21) will follow from Theorem 19.2 if we can prove

$$D_x \textcircled{J}^I\{(1-y^2)^{\frac{1}{2}\alpha}|u-y|^{-1}\}(x) = \Gamma(\alpha+1)(u^2-1)^{\frac{1}{2}\alpha}|x-u|^{-1}. \tag{19.22}$$

But if we apply the technique of Theorem 3.2 in [52] to

$$f(z) = (z^2-1)^{\frac{1}{2}\alpha}(z-x)^{-\alpha}(u-z)^{-1}$$

exactly as in Section 18, then we get for $u > 1$

$$\sin(\tfrac{1}{2}\pi\alpha)\int_{-1}^{+1}dy(1-y^2)^{\frac{1}{2}\alpha}(u-y)^{-1}|x-y|^{-\alpha}\operatorname{sgn}(x-y) = \pi(u^2-1)^{\frac{1}{2}\alpha}(u-x)^{-\alpha}-\pi$$

and (19.22) follows.

We return now to our discussion of the reflected Dirichlet space $(\underline{\underline{C}}^{ref}, C)$ for the full line. For $R > 0$ we denote by $H^{\sim R}$ the hitting operator for the complement of the open interval $(-R, +R)$. The identity (19.21) is equivalent to

$$(19.23) \qquad H^{\sim 1}f(x) = \pi^{-1}\sin(\tfrac{1}{2}\pi\alpha)(1-x^2)^{\frac{1}{2}\alpha}\int du I(|u| \geq 1)f(u)(u^2-1)^{-\frac{1}{2}\alpha}|x-u|^{-1}$$

for $-1 < x < +1$. More generally we have for $R > 0$ and for $-R < x < +R$

$$(19.24) \quad H^{\sim R}f(x) = \pi^{-1}\sin(\tfrac{1}{2}\pi\alpha)(R^2-x^2)^{\frac{1}{2}\alpha}\int du I(|u| \geq R)f(u)(u^2-R^2)^{-\frac{1}{2}\alpha}|x-u|^{-1}.$$

To prove (19.24) we use stability. If $\{X_t\}_{t\geq 0}$ is a right continuous version of the process starting at $x = 0$, then so is $RX_{tR^{-\alpha}}$. This is clear from (19.1). Let $g(x) = f(Rx)$ and $g^{\sim}(x) = H^{\sim R}f(Rx)$ and let τ be the first time that the process X_t exits from $(-R, +R)$. We can assume f is bounded so that the random variables $H^{\sim R}f(X_{\tau \wedge t})$ form a uniformly integrable martingale and therefore so do

$$H^{\sim R}f(RX_{\sigma \wedge tR^{-\alpha}}) = g^{\sim}(X_{\sigma \wedge tR^{-\alpha}})$$

where σ is the first exit time from $(-1, +1)$. Since the terminal variable for this martingale is

$$f(RX_\sigma) = g(X_\sigma),$$

it follows that $g^{\sim} = H^{\sim 1}g$ and so for $-R < x < +R$

$$H^{\sim R}f(x) = g^{\sim}(x/R)$$

$$= \pi^{-1}\sin(\tfrac{1}{2}\pi\alpha)(1-x^2/R^2)^{\frac{1}{2}\alpha}\int dv I(|v|\geq 1)f(Rv) \quad (v^2-1)^{-\frac{1}{2}\alpha}|(x/R)-v|^{-1}$$

and (19.24) follows after applying the substitution $u = vR$.

It follows from (7.4') that for $f \in \underline{\underline{C}}^{ref}$ and for quasi-every x (and therefore almost every x) the boundary function $\gamma f(\infty)$ can be computed from

(19.25) $$\gamma f(\infty) = \operatorname{Lim}_{R\uparrow\infty} H^{\sim R}f(x).$$

From the form of (19.24) it is clear that (19.25) must be true for all x and in particular for $x = 0$. Thus

(19.26) $$\gamma f(\infty) = \operatorname{Lim}_{R\uparrow\infty}\pi^{-1}\sin(\tfrac{1}{2}\pi\alpha)R^{\alpha}\int du I(|u|\geq R)f(u)(u^2-R^2)^{-\frac{1}{2}\alpha}|u|^{-1}.$$

In order to give an explicit formula for the normal derivative at infinity we must identify the equilibrium distribution $\ell_R(dx)$ for the interval $(-R,+R)$. The change of variables $v = Ry$ in (19.14) for $m = 0$ gives

(19.27) $$\int_{-R}^{+R} dv(R^2-v^2)^{-\frac{1}{2}\alpha}|x-v|^{\alpha-1} = \pi/\sin(\tfrac{1}{2}\pi\alpha).$$

This together with (19.3) gives for $-R \leq x \leq +R$

$$\Gamma(1-\alpha)^{-1}\int_{-R}^{+R} dy(R^2-y^2)^{-\frac{1}{2}\alpha}N(x-y) = 1$$

and we conclude that

(19.28) $$\ell_R(dx) = I(|x|\leq R)\Gamma(1-\alpha)^{-1}(R^2-x^2)^{-\frac{1}{2}\alpha}dx.$$

We conclude from Theorem 7.1 that if $f \in \underline{\underline{C}}^{ref}$ and if $D_x \text{Ⓙ}^m f$ is an integrable function (or even a signed measure in bounded variation), then

(19.29) $$(\partial f/\partial n)(\infty) = \operatorname{Lim}_{R\uparrow\infty}\Gamma(1-\alpha)^{-1}\int_{-R}^{+R} dx(R^2-x^2)^{-\frac{1}{2}\alpha}\{f(x)-\gamma f(\infty)\}$$

exists and

(19.30) $$(\partial f/\partial n)(\infty) = \int dx D_x \textcircled{J}^m f = \textcircled{J}^m f(\infty) - \textcircled{J}^m f(-\infty)$$

Just as in Section 14 for $d = 3$ we can replace Lebesgue measure by a bounded reference measure, and then the boundary function $\gamma f(\infty)$ and the normal derivative $(\partial f/\partial n)(\infty)$ appear in boundary conditions for $(\underline{\underline{C}}, C)$ and its extensions. The details are essentially the same as in Section 14.

In the remainder of this section we study symmetric processes on $\underline{\underline{I}}$ which dominate the absorbed process. The main difference with Section 18 (and also Section 20) is that $\kappa(x)$ is integrable and so $(\underline{\underline{F}}, E)$ has no extensions. Indeed the regular boundary Δ_r is always empty, and so $(\underline{\underline{F}}^{\sim}, E^{\sim})$ always coincides with the perturbed space of Definition 5.3. There are many possibilities for Δ and μ, but just as for the Cauchy process we restrict our attention to the special case

(19.31) $$\Delta = \{1\} \ ; \ \kappa\mu_1(x) = (c/\alpha)(1-x)^{-\alpha} \ ; \ \kappa r(x) = (c/\alpha)(1+x)^{-\alpha}.$$

This will permit direct comparison with Section 5 in [52].

Of course $\underline{\underline{H}}$ is the one-dimensional vector space of functions on Δ and the hitting operator H is given by

$$H\varphi(x) = \int_{-1}^{+1} dy N^I(x,y)(c/\alpha)(1-y)^{-\alpha}\varphi(1).$$

The most general Q has the form

(19.32) $$Q(\varphi,\varphi) = \{\int_{-1}^{+1} dx \int_{-1}^{+1} dy N(x,y)(c/\alpha)^2(1-y)^{-\alpha}(1+x)^{-\alpha} + a\}\varphi^2(0)$$

with $a \geq 0$. We have already mentioned that $(\underline{\underline{F}}^{\sim}, E^{\sim})$ cannot be an extension of $(\underline{\underline{F}}, E)$, and therefore the description of $\underline{\underline{F}}^{\sim}$ in Section 18 is not applicable. Indeed it doesn't make sense. To obtain information we follow our usual procedure and look first at the expanded space $(\underline{\underline{F}}^e, E^e)$ introduced near the end of Section 5. By

(15.30) and Theorem 3.2, we have

(19.33) $$E^e(f,f) = \tfrac{1}{2}J\langle f,f\rangle + (c/\alpha)\int_{-1}^{+1}dy(1+y)^{-\alpha}f^2(y)$$

$$+ (c/\alpha)\int_{-1}^{+1}dy(1-y)^{-\alpha}\{f(y)-f(1)\}^2 + af^2(1),$$

and it follows that

$$\kappa^e(\{1\}) = a$$

$$J^e(1,dx) = (c/\alpha)(1-x)^{-\alpha}$$

with κ^e, J^e the killing measure and Lévy measure for $(\underline{\underline{F}}^e, E^e)$. Just as for Case 4 in Section 12, this implies that the operator M of (4.3) is given by

(19.34) $$Mf(1) = q^{-1}\int_{-1}^{1}dx(c_\alpha/\alpha)(1-x)^{-\alpha}f(x)$$

where q is the constant

(19.35) $$q = a+(c_\alpha/\alpha)\int_{-1}^{+1}dx(1-x)^{-\alpha}.$$

Plugging (19.34) into the formula for E^0 in Definition 5.3, we get

(19.36) $$E^\sim(f,f) = \tfrac{1}{2}J\langle f,f\rangle$$

$$+ \tfrac{1}{2}(c\ /\alpha)^2q^{-1}\int_{-1}^{+1}dx\int_{-1}^{+1}dy(1-x)^{-\alpha}(1-y)^{-\alpha}\{f(x)-f(y)\}^2$$

$$+ (c\ /\alpha)\int_{-1}^{+1}dx\{(1+x)^{-\alpha}+(a/q)(1-x)^{-\alpha}\}f^2(x).$$

Also $\underline{\underline{F}}^\sim$ is the $E_1^\sim$ closure of $\underline{\underline{F}}$. Thus $\underline{\underline{F}}^\sim$ is "essentially the same" as $\underline{\underline{F}}$. Indeed, it is easy to see that

(19.37) $$\underline{\underline{F}}_b^\sim = \underline{\underline{F}}_b.$$

<u>Theorem 19.4.</u> <u>The generator</u> $A^\sim$ <u>is defined by</u>

(19.38) $$A^{\sim}f(x) = \text{Ⓐ}f(x)+(c/\alpha)(1-x)^{-\alpha}Mf(1)$$

supplemented by the requirement that (19.34) must converge. ///

Proof. The only thing to be proved is that additional restrictions need not be imposed. It suffices to show that if (19.34) converges and the right side of (19.36) is square integrable, then $f \in \underline{\underline{F}}^{\sim}$. For $\alpha < \frac{1}{2}$ the function $(1-x)^{-\alpha}$ is square integrable, and we need only apply Theorem 19.2. (Also convergence in (19.34) is automatic.) For $\frac{1}{2} \leq \alpha < 1$ we introduce the function

(19.39) $$\xi_1(x) = \int_{-1}^{+1} dy N^I(x,y)(c\ /\alpha)(1-y)^{1-\alpha}.$$

This function is also introduced on p. 183 in [52] where it is represented as an integral. We need only observe here that ξ_1 belongs to $\underline{\underline{F}}$ since ξ_1 is bounded and therefore

$$(c/\alpha)\int_{-1}^{+1} dy\,(1-y)^{1-\alpha}\xi_1(y) < +\infty$$

and also that by (19.12)

$$D_x \text{Ⓙ}^I \xi_1(x) = (c/\alpha)(1-x)^{-\alpha}.$$

Now it suffices to apply Theorem 19.2 to $f(x)-Mf(1)\xi_1(x)$. ///

We emphasize that with the formulation in Theorem 19.4 there are no boundary conditions. The constant a enters directly into the definition of the generator in the interior of $\underline{\underline{I}}$.

Just as for Case 4 in Section 12 the direct application of the technique of Theorem 11.4 leads to the same result. However, if we restrict attention to the case when $A^{\sim}f$ is bounded, then we are naturally led to a second quite different formulation. The point is that such f can be represented

$$f(x) = -N^I A^{\sim} f(x) + N^I \kappa \mu_1(x) M f(1),$$

and therefore f is continuous on the closed interval [-1, +1] and satisfies

(19.40) $$f(-1) = 0 \; ; \; f(+1) = Mf(1).$$

We define the operator

(19.41) $$Ⓐ^{\sim} f(x) = Ⓐ f(x) + (c/\alpha)\{f(-1)(1+x)^{-\alpha} + f(1)(1-x)^{-\alpha}\}$$

acting on functions f defined and continuous on the closed interval [-1, +1]. If f is absolutely continuous on [-1, +1] then the proof of Lemma 18.6 shows that also

(19.42) $$Ⓐ^{\sim} f(x) = -Ⓙ D_x f(x).$$

Now it is easy to prove

<u>Theorem 19.5.</u> <u>When restricted to function</u> f <u>for which</u> $A^{\sim} f$ <u>is bounded,</u> $A^{\sim}$ <u>is defined by</u>

(19.43) $$A^{\sim} f = Ⓐ^{\sim} f$$

<u>supplemented by the boundary conditions</u>

(19.44) $$f(-1) = 0 \; ; \; af(1) = (c/\alpha)\int_{-1}^{+1} dx (1-x)^{-\alpha}\{f(x) - f(1)\}.$$ ///

Theorem 19.5 agrees with Theorem 5.1 in [52] for the choice

(19.45) $$\sigma = \gamma = 0 \; ; \; p = a \; ; \; n(dx) = (c/\alpha)(1-x)^{-\alpha} dx.$$

As in Section 18, Theorem 19.5 can be used to establish the strong Feller property for the resolvent operators and transition operators.

20. One Dimensional Stable Process, $1 < \alpha < 2$.

In this section we assume that $1 < \alpha < 2$. Again our guide is the paper of S. Watanabe [52].

The transition operators C_t and the densities c_t are defined exactly as in Section 19, but a different argument is needed to show that $c_t(x) \geq 0$. We will give an indirect proof which exploits symmetry, but first we take care of some preliminaries.

The analogue of Ⓙ in Section 18 would be quite singular, and so we work instead with

$$\text{(20.1)} \qquad Ⓚf(x) = \{c/\alpha(\alpha-1)\}\int dy\,|x-y|^{1-\alpha}f(y).$$

Just as for Ⓙ in Section 19 this makes sense when f is integrable, but for more general f belonging to any $L^p(\underline{\underline{R}}, dx)$, $p \geq 1$, we work instead with

$$\text{(20.1')} \qquad Ⓚ^m f(x) = \{c/\alpha(\alpha-1)\}\int_{-1}^{+1} dy\,|x-y|^{1-\alpha}f(y)$$

$$+ \{c/\alpha(\alpha-1)\}\int dy I(|y| \geq 1)\{|x-y|^{1-\alpha} - (y)^{1-\alpha}\}f(y).$$

Lemma 20.1. For $f \in L^1(\underline{\underline{R}}, dx)$

$$ⒻⓀf(\lambda) = |\lambda|^{\alpha-2}Ⓕf(\lambda). \qquad ///$$

Proof. The point here is to show that

$$\text{(20.2)} \qquad \{c/\alpha(\alpha-1)\}\operatorname{Lim}_{R\uparrow\infty}\int_{-R}^{+R} dx e^{-i\lambda x}|x|^{1-\alpha} = |\lambda|^{\alpha-2}$$

with the left side suitably controlled and again we only verify (20.2) itself. Arguing more or less as in the proof of Lemma 19.1 we get for $\lambda > 0$

$$\mathrm{Lim}_{R\uparrow\infty}\int_{-R}^{+R}dxe^{-i\lambda x}|x|^{1-\alpha}$$

$$= 2\mathrm{Re.}\int_0^{\infty}dxe^{i\lambda x}|x|^{1-\alpha}$$

$$= 2\mathrm{Re.}\int_0^{\infty}id\tau e^{-\tau\lambda}\tau^{1-\alpha}e^{\frac{1}{2}i\pi(1-\alpha)}$$

$$= -2\lambda^{\alpha-2}\Gamma(2-\alpha)\sin\tfrac{1}{2}\pi(1-\alpha)$$

$$= \alpha(\alpha-1)/c. \qquad ///$$

Now for $f \in$ Ⓢ we have

(20.3) $$(2\pi)^{-1}\int_{-\infty}^{+\infty}d\lambda|\lambda|^{\alpha}|\text{Ⓕ}f(\lambda)|^2$$

$$= -\int_{-\infty}^{+\infty}dxf(x)D_xD_x\text{Ⓚ}f(x)$$

$$= -\{c/\alpha(\alpha-1)\}\int_{-\infty}^{+\infty}dxD_x\{D_xf(x)-D_xf(y)\}\int_{-\infty}^{+\infty}dy|x-y|^{1-\alpha}f(y)$$

$$= -(c/\alpha)\int_{-\infty}^{+\infty}dx\{D_xf(x)-D_xf(y)\}\int_{-\infty}^{+\infty}dy|x-y|^{-\alpha}\mathrm{sgn}(x-y)f(y)$$

$$= -(c/\alpha)\int_{-\infty}^{+\infty}dxD_x\{f(x)-f(y)-(x-y)D_xf(y)\}\int_{-\infty}^{+\infty}dy|x-y|^{-\alpha}\mathrm{sgn}(x-y)f(y)$$

$$= -c\int_{-\infty}^{+\infty}dx\{f(x)-f(y)-(x-y)D_xf(y)\}\int_{-\infty}^{+\infty}dy|x-y|^{-\alpha-1}f(y)$$

$$= -c\mathrm{Lim}_{\varepsilon\downarrow 0}\int_{-\infty}^{+\infty}dx\int_{-\infty}^{+\infty}dyI(|x-y|\geq\varepsilon)|x-y|^{-\alpha-1}\{f(x)-f(y)-(x-y)D_xf(y)\}f(y)$$

$$= \tfrac{1}{2}c\int_{-\infty}^{+\infty}dx\int_{-\infty}^{+\infty}dy|x-y|^{-\alpha-1}\{f(x)-f(y)\}^2,$$

and this extends to general $f \in L^2(\underline{\underline{R}}, dx)$. Let $\underline{\underline{C}}$ be the subspace of $f \in L^2(\underline{\underline{R}}, dx)$ for which the equivalent expressions

(20.4) $$C(f,f) = (2\pi)^{-1}\int_{-\infty}^{+\infty}d\lambda|\lambda|^{\alpha}|\text{Ⓕ}f(\lambda)|^2$$

(20.4') $$C(f,f) = \tfrac{1}{2}c\int_{-\infty}^{+\infty}dx\int_{-\infty}^{+\infty}dy|x-y|^{-1-\alpha}\{f(x)-f(y)\}^2$$

converge. It is easy to check now that the pair $(\underline{\underline{C}}, C)$ is a Dirichlet space on $L^2(\underline{\underline{R}}, dx)$, and so by the results in SMP, Section 1, the semigroup C_t, $t > 0$, is Markovian. Equivalently, the density $c_t(x)$ is nonnegative. The first criterion of the second remark following Lemma 18.1 shows that $(\underline{\underline{C}}, C)$ is recurrent. Thus $\underline{\underline{C}}_{(e)} = \underline{\underline{C}}^{ref}$, and the boundary function $\gamma f(\infty)$ and normal derivative $(\partial f/\partial n)(\infty)$ do not exist. Also $(\underline{\underline{C}}, C)$ has no extensions, even if we replace dx by a bounded reference measure. The second criterion shows that singletons are nonpolar for $(\underline{\underline{C}}, C)$. Moreover, with the help of Fourier transforms it is easy to check directly that every $f \in \underline{\underline{C}}$ is uniformly continuous. If $f \in \underline{\underline{C}}_{(e)}$ is bounded and if $\varphi \in C^\infty_{com}(\underline{\underline{R}})$, then certainly $f\varphi \in \underline{\underline{C}}$, and therefore $f\varphi$ is continuous. This shows that every bounded function and therefore every function in $\underline{\underline{C}}_{(e)}$ is continuous. As in Section 18

$$A^R f = D_x D_x \circledK^m f, \tag{20.5}$$

and if f is integrable

$$A^R f = D_x D_x \circledK f. \tag{20.5'}$$

Again we denote by $(\underline{\underline{F}}, E)$ the absorbed Dirichlet space for $\underline{\underline{I}}$, and we introduce the restricted operator

$$\circledK^I f(x) = \{c/\alpha(\alpha-1)\} \int_{-1}^{+1} dy\, |x-y|^{1-\alpha} f(y).$$

The space $(\underline{\underline{F}}, E)$ can be described explicitly exactly as in Section 19, and again it is true that a potential density $N^I(x,y)$ exists for $(\underline{\underline{F}}, E)$. The significant difference is that $\kappa(x)$ is not integrable. The local generator is defined by

$$\circledA f(x) = D_x D_x \circledK^I f(x) \tag{20.6}$$

and exactly as in Section 19,

Theorem 20.2. The generator A is defined by

$$Af(x) = \textcircled{A}f. \tag{20.7}$$ ///

Remark. Every $f \in$ domain A and indeed every $f \in \underline{\underline{F}}$ belongs to $C_0(\underline{\underline{I}})$. In Section 19 this was true only for Af bounded. ///

The formula (19.21) is also valid here, but the proof must be modified slightly since $|x-y|^{-\alpha}$ is not integrable. The basic idea is to consider f(z) as in Section 19 for a large contour but to look instead at an anti-derivative with respect to x when considering contours which "shrink to the cut". Also (19.23) and (19.24) are valid and are proved exactly as before. However, there are no equilibrium distributions since $(\underline{\underline{C}}, C)$ is recurrent.

We look now at symmetric processes on $\underline{\underline{I}}$ which dominate the absorbed process. Since $\kappa(x)$ is not integrable, extensions $(\underline{\underline{F}}^{\sim}, E^{\sim})$ of $(\underline{\underline{F}}, E)$ exist. However, every function in $\underline{\underline{F}}^{env}$ is uniformly continuous, and it follows just as in Section 12 that the regular part of the boundary can have at most two points. We will restrict our attention to the choices (19.31), (19.32) when the situation is almost exactly the same as for the Cauchy process. The space $\underline{\underline{F}}^{\sim}$ is the set of continuous functions f on the closed interval [-1, +1] for which

$$\tfrac{1}{2}J\langle f, f\rangle + (c\ /\alpha)\int_{-1}^{+1} dx\,(1+x)^{-\alpha} f^2(x) + (c_\alpha/\alpha)\int_{-1}^{+1} dx\,(1-x)^{-\alpha}\{f(x)-f(1)\}^2$$

converges and for such f

$$E^{\sim}(f,f) = \tfrac{1}{2}J\langle f, f\rangle + (c\ /\alpha)\int_{-1}^{+1} dx\,(1+x)^{-\alpha} f^2(x)$$

$$+ (c\ /\alpha)\int_{-1}^{+1} dx\,(1-x)^{-\alpha}\{f(x)-f(1)\}^2 + af^2(1).$$

The generator $A^{\sim}$ is defined by

(20.8) $$A^{\sim}f(x) = Ⓐf(x) + f(1)(c/\alpha)(1-x)^{-\alpha}$$

supplemented by the boundary condition

(20.9) $$af(1) = (c/\alpha)\int_{-1}^{+1}dx(1-x)^{-\alpha}\{f(x)-f(1)\}.$$

There is no need to supplement (20.9) with the requirement $f(-1) = 0$. Of course, this is automatically true for $f \in$ domain $A^{\sim}$ and indeed for $f \in \underline{\underline{F}}^{\sim}$. Thus the direct application of our general theory yields results essentially identical with Theorem 5.1 in [52]. We saw that this was not the case for $0 < \alpha < 1$.

Remark. Once we restrict ourselves to the case (19.31) the situation is pretty much the same as for the Cauchy process. However, the fact that singletons are nonpolar here does make a difference in the general context. For example we have already mentioned above that every f in $\underline{\underline{F}}^{env}$ is uniformly continuous, and therefore the regular boundary can have at most two points. Another interesting distinction becomes apparent if we consider the case where killing is simply suppressed. This does not fit into the general setup of Chapter II, but it is covered by Section 21 in SMP. For the Cauchy process there is only one possibility, and it is recurrent. For $\alpha > 1$ the terminal boundary is nonempty, and there is a classification more or less as for standard Brownian motion without killing on a bounded interval. ///

21. Multi-dimensional Stable Process, $0 < \alpha < 2$.

In this section we consider d-dimensional isotropic stable processes of index α. We assume always that $d \geq 2$ and that $0 < \alpha < 2$. Our guide is the two papers [12] of J. Elliott.

The transition density is defined by Fourier inversion

$$c_t(x) = (2\pi)^{-d} \int_{\mathbb{R}^d} d\lambda e^{i\lambda \cdot x} e^{-t|\lambda|^\alpha}.$$

Here and below in this section integrals are understood to be over $\mathbb{R}^d$ unless otherwise specified. Just as in Section 20 we will show indirectly that $c_t(x) \geq 0$ and therefore the transition operators

$$C_t f(x) = \int dy c_t(x-y) f(y)$$

form a symmetric Markovian semigroup on $L^2(\mathbb{R}^d, dx)$. We begin by introducing the auxiliary operator

$$\text{(21.1)} \qquad Ⓚf(x) = \{c/\alpha(\alpha+d-2)\} \int dy |x-y|^{2-d-\alpha} f(y)$$

where now

$$\text{(21.2)} \qquad c/\alpha(\alpha+d-2) = \pi^{-\frac{1}{2}d} 2^{\alpha-2} \Gamma(\tfrac{1}{2}[d+\alpha-2])/\Gamma(1-\tfrac{1}{2}\alpha).$$

The expression (21.1) converges absolutely for almost every x when f is integrable. For nonintegrable f the operator Ⓚ must be replaced by a modified operator $Ⓚ^m$ as in Sections 19 and 20.

Lemma 21.1. *If f is integrable, then*

$$ⒻⓀf(\lambda) = |\lambda|^{\alpha-2} Ⓕf(\lambda). \qquad ///$$

Proof. The point is to show that

(21.3) $$\int dx e^{-i\lambda\cdot x}|x|^{2-d-\alpha} = \{c/\alpha(\alpha+d-2)\}^{-1}|\lambda|^{\alpha-2}$$

with the integral defined in the appropriate Cauchy-Riemann sense. But everything is clear except for the constant, and this can be checked by multiplying both sides by $e^{-\frac{1}{2}|\lambda|^2}$ and then integrating with respect to λ. ///

Next for $f \in$ Ⓢ we reason as in (20.3) and use the identity

(21.4) $$\Delta_x|x-y|^{2-d-\alpha} = \alpha(\alpha+d-2)|x-y|^{-d-\alpha}$$

to get

(21.5) $$(2\pi)^{-d}\int d\lambda|\lambda|^{\alpha}|Ⓕf(\lambda)|^2 = \tfrac{1}{2}c\int dx\int dy|x-y|^{-d-\alpha}\{f(x)-f(y)\}^2.$$

Now it follows that the C_t are Markovian, that the associated Dirichlet norm is

(21.6) $$C(f,f) = \tfrac{1}{2}c\int dx\int dy|x-y|^{-d-\alpha}\{f(x)-f(y)\}^2$$

and that the Dirichlet space $\underline{\underline{C}}$ is the subspace of $f \in L^2(\underline{\underline{R}}^d, dx)$ for which (21.6) converges. By the criterion in the second remark following Lemma 18.1 the space $(\underline{\underline{C}}, C)$ is transient and just as in Section 19 the extended space $\underline{\underline{C}}_{(e)}$ has codimension one in the reflected space $\underline{\underline{C}}^{ref}$. There exists a potential density $N(x)$, and it can be computed explicitly

(21.7) $$N(x) = |x|^{\alpha-d}\pi^{-\frac{1}{2}d}2^{-\alpha}\Gamma(\tfrac{1}{2}[d-\alpha])/\Gamma(\tfrac{1}{2}\alpha).$$

This is clear except for the constant and the latter can be determined from (21.3) with $2-d-\alpha$ replaced by $-\alpha$. At least for f integrable

(21.8) $$A^R f(x) = \Delta Ⓚf(x).$$

Continuous passage across the boundary of any ball is not possible.

That is, if for $R \geq 0$ we let σ_R be the hitting time

$$\sigma_R = \inf\{t > 0 : |X_t| \geq R\},$$

then

$$P_x\left[|X_{\sigma_R}| = R\right] = 0$$

whenever $|x| < R$. This follows as a simple special case of P. W. Millar's results [40]. It follows that the hitting operator $H^{\sim R}$ for the complement of

$$D_R = \{x \in \underline{\underline{R}}^d : |x| < R\}$$

can be represented

$$H^{\sim R}f(x) = c\int dy I(|y| < R) N^R(x, y) \int dz I(|z| \geq R) |y-z|^{-d-\alpha} f(z)$$

where $N^R(x, y)$ is the potential density for the absorbing Dirichlet space for D_R. In particular

$$H^{\sim R}f(x) = \int dz I(|z| > R) h^R(x, y) f(z)$$

for some density $h^R(x, z)$. To our knowledge it is not possible to compute $h^R(x, y)$ explicitly as was the case for $d = 1$. It follows from our general theory that

(21.10) $$\gamma f(\infty) = \operatorname{Lim}_{R \uparrow \infty} \int dz h^R(0, z) f(z)$$

exists and is finite for general $f \in \underline{\underline{C}}^{ref}$ and that $f(x) - \gamma f(\infty) 1 \in \underline{\underline{C}}_{(e)}$. If $f \in \underline{\underline{C}}^{ref}$ and if $\Delta \textcircled{K}^m f$ is integrable, then

(21.11) $$(\partial f / \partial n)(\infty) = \operatorname{Lim}_{R \uparrow \infty} \int dx I(|x| < R) \ell_R(x) \{f(x) - \gamma f(\infty)\}$$

exists where $\ell_R(x)$ is the equilibrium density for the ball D_R and

(21.12) $$(\partial f/\partial n)(\infty) = -\int dx \Delta \textcircled{K}^m f(x).$$

In particular if f is integrable, then

(21.12') $$(\partial f/\partial n)(\infty) = -\int dx \Delta \textcircled{K} f(x).$$

All this plays a crucial role in the classification of extensions of $(\underline{\underline{C}}, C)$ when dx is replaced by a bounded reference measure. The equilibrium density is determined by the condition

(21.13) $$\int dy I(|y| \leq R) \ell_R(y) N(x-y) = 1$$

for $|x| \leq R$, and it can be explicitly determined

(21.14) $$\ell_R(x) = (R^2 - |x|^2)^{-\frac{1}{2}\alpha} 2^\alpha \Gamma(\tfrac{1}{2}d)/\Gamma(\tfrac{1}{2}[d-\alpha])\Gamma(1-\tfrac{1}{2}\alpha).$$

We refer to the appendix in N. S. Landkof's book [35] for a derivation.

We look next at the absorbed Dirichlet space $(\underline{\underline{F}}, E)$ for $\underline{\underline{I}} = D_1$, the unit ball. Clearly

(21.15) $$E(f, f) = \tfrac{1}{2} J\langle f, f\rangle + \int dx I(|x| < 1)\kappa(x) f^2(x)$$

where now

$$J\langle f, f\rangle = c\int dx \int dy I(|x| < 1, |y| < 1)|x-y|^{-d-\alpha}\{f(x)-f(y)\}^2$$

$$\kappa(x) = c\int dy I(|y| \geq 1)|x-y|^{-d-\alpha}$$

and $\underline{\underline{F}}$ is the subspace of functions f such that (21.14) converges. The restricted operator $\textcircled{K}^I$ is defined exactly as in one dimension. The next theorem gives a partial description of the generator A for $(\underline{\underline{F}}, E)$.

<u>Theorem 21.2.</u> <u>The generator</u> A <u>is defined by</u>

(21.16) $$Af(x) = \Delta \textcircled{K}^I f(x)$$

<u>supplemented by either one of the following two conditions.</u>

(21.17) f <u>is bounded and</u> $f \in L^1(\underline{\underline{I}}, \kappa)$

(21.17') f <u>extends to a function in</u> $\underline{\underline{C}}_b$. ///

<u>Proof.</u> The point is to show that if the right side of (21.16) is square integrable, then either side condition implies $f \in \underline{\underline{C}}$. Suppose first that (21.17) is satisfied. Our previous arguments show that for $\varphi \in C^\infty_{com}(\underline{\underline{R}}^d)$

(21.18) $$(2\pi)^{-d}\int d\lambda |\lambda|^\alpha \textcircled{F}f(\lambda)\textcircled{F}\varphi(\lambda)^* = -\int dx \Delta \textcircled{K}^I f(x)\varphi(x).$$

We can write

(21.19) $$\int dx \Delta \textcircled{K}^I f(x)\varphi(x) = \int_{\underline{\underline{I}}} dx \Delta \textcircled{K}^I f(x)\varphi(x)$$

$$+ \int dx I(|x|>1)\varphi(x)\int_{\underline{\underline{I}}} dy \alpha(\alpha+d-2)|x-y|^{-d-\alpha} f(y).$$

This is clear when f has compact support contained in $\underline{\underline{I}}$ and follows in general by passage to the limit in f since the kernel for $\textcircled{K}^I$ is integrable. Since $\Delta \textcircled{K}^I f(x)$ is square integrable and therefore integrable on $\underline{\underline{I}}$ and since f is in $L^1(\underline{\underline{I}}, \kappa)$ the right side of (21.19) remains bounded as φ runs over a set of uniformly bounded functions. Finally, we let $\varphi = f*\psi$ where $\psi \in C^\infty_{com}$, $\psi \geq 0$, $\int dx \psi(x) = 1$, and $\textcircled{F}\psi \geq 0$, and let support ψ shrink to 0 to conclude that the left side of (21.5) is finite and therefore $f \in \underline{\underline{C}}$. Our proof assuming (21.17') depends on the identity

(21.20) $$\int_{\underline{\underline{I}}} dx g(x)f(x) = -\int_{\underline{\underline{I}}} dx N^I g(x)\Delta \textcircled{K}^I f(x)$$

valid at least for g bounded. Of course N^I is the potential operator for the absorbed space $(\underline{\underline{F}}, E)$. This identity is a slight extension of (4.14) in [12, I]. To prove (21.20) we use the same symbol f for an integrable extension in $\underline{\underline{C}}_b$ and note first that for $\psi \in C^\infty_{com}(\underline{\underline{I}})$

(21.21) $$\tfrac{1}{2}J\langle\psi, f\rangle + \int_{\underline{\underline{I}}} dx\kappa(x)\psi(x)f(x) = -\int_{\underline{\underline{I}}} dx\psi(x)\Delta\textcircled{K}^I f(x).$$

This follows since the left side of (21.21)

$$= C(\psi, f) + c\int dx\int dy I(|x| < 1, |y| \geq 1)|x-y|^{-d-\alpha}\psi(x)f(y)$$

$$= -\int_{\underline{\underline{I}}} dx\psi(x)\Delta\textcircled{K}f(x) + c\int dx\int dy I(|x| < 1, |y| \geq 1)|x-y|^{-d-\alpha}\psi(x)f(y)$$

which agrees with the right side. (See (21.4).) With the help of the estimate

(21.22) $$|N^I g(x)| \leq \text{const.}\, (1-|x|^2)^{\frac{1}{2}\alpha}$$

(see Theorem 5.2 in [12, I]) it is then easy to approximate N^I_g with such ψ and establish (21.21) with ψ replaced by $N^I g$. Finally, the left side of (21.21) with $N^I g$

$$= C(N^I g, f) + c\int dx\int dy I(|x| < 1, |y| \geq 1)|x-y|^{-d-\alpha}N^I g(x)f(y)$$

$$= \int_{\underline{\underline{I}}} dx g(x)\{f(x) - \tilde{H}^I f(x)\} + c\int dx\int dy I(|x| < 1, |y| \geq 1)|x-y|^{-d-\alpha}N^I g(x)f(y)$$

which establishes (21.20) since the last two terms cancel. We are done since also (21.20) is valid with f replaced by $N^I\Delta\textcircled{K}^I f$. ///

<u>Remark.</u> For d = 1 we were able to describe the generator A without any side conditions. Our main tool was the identity (19.14) for m = 0, 1 which gave us direct control over the annihilators of $D_x\textcircled{J}^I$ or $D_x D_x\textcircled{K}^I$. Possibly a d-dimensional analogue would allow us to remove the side conditions here. ///

As in one dimension, it is possible to prove the strong Feller property with the help of the identity

(21.23) $$N^{I}g(x) = Ng(x) - \int dz I(|z| \geq 1) N(x-z) \int_{\underline{\underline{I}}} dy\, c\,|z-y|^{-d-\alpha} N^{I} g(y)$$

which is a special case of the duality relation (7.4) in SMP. We refer to Lemma 4.6 in [12, I] for a direct proof of (21.23).

We look now at symmetric processes on $\underline{\underline{I}}$ which dominate the absorbing one. We have already noted that the terminal boundary is empty, and it is clear from our study of simpler examples in earlier sections that we have little "a priori" control over Δ. However, there are good reasons for choosing

(21.24) $$\Delta = S, \text{ the unit sphere;}$$

$$\nu(ds) = \text{surface area;}$$

$$\kappa(x)\mu(x, ds) = (c/\alpha)|x-s|^{-d-\alpha}\{(s-x)\cdot s\}\nu(ds).$$

To simplify the notation we will usually write

(21.25) $$\kappa(x)\mu(x, s) = (c/\alpha)|x-s|^{-d-\alpha}\{(s-x)\cdot s\}$$

for $x \in \underline{\underline{I}}$ and $s \in S$. The special properties of this choice are due to the relation

(21.26) $$(\partial/\partial n)_{s}\{c/\alpha(\alpha+d-2)\}|x-s|^{2-d-\alpha} = \kappa(x)\mu(x, s)$$

which implies

(21.27) $$\Delta \textcircled{K}^{I} f(x) = \operatorname{div}_{x}\{c/\alpha(\alpha+d-2)\}\int_{\underline{\underline{I}}} dy \operatorname{grad} f(y)|x-y|^{2-d-\alpha}$$

$$- \int_{S}\nu(ds)\kappa(x)\mu(x, s)f(s)$$

at least for f having a C^2 extension to a neighborhood of the closure $cl(\underline{\underline{I}})$. We refer to Section 1 in [12, 2] for a proof. Thus our choice is analogous to the one made for $d = 1$ and (21.27) corresponds to (18.38). After approximating κ from below by $\kappa\varphi_n$ where φ_n is the indicator of a compact subset of $\underline{I}$ and applying Theorem 21.2 we see that

(21.28) $$\Delta \textcircled{K}^{I} 1(x) = \kappa(x),$$

and so the special case $f = 1$ in (21.26) gives

(21.29) $$\int_S \nu(ds)\mu(x, s) = 1,$$

which shows that our choice (21.25) is legitimate and also that $r(x) = 0$. The main significance of (21.25) is that it permits

Lemma 21.3. If φ is continuous on $\underline{S}$, then the function

(21.30) $$H\varphi(x) = N^I \kappa\mu\varphi(x)$$
$$= \int_{\underline{I}} dy N^I(x, y)\kappa(y)\int_{\underline{S}}\nu(ds)\mu(y, s)\varphi(s)$$

belongs to $C(\underline{I}^-)$ and $f = \varphi$ on $\underline{S}$. ///

Proof. We need only consider the special case when φ is C^2. Let f be a C^2 function defined in a neighborhood of the closure $\underline{I}^-$ such that

$$f = \varphi \text{ on } S \text{ ; } \partial f/\partial n = 0 \text{ on } \underline{S}.$$

It is clear that such f exists. We refer to Lemma 3.3 in [12] for an explicit construction in a more general context. Let $(\underline{F}^e, E^e)$ be the expanded Dirichlet space on $L^2(\underline{I} \cup \underline{S}; dx+\nu)$ which corresponds to the excursion space $(\underline{N}, N)$. That is,

(21.31) $$E^e(g, g) = \tfrac{1}{2}J<g, g> + \int_{\underline{I}} dx \int_{\underline{S}}\nu(ds)\kappa(x)\mu(x, s)\{g(x)-g(s)\}^2$$

and $\underline{\underline{F}}^e$ is the set of $g \in L^2(\underline{\underline{I}} \cup \underline{\underline{S}}; dx+\nu)$ for which (21.31) converges. Then $\underline{\underline{F}}$ is contained in $\underline{\underline{F}}^e$, and (21.30) is the E^e orthogonal projector onto the complement of $\underline{\underline{F}}$ in the sense of Lemma 8.3 in SMP. The function f specified above belongs to $\underline{\underline{F}}^e$ and for $g \in \underline{\underline{F}} \cap C_{com}(\underline{\underline{I}})$ we have

$$E^e(g,f) = \tfrac{1}{2} J\langle g,f\rangle + \int_{\underline{\underline{I}}} dx\kappa(x)g(x)f(x) - \int_{\underline{\underline{I}}} dxg(x)\kappa\mu f(x).$$

$$= - \int_{\underline{\underline{I}}} dxg(x)\Delta \textcircled{K}^I f(x) - \int_{\underline{\underline{I}}} dxg(x)\kappa\mu f(x).$$

which by (21.27)

$$= -\{c/\alpha(\alpha+d-2)\}\int_{\underline{\underline{I}}} dxg(x)\mathrm{div}_x \int_{\underline{\underline{I}}} dy\,|x-y|^{2-d-\alpha} \mathrm{grad}\, f(y),$$

and after application of the divergence theorem

$$E^e(g,f) = -\int_{\underline{\underline{I}}} dxg(x)\textcircled{K}^I \Delta f(x).$$

In the last step we used the condition $\partial f/\partial n = 0$. From this it follows that

$$f(x) - N^I\kappa\mu\varphi(x) = N^I \textcircled{K}^I \Delta f(x), \tag{21.32}$$

and we are done since $\textcircled{K}^I \Delta f(x)$ is bounded. ///

<u>Remark.</u> In putting together our proof of Lemma 21.3 we were strongly influenced by Section 3 in [12, 2] where a more general result is proved using different tools. ///

It is easy to establish for the killing density $\kappa(x)$ an estimate

$$m(1-|x|)^{-\alpha} \le \kappa(x) \le M(1-|x|)^{-\alpha} \tag{21.33}$$

where m, M are positive constants depending on d and α. For $0 < \alpha < 1$ the density $\kappa(x)$ is integrable, and therefore $\Delta = \Delta_s$. For $1 \le \alpha < 2$ we can apply spherical symmetry to conclude that

(21.34) $$\int_{\underline{\underline{I}}} dx\kappa(x)\int_{\underline{\underline{S}}}\nu(ds)\mu(x,s)\varphi(s) = \nu(\underline{\underline{S}})^{-1}\int_{\underline{\underline{I}}} dx\kappa(x)\int_{\underline{\underline{S}}}\nu(ds)\varphi(s) = +\infty$$

whenever $\varphi \geq 0$ is not ν-null. This guarantees that $\Delta = \Delta_r$ for $1 \leq \alpha < 2$. We have already noted that the excursion space $\underline{\underline{N}}$ contains all C^2 functions on $\underline{\underline{S}}$.

Let $(\underline{\underline{H}}, Q)$ be a pair satisfying the conditions of paragraph 3.4. Just as in Section 15 the verification of 3.4.2 in practice can be quite difficult. However, it is relatively easy to give examples. The situation considered in |12, 2| would correspond to the choice $\underline{\underline{H}} = \underline{\underline{N}}$ and

(21.35) $$Q(\varphi,\varphi) - N(\varphi,\varphi)$$
$$= \int_{\underline{\underline{S}}}\nu(ds)\int_{\underline{\underline{S}}}\nu(dt)b(s,t)\{\varphi(s)-\varphi(t)\}^2 + \int_{\underline{\underline{S}}}\nu(ds)a(s)\varphi^2(s)$$

with $a, b \geq 0$.

We consider first the range $0 < \alpha < 1$ and begin by examining the expanded Dirichlet space $(\underline{\underline{F}}^e, E^e)$, already introduced in the proof of Lemma 21.3. Let $\underline{\underline{F}}_a^{env}$ be the set of functions $f \in L^2(\underline{\underline{I}}, dx)$ for which

(21.36) $$J<f,f> < +\infty.$$

The function space $\underline{\underline{F}}^e$ consists of pairs (f,φ) with $f \in \underline{\underline{F}}_a^{env}$ and $\varphi \in \underline{\underline{H}}$ such

(21.37) $$\int_{\underline{\underline{I}}} dx\kappa(x)\int_{\underline{\underline{S}}}\nu(ds)\mu(x,s)\{f(x)-\varphi(s)\}^2 < +\infty.$$

The norm E^e is given by

(21.38) $$E^e(\{f,\varphi\},\{f,\varphi\})$$
$$= \tfrac{1}{2}J<f,f> + \int_{\underline{\underline{I}}} dx\kappa(x)\int_{\underline{\underline{S}}}\nu(ds)\mu(x,s)\{f(x)-\varphi(s)\}^2$$
$$+ \{Q-N\}(\varphi,\varphi).$$

If $\{f, \varphi\}$ belongs to the domain of the expanded generator A^e, then for $g \in C^{\infty}_{com}(\underline{\underline{I}})$

(21.39)
$$\tfrac{1}{2} J<f, g> + \int_{\underline{\underline{I}}} dx \kappa(x) g(x) \int_{\underline{\underline{S}}} \nu(ds) \mu(x, s)\{f(x) - \varphi(s)\}$$
$$= - \int_{\underline{\underline{I}}} dx g(x) A^e\{f, \varphi\}(x),$$

and for $\psi \in \underline{\underline{H}}_b$,

(21.40)
$$\tfrac{1}{2} J< f, H\psi> + \int_{\underline{\underline{I}}} dx \kappa(x) \int_{\underline{\underline{S}}} \nu(ds) \mu(x, s)\{f(x) - \varphi(s)\}\{H\psi(x) - \psi(s)\}$$
$$+ \{Q-N\}(\varphi, \psi)$$
$$= - \int_{\underline{\underline{I}}} dx H\psi(x) A^e\{f, \varphi\}(x) - \int_{\underline{\underline{S}}} \nu(ds) \psi(s) A^e\{f, \varphi\}(s).$$

The left side of (21.39)

$$= \tfrac{1}{2} J<f, g> + \int_{\underline{\underline{I}}} dx \kappa(x) g(x) f(x) - \int_{\underline{\underline{I}}} dx \kappa(x) g(x) \mu\varphi(x)$$
$$= - \int_{\underline{\underline{I}}} dx g(x) \Delta \circledR^{I} f(x) - \int_{\underline{\underline{I}}} dx \kappa(x) g(x) \mu\varphi(x),$$

and so (21.39) is equivalent to

(21.41)
$$A^e\{f, \varphi\}(x) = \Delta \textcircled{K}^{I} f(x) + \kappa(x) \mu\varphi(x),$$

of course in the distribution sense on $\underline{\underline{I}}$. Applying Theorem 7.1, we see that (21.40) is equivalent to

(21.42)
$$(\partial f/\partial n)(\psi) - \{Q-N\}(\varphi, \psi) = \int \nu(ds) \psi(s) A^e\{f, \varphi\}(s)$$

where $(\partial f/\partial n)(\psi)$ is defined by (7.9) and has no "a priori" connection with the normal derivative. Conversely, a pair $\{f, \varphi\} \in \underline{\underline{F}}^e$ belongs to domain A^e if the right side of (21.41) is square integrable and if there exists $\theta(s) \in L^2(\underline{\underline{S}}, \nu)$ such that for $\psi \in \underline{\underline{H}}_b$ the left side of (21.42) agrees with the right side with

$A^e\{f,\varphi\}(s)$ replaced by $\theta(s)$. This can be simplified somewhat if $\{f,\varphi\}$ is known to be bounded. Since the terminal boundary is empty, it is clear that

$$P^{ex}(\Omega^{ex}) = \int_{\underline{\underline{I}}} dx\kappa(x) \tag{21.43}$$

which converges for $0 < \alpha < 1$. By the dominated convergence theorem

$$(\partial f/\partial n)(\psi) = \int_{\underline{\underline{I}}} dx\kappa(x)\int_{\underline{\underline{S}}}\nu(ds)\mu(x,s)\psi(s)\{f(x)-\varphi(s)\},$$

and so (21.42) is equivalent to

$$\int_{\underline{\underline{I}}} dx\kappa(x)\int_{\underline{\underline{S}}}\nu(ds)\mu(x,s)\psi(s)\{f(x)-\varphi(s)\} - \{Q-N\}(\varphi,\psi) = \int\nu(ds)\psi(s)A^e\{f,\varphi\}(s). \tag{21.42'}$$

For the special case (21.35) this becomes

$$\begin{aligned}&\int_{\underline{\underline{I}}} dx\kappa(x)\int_{\underline{\underline{S}}}\nu(ds)\mu(x,s)\psi(s)\{f(x)-\varphi(s)\}\\ &\quad - \int_{\underline{\underline{S}}}\nu(ds)\int_{\underline{\underline{S}}}\nu(dt)b(s,t)\{\varphi(s)-\varphi(t)\}\{\psi(s)-\psi(t)\}\\ &\quad - \int_{\underline{\underline{S}}}\nu(ds)a(s)\varphi(s)\psi(s)\\ &= \int_{\underline{\underline{S}}}\nu(ds)\psi(s)A^e\{f,\varphi\}(s).\end{aligned} \tag{21.44}$$

At least when $b(s,t)$ is integrable with respect to $\nu(ds)\times\nu(dt)$, we conclude that

$$\begin{aligned}A^e\{f,\varphi\}(s) &= \int_{\underline{\underline{I}}} dx\kappa(x)\mu(x,s)\{f(x)-\varphi(s)\}\\ &\quad + 2\int_{\underline{\underline{S}}}\nu(dt)b(s,t)\{\varphi(t)-\varphi(s)\} - a(s)\varphi(s),\end{aligned} \tag{21.45}$$

which is consistent with (0.4) in [12, 2].

We consider next the Dirichlet space $(\underline{\underline{F}}^{\sim}, E^{\sim})$ obtained from $(\underline{\underline{F}}^e, E^e)$ by collapsing the time scale. A given $f \in L^2(\underline{\underline{I}}, dx)$ belongs to $\underline{\underline{F}}^{\sim}$ if and only if there exists at least one $\varphi \in \underline{\underline{H}}$ such that the pair $\{f,\varphi\}$ belongs to $\underline{\underline{F}}^e$. In this case there is a unique function $Mf \in \underline{\underline{H}}$ which minimizes (21.38), and the Dirichlet norm $E^{\sim}$ is defined by

$$E^{\sim}(f, f) = E^{e}(\{f, Mf\}, \{f, Mf\}). \tag{21.46}$$

The absorbed space $\underline{\underline{F}}$ is dense in $\underline{\underline{F}}^{\sim}$ and indeed

$$\underline{\underline{F}}_b^{\sim} = \underline{\underline{F}}_b = \underline{\underline{F}}_b^{env}. \tag{21.47}$$

All this follows from our general results in Section 5 with the simplification that $\Delta_r = \phi$. The operator M is a hitting operator for the expanded process. It is not given directly when $(\underline{\underline{H}}, Q)$ is specified, but rather must be computed. This was easily done in Section 19 but could be quite difficult here. The Dirichlet form can also be written

$$E^{\sim}(f, f) = \tfrac{1}{2}J\langle f, f\rangle + \int_{\underline{\underline{I}}} dx\kappa(x)\{1-\mu M1(x)\}f^2(x) + \tfrac{1}{2}\kappa\mu M\langle f, f\rangle \tag{21.48}$$

with conventions as in Section 5. Thus

$$\mu M1(x) = \int \nu(ds)\mu(x, s)M1(s)$$

$$\kappa\mu M\langle f, f\rangle = \int dx\kappa(x)\int \nu(ds)\mu(x, s)\int M(s, dy)\{f(x)-f(y)\}^2.$$

If $f \in$ domain $A^{\sim}$, then for $g \in C^{\infty}_{com}(\underline{\underline{I}})$

$$-\int dx A^{\sim}f(x)g(x) = E^{\sim}(f, g),$$

which is equivalent to

$$\begin{aligned}
-\int dx A^{\sim}f(x)g(x) &= \tfrac{1}{2}J\langle f, g\rangle + \int dx\kappa(x)f(x)g(x) \\
&\quad - \int dx\kappa(x)f(x)g(x)\mu M1(x) \\
&\quad + \int dx\kappa(x)g(x)\int \mu M(x, dy)\{f(x)-f(y)\} \\
&= -\int dx\Delta \textcircled{K}^{I} f(x)g(x) - \int dx\kappa(x)g(x)\mu Mf(x),
\end{aligned}$$

and therefore

$$A^{\sim}f(x) = \Delta \textcircled{K}^{I} f(x) + \kappa(x)\mu M f(x). \tag{21.49}$$

Suppose conversely that f is bounded and that the right side of (21.49) is square integrable on $\underline{\underline{I}}$. Then $\Delta \textcircled{K}^{I} f(x)$ is integrable on $\underline{\underline{I}}$, and our proof of Theorem 21.2 shows that $f \in \underline{\underline{C}}$, and therefore $f \in \underline{\underline{F}}^{\sim}$. This is enough to guarantee that $f \in$ domain $A^{\sim}$. We conclude that a bounded function f is in domain $A^{\sim}$ if and only if the right side of (21.49) is square integrable on $\underline{\underline{I}}$. There are no boundary conditions, but our choice of the pair $(\underline{\underline{H}}, Q)$ does effect the hitting operator M.

We cannot hope to establish strong Feller properties without imposing some restrictions on the pair $(\underline{\underline{H}}, Q)$. (See for example Lemma 3.6 in [12].) A condition which seems to be reasonable in the present context is

__21.1. Regularity Condition.__ The associated Green's operator $R_{(1)}$ defined by (3.27) maps bounded functions on $\underline{\underline{S}}$ into continuous functions. ///

__Theorem 21.4. Assume that the regularity condition__ 21.1 __is satisfied, and let__ $C^{\sim}(\underline{\underline{I}}^{-})$ __be the uniform closure of functions__ $f \in \underline{\underline{C}}(\underline{\underline{I}}^{-})$ __whose restrictions__ γf __to__ $\underline{\underline{S}}$ __can be represented__

$$\gamma f(s) = R_{(1)}\{\textstyle\int_{\underline{\underline{I}}} dx N^{I} g(x)\kappa\mu(x,\cdot)\}(s) = R_{(1)}\pi g(x) \tag{21.50}$$

__with__ g __bounded on__ $\underline{\underline{I}}$.

(i) __The resolvent operators__ $G_{\lambda}^{\sim}, \lambda > 0$ __map bounded functions into__ $C^{\sim}(\underline{\underline{I}})$.

(ii) __The transition operators__ $P_t^{\sim}$, $t > 0$, __form a strongly continuous semi-group on the Banach space__ $C^{\sim}(\underline{\underline{I}})$. __A function__ $f \in C(\underline{\underline{I}}^{-})$ __belongs to the domain of the generator__ $A_c^{\sim}$ __for this semigroup if and only if__

$$\textcircled{A}^{\sim} f(x) = \Delta \textcircled{K}^{I} f(x) + \kappa(x)\mu f(x) \tag{21.51}$$

<u>belongs to</u> $C^{\sim}(\underline{\underline{I}})$, <u>the restriction</u> γf <u>belongs to</u> $\underline{\underline{H}}$, <u>and</u> f <u>satisfies the boundary condition</u>

(21.52) $$\int_{\underline{\underline{I}}} dx\kappa(x)\int_{\underline{\underline{S}}}\nu(ds)\mu(x,s)\psi(s)\{f(x)-f(s)\} = \{Q-N\}(\gamma f,\psi)$$

<u>for</u> $\psi \in \underline{\underline{H}}_b$. <u>In this case</u>

(21.53) $$A_c^{\sim}f(x) = Ⓐ^{\sim}f(x). \qquad ///$$

<u>Proof.</u> If $f = G_1^{\sim}g$ with g bounded, then by (3.41)

(21.54) $$\begin{aligned} f &= N_1^I g + N_1^I\kappa\mu R_{(1)}\pi_1 g \\ &= N^I(g-f) + N^I\kappa\mu R_{(1)}\pi_1 g, \end{aligned}$$

where the operator π_1 is defined by

$$\pi_1 g(s) = \int_{\underline{\underline{I}}} dx N_1^I g(x)\kappa(x)\mu(x,s).$$

Since

$$\int_{\underline{\underline{I}}} dx N^I 1(x)\int_{\underline{\underline{S}}}\nu(ds)\kappa(x)\mu(x,s) = \int_{\underline{\underline{I}}} dx$$

converges, it is clear from spherical symmetry that π, and therefore π_1 maps bounded functions on $\underline{\underline{I}}$ into bounded functions on $\underline{\underline{S}}$. Our regularity condition 21.1 guarantees that $R_{(1)}\pi_1 g$ is continuous on $\underline{\underline{S}}$, and from Lemma 21.3 and (21.23) we conclude that $f \in C(\underline{\underline{I}}^-)$ and

(21.55) $$\gamma f = R_{(1)}\pi_1 g.$$

This establishes everything except the characterization of domain $A_c^{\sim}$. If $f \in$ domain $A_c^{\sim}$, then (21.54) and (21.55) imply (21.53), and then (21.49) implies

(21.56) $$Mf = \gamma f,$$

and the boundary condition (21.52) follows from the properties of M as an orthogonal projector on the expanded space $(\underline{\underline{F}}^e, E^e)$. This establishes the direct part of the characterization, and the converse follows routinely. ///

This theorem is closely related to the results contained in Section 4 of [12, 2].

We turn now to the case $1 \leq \alpha < 2$. Since $\Delta = \Delta_r$, a consequence of (21.34), there is little point to considering separately the expanded space $(\underline{\underline{F}}^e, E^e)$. A given function $f \in \underline{\underline{F}}_a^{env}$ belongs to the function space $\underline{\underline{F}}^\sim$ if and only if there exists $\varphi \in \underline{\underline{H}}$ such that (21.37) converges. In this case such φ is unique, and we denote it by γf. The norm $E^\sim$ is then given by

$$E^\sim(f,f) = \tfrac{1}{2}J\langle f,f\rangle + \int_{\underline{\underline{I}}} dx\kappa(x) \int_{\underline{\underline{S}}} \nu(ds)\mu(x,s)\{f(x)-\gamma f(s)\}^2 + \{Q-N\}(\gamma f, \gamma f).$$

The absorbed space $(\underline{\underline{F}}, E)$ is a proper closed subspace of $(\underline{\underline{F}}^\sim, E^\sim)$, and the operator (21.30) is $E^\sim$ orthogonal projection of $\underline{\underline{F}}^\sim$ onto the complement of $\underline{\underline{F}}$. If $f \in$ domain $A^\sim$, then for $g \in C^\infty_{com}(\underline{\underline{I}})$,

$$-\int dx g(x) A^\sim f(x) = -\int dx g(x) \Delta Ⓚ^I f(x) - \int dx \kappa(x) g(x) \mu\gamma f(x), \tag{21.57}$$

and therefore

$$A^\sim f(x) = Ⓐ^\sim f(x). \tag{21.58}$$

For $\psi \in \underline{\underline{H}}_b$ we get

$$(\partial f/\partial n)(\psi) = \{Q-N\}(\gamma f, \psi) \tag{21.59}$$

with the left side again defined by (7.9). We can regard (21.59) as a rigorous formulation of (21.52), which is equivalent to (21.59) at least when f is differentiable. (See the discussion on p. 86 in [12].) Conversely, if f is known to be in $\underline{\underline{F}}^{\sim}$, if Ⓐ$^{\sim}$f(x) is square integrable and if f satisfies the boundary condition (21.59), then $f \in$ domain $A^{\sim}$. It is much more difficult here to remove the "a priori" condition that $f \in \underline{\underline{F}}^{\sim}$ when characterizing domain $A^{\sim}$. The best we can do at present is the following If $f \in \underline{\underline{F}}_b^{env}$ and if there exists $\varphi \in \underline{\underline{H}}$ such that ΔⓀ^{I}f(x) + $\kappa(x)\mu\varphi(x)$ is square integrable on $\underline{\underline{I}}$, then $f \in \underline{\underline{F}}^{\sim}$ and $\gamma f = \varphi$. If also f satisfies the boundary condition (21.59), then $f \in$ domain $A^{\sim}$. To prove this it suffices to observe that $f - N\kappa\mu\varphi \in \underline{\underline{F}}$ by Theorem 21.2. Also Theorem 21.4 is valid for the range $1 \leq \alpha < 2$, except that when characterizing domain $A_c^{\sim}$ we must impose the "a priori" requirement $f \in \underline{\underline{F}}^{env}$.

Chapter V. Hyperbolic Half Plane

As in Section 15 the basic state space $\underline{\underline{X}}$ is the upper half plane, but now we view it as a symmetric space $\underline{\underline{X}} = SL_2(R)/SO(2)$, and we focus our attention on Dirichlet spaces on $\underline{\underline{X}}$ which are invariant in an appropriate sense. Every invariant symmetric Markov process on $\underline{\underline{X}}$ certainly determines an invariant Dirichlet space, but the converse is false. For example, the Dirichlet space for absorbing Brownian motion is invariant, and it corresponds to a time changed Brownian motion which is also invariant. But the Dirichlet space for reflected Brownian motion is invariant and does not correspond to an invariant Markov process. This means that there exist invariant Dirichlet spaces not covered by the results in [27], [22], and [20]. In Section 25 we describe at least some of them.

The necessary preliminaries are collected and presented from an elementary point of view in Section 22. In Section 23 we rederive the well-known classification of invariant semigroups on $\underline{\underline{X}}$ using Dirichlet space techniqes. The harmonic functions for any one of these semigroups are precisely the classical harmonic functions. This follows from a general result of H. Furstenberg [19]. In Section 24 we reproduce Furstenberg's proof.

One interesting consequence of our analysis in Section 25 is that if there are "too many large jumps", then extensions can be constructed only if the boundary is "collapsed to a point". This is the meaning of Theorem 25.1.

22. Preliminaries.

Our presentation here was strongly influenced by the first part of the book of S. Lang [36].

Throughout this chapter $\mathbf{G} = SL_2(R)$, the group of real 2×2 matrices with determinant one. The following notation will be used for certain special elements in $\mathbf{G}$,

(22.1) $$r_\theta = \begin{bmatrix} \cos\theta & \sin\theta \\ -\sin\theta & \cos\theta \end{bmatrix}, \quad \theta \in T = (-\pi, +\pi].$$

(22.2) $$h_a = \begin{bmatrix} a & 0 \\ 0 & a^{-1} \end{bmatrix}, \quad a \text{ real and } a \neq 0.$$

(22.3) $$n_b = \begin{bmatrix} 1 & b \\ 0 & 1 \end{bmatrix}, \quad b \text{ real.}$$

These generate the following subgroups of $\mathbf{G}$,

(22.4) $$\mathbf{K} = \{\Lambda_\theta : \theta \in T\}$$

$$\mathbf{A} = \{h_a : a > 0\}$$

$$\mathbf{N} = \{n_b : b \text{ real}\}$$

$$\mathbf{P} = \mathbf{NA} = \mathbf{AN}.$$

Typical members of $\mathbf{G}$ will be denoted by Greek letters σ, τ, ρ, etc. The functions $\underline{\alpha}, \underline{\beta}, \underline{\gamma}, \underline{\delta}$, are defined on $\mathbf{G}$ by

(22.5) $$\sigma = \begin{bmatrix} \underline{\alpha}(\sigma) & \underline{\beta}(\sigma) \\ \underline{\gamma}(\sigma) & \underline{\delta}(\sigma) \end{bmatrix}.$$

Of course these do not form a coordinate system on $\mathbf{G}$. The following decomposition theorem for $\mathbf{G}$ permits us to introduce two coordinate systems which will turn out to be quite convenient.

Theorem 22.1. (Iwasawa Decomposition). Every $\sigma \in \underline{\underline{G}}$ has a unique representation

$$\sigma = h_a n_b r_\theta \tag{22.6}$$

and also

$$\sigma = n_{b^{\#}} h_a r_\theta \tag{22.7}$$

with $a > 0$. Moreover

$$b^{\#} = a^2 b. \tag{22.8}$$

///

Proof. We note first that if $\tau \in \underline{\underline{G}}$ with $\underline{\delta}(\tau) \neq 0$, then τ has a unique representation

$$\tau = \mu\lambda\nu \tag{22.9}$$

with $\mu \in N$, $\lambda \in A \cup (-A)$, and $\nu \in N^t$. (Here and below we use the superscript t to denote the transpose operation.) The factorization (22.9) is just the classical Gaussian factorization in inverted form. To prove it we note first that the equation

$$\underline{\beta}(n_x \sigma) = 0$$

has the unique solution $x = -\underline{\beta}(\sigma)/\delta(\sigma)$, and therefore $\sigma = n_{-x}\tau$ with $\underline{\beta}(\tau) = 0$. But then

$$\tau = \begin{bmatrix} y & 0 \\ z & y^{-1} \end{bmatrix} = h_y n^t_{yz},$$

and (22.9) is established. To establish (22.7) we apply (22.9) to $\rho = \sigma\sigma^t$ which is permissible since ρ is positive definite and therefore $\underline{\delta}(\rho) > 0$. In this case it is easy to check that (22.9) specializes to

(22.9') $$\rho = n_{b^{\#}} h_a h_a n_{b^{\#}}^t$$

with $a > 0$. Since $(h_a^{-1} n_{b^{\#}}^{-1} \sigma)(h_a^{-1} n_{b^{\#}}^{-1} \sigma)^t = 1$ we have $h_a^{-1} n_{b^{\#}}^{-1} \sigma = r_\theta$ which implies the existence of a representation (22.7). Since (22.7) implies (22.9'), uniqueness in (22.9') implies uniqueness in (22.7). Finally (22.6) follows from (22.7) since

(22.10) $$h_a n_b = n_{a^2 b} h_a . \qquad ///$$

The functions $\underline{a}$, $\underline{b}$, $\underline{b}^{\#}$, $\underline{\theta}$ are defined on $\underline{\underline{G}}$ by the decompositions (22.6) and (22.7). Also we view $\underline{a}$, $\underline{b}$, $\underline{b}^{\#}$, $\underline{\theta}$ as functions defined on the appropriate subgroups in (22.4). Clearly $(\underline{a}, \underline{b}, \underline{\theta})$ and $(\underline{a}, \underline{b}^{\#}, \underline{\theta})$ are global coordinate systems for $\underline{\underline{G}}$ while $(\underline{a}, \underline{b})$ and $(\underline{a}, \underline{b}^{\#})$ are coordinate systems on $\underline{\underline{P}}$. Now it is fairly easy to give explicit formulae for the Haar measures on $\underline{\underline{G}}$ and also its subgroups (22.4).

Theorem 22.2. (i) *$\underline{a}^{-1}d\underline{a}$ is a 2-sided Haar measure for A.*

(ii) $d\underline{b}$ is 2-sided Haar measure for $\underline{\underline{N}}$.

(iii) $(2\pi)^{-1}d\underline{\theta}$ is normalized 2-sided Haar measure for K.

(iv) $\underline{a}^{-1}d\underline{a}d\underline{b} = \underline{a}^{-3}d\underline{a}d\underline{b}^{\#}$ is left invariant Haar measure for $\underline{\underline{P}}$.

(v) $\underline{a}^{-1}d\underline{a}d\underline{b}^{\#} = \underline{a}d\underline{a}d\underline{b}$ is right invariant Haar measure for $\underline{\underline{P}}$.

(vi) $d\sigma = \underline{a}^{-1}d\underline{a}d\underline{b}d\underline{\theta} = \underline{a}^{-3}d\underline{a}d\underline{b}^{\#}\, d\underline{\theta}$ is 2-sided Haar measure for $\underline{\underline{G}}$. ///

Proof. (i) through (iii) are entirely elementary. For (iv) and (v) the main point is to observe that

$$\int_0^\infty a^{-1}da \int_{-\infty}^{+\infty} db f(n_d h_a n_b) = \int_0^\infty a^{-1}da \int_{-\infty}^{+\infty} db f(h_a n_{b+d/a^2})$$

$$\int_0^\infty a^{-1}da \int_{-\infty}^{+\infty} db^{\#} f(n_{b^{\#}} h_a n_d) = \int_0^\infty a^{-1}da \int_{-\infty}^{+\infty} db^{\#} f(n_{b^{\#}+a^2 d} h_a)$$

and to apply translation invariance of db and $db^{\#}$. In proving (vi) we use the fact that 2-sided Haar measure $d\sigma$ exists for $\underline{\underline{G}}$. (See for example [27, p. 366].) We observe first that if $F \in C_{com}(\underline{\underline{G}})$ and if

(22.11) $$\int_{-\pi}^{+\pi} d\theta F(\sigma r_\theta) = 0$$

for every $\sigma \in \underline{\underline{G}}$, then also

(22.12) $$\int_{\underline{\underline{G}}} d\sigma F(\sigma) = 0.$$

This follows since if $\psi \in C_{com}(\underline{\underline{P}})$ satisfies

$$\psi(\sigma) = 1 \text{ whenever } F(\sigma r_\theta) \neq 0 \text{ for some } \theta \in \underline{\underline{T}},$$

then

$$\int_{\underline{\underline{G}}} d\sigma F(\sigma) = \int_{\underline{\underline{G}}} d\sigma F(\sigma)(2\pi)^{-1}\int_{-\pi}^{+\pi} d\theta\psi(\sigma r_\theta)$$

$$= \int_{\underline{\underline{G}}} d\sigma\psi(\sigma)(2\pi)^{-1}\int_{-\pi}^{+\pi} d\theta F(\sigma r_{-\theta}).$$

Now for $f \in C_{com}(\underline{\underline{P}})$ define $f^{\#}$ on $\underline{\underline{G}}$ by

$$f^{\#}(pr_\theta) = f(p),$$

and then define a linear functional L on $C_{com}(\underline{\underline{P}})$ by

$$L(f) = \int_{\underline{\underline{G}}} d\sigma f^{\#}(\sigma).$$

It is easy to check that L is left (but not right) invariant on $\underline{\underline{P}}$, and therefore it must be that

$$L(f) = \int_{\underline{\underline{P}}} dp f(p)$$

where dp is left invariant Haar Measure on $\underline{\underline{P}}$. For general $F \in C_{com}(\underline{\underline{G}})$ define F^b on $\underline{\underline{P}}$ by

$$F^b(p) = (2\pi)^{-1}\int_{-\pi}^{+\pi} d\theta F(pr_\theta).$$

Now it is clear that $F-F^{b\#}$ satisfies (22.11), and therefore by our preliminary observation

$$\int_{\underline{\underline{G}}} d\sigma F(\sigma) = \int_{\underline{\underline{G}}} d\sigma F^{b\#}(\sigma)$$

$$= \int_{\underline{\underline{P}}} dp F^b(p),$$

and (vi) follows directly from (iv). ///

From now on we use the symbol $d\sigma$ (or possibly $d\rho$, $d\tau$, etc.) to denote the Haar measure on $\underline{\underline{G}}$ determined in Theorem 22.2(vi). The next theorem gives an important alternative formula for $d\sigma$.

Theorem 22.3. For $f \geq 0$ on $\underline{\underline{G}}$

(22.12)
$$\int_{\underline{\underline{G}}} d\sigma f(\sigma) = \int_{-\pi}^{+\pi} d\theta_1 \int_{-\pi}^{+\pi} d\theta_2 \int_1^\infty a^{-1} da \tfrac{1}{2}\{a^2 - a^{-2}\} f(r_{\theta_1} h_a r_{\theta_2}).$$
///

Our proof of Theorem 22.3, which is essentially the one given in [36, p. 139] depends on some analysis of the adjoint action of $\underline{\underline{G}}$ on the Lie algebra $sl_2(R)$. We introduce the matrices

(22.13)
$$X^+ = \begin{bmatrix} 0 & 1 \\ 0 & 0 \end{bmatrix}; \; W = \begin{bmatrix} 0 & 1 \\ -1 & 0 \end{bmatrix}; \; V = \begin{bmatrix} 0 & 1 \\ 1 & 0 \end{bmatrix}$$
$$H = \begin{bmatrix} 1 & 0 \\ 0 & -1 \end{bmatrix}; \; X^- = \begin{bmatrix} 0 & 0 \\ 1 & 0 \end{bmatrix}$$

which all belong to $sl_2(R)$, and we note the identities

(22.14) $$\mathrm{ad}(h_a)W = h_a W h_a^{-1} = a^2 X^+ - a^{-2} X^-$$

(22.15) $$\mathrm{ad}(n_b)H = n_b H n_b^{-1} = H - 2bX^+,$$

which can be established by direct computation. Also we introduce the subset

(22.16) $$A^+ = \{h_a : a > 1\}.$$

Consider first the mapping S from $\underline{\underline{K}} \times \underline{\underline{A}}^+ \times \underline{\underline{K}}$ to $\underline{\underline{G}}$ given by

(22.17) $$S(r_{\theta_1}, h_a, r_{\theta_2}) = r_{\theta_1} h_a r_{\theta_2}.$$

Any $\sigma \in \underline{\underline{G}}$ has a unique polar decomposition $\sigma = r_\theta \rho$ with ρ positive definite. If $\rho \neq 1$, then ρ can be represented $\rho = r_\varphi h_a r_{\varphi'}$ with $a > 1$ in exactly two ways. (If φ works, then so does $\varphi + \pi$.) It follows that except for a null set the map S is surjective and of degree 2. Here and below we identify a matrix $X \in sl_2(R)$ with the left invariant vector field on $\underline{\underline{G}}$ which corresponds to the differential operator

(22.18) $$Ⓓ_X f(\sigma) = (d/dt) f(\sigma e^{tX})\big|_{t=0}.$$

By the usual conventions the mapping S induces a mapping of tangent spaces at corresponding points which we denote simply by S_*. Let w_1, h, w_2 denote the vector fields on $\underline{\underline{K}} \times \underline{\underline{A}}_+ \times \underline{\underline{K}}$ which are just the partial derivatives with respect to the obvious product coordinate system. Also let h, x^+, w be the corresponding vector fields on $\underline{\underline{A}} \times \underline{\underline{N}} \times \underline{\underline{K}}$, and let T be the mapping into $\underline{\underline{G}}$ determined by (22.6). Then it is clear from the classical change of variables formula for multiple integrals that the left side of (22.12)

$$= \tfrac{1}{2}\int_{-\pi}^{+\pi} d\theta_1 \int_{-\pi}^{+\pi} d\theta_2 \int_1^\infty a^{-1} da \,|\omega(S_* w_1, S_* h, S_* w_2)/\omega(T_* h, T_* x^+, T_* w)|\, f(r_{\theta_1} h_a r_{\theta_2}).$$

Here ω is any nonzero 3-form on $\underline{\underline{G}}$. The ratio

$$(22.19) \qquad |\omega(S_* w_1, S_* h, S_* w_2)/\omega(T_* h, T_* x^+, T_* w)|$$

is pointwise independent of the choice of ω and can be computed by looking at determinants in a local coordinate system. Taking into account the identification (22.18), we have

$$S_* w_1 = \mathrm{ad}(r_{\theta_2}^{-1})\mathrm{ad}(h_a^{-1})W = \mathrm{ad}(r_{\theta_2}^{-1})\{a^{-2}X^+ - a^2 X^-\}$$

$$S_* h = \mathrm{ad}(r_{\theta_2}^{-1})H$$

$$S_* w_2 = W = \mathrm{ad}(r_{\theta_2}^{-1})W.$$

Since 1 is the only nonnegative character on $\underline{\underline{T}}$, we can ignore the operator $\mathrm{ad}(r_{\theta_2}^{-1})$ and compute

$$|\omega(S_* w_1, S_* h, S_* w_2)| = |a^{-2}\omega(X^+, H, -X^-) - a^2\omega(X^-, H, X^+)|$$

$$= \{a^2 - a^{-2}\}|\omega(X^+, H, X^-)|.$$

Similarly

$$T_* h = \mathrm{ad}(r_\theta^{-1})\mathrm{ad}(n_b^{-1})H = \mathrm{ad}(r_\theta^{-1})\{H + 2bX^+\}$$

$$T_* x^+ = \mathrm{ad}(r_\theta^{-1})X^+$$

$$T_* w = W = \mathrm{ad}(r_\theta^{-1})W,$$

and therefore

$$|\omega(T_* h, T_* x^+, T_* w)| = |\omega(X^+, H, X^-)|,$$

which completes the proof. ///

Notice that (22.12) can also be written

(22.12') $$\int_{\underline{\underline{G}}} d\sigma f(\sigma) = \tfrac{1}{2}\int_{-\pi}^{+\pi} d\theta_1 \int_{-\pi}^{+\pi} d\theta_2 \int_0^{\infty} dt\{\sinh t\} f(r_{\theta_1} e^{\frac{1}{2}tH} r_{\theta_2}).$$

Now let $\underline{\underline{X}}$ be the open upper half plane $\underline{\underline{R}}^{2,+}$ as in Section 15. It will be convenient here to view points in $\underline{\underline{X}}$ as complex numbers, so that also

(22.20) $$\underline{\underline{X}} = \{z : \mathrm{Im}\, z > 0\}.$$

For $\sigma \in \underline{\underline{G}}$ we define the transformation

(22.21) $$T_\sigma z = \{\underline{\alpha}(\sigma)z+\underline{\beta}(\sigma)\}/\{\gamma(\sigma)z+\underline{\delta}(\sigma)\}.$$

Of course each T_σ is a diffeomorphism on $\underline{\underline{X}}$ and

(22.22) $$T_\sigma T_\tau = T_{\sigma\tau}.$$

It is easy to check that

$$T_\sigma i = i$$

if and only if $\sigma \in \underline{\underline{K}}$. Therefore we can identify $\underline{\underline{X}}$ with the coset space $\underline{\underline{G}}/\underline{\underline{K}}$ according to the usual conventions (see for example [27, p. 114]), and the canonical projection of $\underline{\underline{G}}$ onto $\underline{\underline{X}}$ is given by

(22.23) $$\Pi\sigma = T_\sigma i.$$

It is convenient now to introduce functions $\underline{x}$, $\underline{y}$ on $\underline{\underline{G}}$ by the formula

(22.24) $$\Pi\sigma = \underline{x}+i\underline{y},$$

and then it is easy to check that

(22.25) $$\underline{x} = \underline{b}^{\#} \;;\; \underline{y} = \underline{a}^2$$

and therefore $(\underline{x}, \underline{y}, \underline{\theta})$ is a global coordinate system for $\underline{\underline{G}}$. The restriction of T to the subgroup $\underline{\underline{P}}$ corresponds to the left action of $\underline{\underline{P}}$ on itself, and it follows from (22.25) and Theorem 22.2(iv) that the measure $d_i z$ determined on $\underline{\underline{X}}$ by

$$\int_{\underline{\underline{X}}} d_i z f(z) = \tfrac{1}{2}\int_{-\infty}^{+\infty} dx \int_0^{\infty} dy\, y^{-2} f(x+iy) \tag{22.26}$$

satisfies

$$\int_{\underline{\underline{X}}} d_i z f(T_\sigma z) = \int_{\underline{\underline{X}}} d_i z f(z) \tag{22.27}$$

whenever $\sigma \in \underline{\underline{P}}$ and is uniquely determined up to a constant multiple by this. In fact (22.27) is valid for general $\sigma \in \underline{\underline{G}}$. This can be established either by checking directly for $\sigma \in \underline{\underline{K}}$ or by applying the criterion of [27, p. 368].

We say that a function F on $\underline{\underline{G}}$ is <u>right invariant</u> if

$$F(\sigma r_\theta) = F(\sigma) \tag{22.28}$$

and <u>left invariant</u> if

$$F(r_\theta \sigma) = F(\sigma) \tag{22.28'}$$

for all $\theta \in \underline{\underline{T}}$ and $\sigma \in \underline{\underline{G}}$. If f is defined on $\underline{\underline{X}}$, then the pull back

$$\Pi^* f(\sigma) = f(\Pi\sigma) \tag{22.29}$$

is right invariant on $\underline{\underline{G}}$. Conversely if F is right invariant on $\underline{\underline{G}}$, then there exists a unique function $\Pi^{*-1}F$ on $\underline{\underline{X}}$ such that

$$F = \Pi^* \Pi^{*-1} F. \tag{22.30}$$

Thus any function space on $\underline{\underline{X}}$ can be identified with a right invariant function space on $\underline{\underline{G}}$. We say that F on $\underline{\underline{G}}$ is bi-invariant if it is both right and left invariant and that a Radon measure μ on $\underline{\underline{G}}$ is bi-invariant if for all $\theta \in \underline{\underline{T}}$ and

$F \in C_{com}(\underline{\underline{G}})$

$$\int_{G}\mu(d\sigma)F(r_\theta\sigma) = \int_{\underline{\underline{G}}}\mu(d\sigma)F(\sigma r_\theta) = \int_{\underline{\underline{G}}}\mu(d\sigma)F(\sigma). \tag{22.31}$$

The significance for us of bi-invariant measures can be traced to the following elementary result.

<u>Lemma 22.4.</u> <u>Let</u> P <u>be an operator mapping bounded measurable functions on</u> $\underline{\underline{X}}$ <u>into bounded measurable functions and satisfying the following three conditions.</u>

(i) $0 \le Pf \le 1$ <u>whenever</u> $0 \le f \le 1$.

(ii) <u>If</u> $f_n \to f$ <u>boundedly, then</u> $Pf_n \to Pf$ <u>boundedly</u>.

(iii) P <u>commutes with the translation operators</u>

$$Ⓛ_\sigma f(z) = f(T_\sigma z). \tag{22.32}$$

<u>Then there exists a unique bi-invariant subprobability</u> p <u>on</u> $\underline{\underline{G}}$ <u>such that for general</u> $\sigma \in \underline{\underline{G}}$

$$Pf(\prod\sigma) = \int_{\underline{\underline{G}}}p(d\tau)f(\prod\sigma\tau). \tag{22.33}$$ ///

<u>Proof.</u> Note first that there exists a unique subprobability p_0 on $\underline{\underline{X}}$ such that

$$Pf(i) = \int_{\underline{\underline{X}}}p_0(dz)f(z)$$

for $f \in C_{com}(\underline{\underline{X}})$ and therefore for bounded measurable f and that p_0 determines a unique right invariant subprobability p on $\underline{\underline{G}}$ such that

$$Pf(i) = \int_{\underline{\underline{G}}}p(d\tau)f(\prod\tau).$$

For general $\sigma \in \underline{\underline{G}}$ we have

$$Pf(\prod\sigma) = ⓛ_\sigma Pf(i) = Pⓛ_\sigma f(i)$$

$$= \int_{\underline{\underline{G}}} p(d\tau) ⓛ_\sigma f(\prod\tau)$$

which implies (22.33), and then the special case $\sigma \in \underline{\underline{K}}$ guarantees that p is left invariant and therefore bi-invariant. ///

The elements σ, σ^{-1} always belong to the same double coset of $\underline{\underline{G}}$ relative to $\underline{\underline{K}}$. To see this observe that if σ has a polar decomposition $\sigma = k\rho$, then $\rho = k_1\rho^{-1}k_1^{-1}$ since ρ, ρ^{-1} have the same eigenvalues and therefore $\sigma = kk_1\rho^{-1}k_1^{-1} = kk_1\sigma^{-1}kk_1^{-1}$. This implies that every bi-invariant function and, more important, every bi-invariant measure on $\underline{\underline{X}}$ is even. It follows that if P can be represented (22.33) with p a bounded bi-invariant measure on $\underline{\underline{G}}$, then also P can be viewed as a bounded symmetric operator on $L^2(\underline{\underline{X}}, d_i z)$. Thus it is reasonable to supplement Lemma 22.4 with an $L^2(\underline{\underline{X}}, d_i z)$ version.

Lemma 22.5. Let P be a bounded symmetric operator on $L^2(\underline{\underline{X}}, d_i z)$ which commutes with the translation operators (21.32) and which satisfies condition (i) of Lemma 22.4 in the almost everywhere sense. Then the conclusion of Lemma 22.4 is valid. ///

For a proof the main point is to observe that there exists a bounded measure $P(dz, dz')$ on $\underline{\underline{X}} \times \underline{\underline{X}}$ such that for $f, g \in C_{com}(\underline{\underline{X}})$

$$\iint P(dz, dz')f(z)g(z') = \int_{\underline{\underline{X}}} d_i z f(z) Pg(z)$$

and that the boundedness of P on $L^2(\underline{\underline{X}}, d_i z)$ guarantees that this identity extends to general pairs f, g in $L^2(\underline{\underline{X}}, d_i z)$. We omit the remainder of the proof which is a special case of the argument given below in Section 23 when we analyze the Lévy measure $J(dz, dz')$.

The convolution of a function F on $\underline{\underline{G}}$ with a measure p on $\underline{\underline{G}}$ is the function $F*p$ defined by

(22.34) $$F*p(\sigma) = \int_{\underline{\underline{G}}} p(d\tau)F(\sigma\tau)$$

when it converges. The convolution $p*p'$ of two measures p, p' is defined by

(22.35) $$\int_{\underline{\underline{G}}} p*p'(d\sigma)F(\sigma) = \int_{\underline{\underline{G}}} p(d\sigma)\int_{\underline{\underline{G}}} p'(d\tau)F(\sigma\tau).$$

It is easy to check associativity

(22.36) $$F*\{p*p'\} = \{F*p\}*p'$$

(22.36') $$p*\{p'*p''\} = \{p*p'\}*p''.$$

Also it is clear that if p, p' are bi-invariant, then $F*p$ is right-invariant, and $p*p'$ is again bi-invariant. It is much deeper, although not particularly important for us that

(22.37) $$p*p' = p'*p,$$

a result first discovered by I. Gelfand [21] in the context of functions. This can be deduced either from the formula (22.43) derived below or from symmetry since the composition of two symmetric operators is symmetric if and only if they commute. For a proof in a more general context we refer to p. 53 in Lang's book [36].

In general, the definitions (22.34) and (22.35) are inconsistent in that if p has density p_0, then the density of $p*p'$ is given not by $p_0*p'(\sigma)$, but instead by $\int p'(d\tau)p_0(\sigma\tau^{-1})$. For bi-invariant measures this inconsistency disappears because of symmetry.

Now we look more closely at (22.34). Suppose first that p is absolutely continuous with density p_0. Then by (22.12') we have

$$Pf(i) = \tfrac{1}{2}\int_0^\infty dt \sinh t\, p_0(e^{\frac{1}{2}tH})\int_{-\pi}^{+\pi} d\theta f(\Pi r_\theta e^{\frac{1}{2}tH})$$

$$= \pi\int_0^\infty dt \sinh t\, p_0(e^{\frac{1}{2}tH}) \text{Ⓐ}_t f$$

where

$$\text{(22.38)} \qquad \text{Ⓐ}_t f = (2\pi)^{-1}\int_{-\pi}^{+\pi} d\theta f(\ r_\theta e^{\frac{1}{2}tH}).$$

More generally for $\sigma \in \underline{\underline{G}}$

$$\text{(22.39)} \qquad Pf(\Pi\sigma) = \pi\int_0^\infty dt \sinh t\, p_0(e^{\frac{1}{2}tH}) \text{Ⓐ}_t \text{Ⓛ}_\sigma f.$$

This together with a routine passage to the limit in p establishes

Theorem 22.6. *Let P be a submarkovian operator which can be represented (22.33) with p bi-invariant. Then there exists a unique subprobability $p^b(dt)$ on $[0,\infty)$ such that*

$$\text{(22.40)} \qquad Pf(\Pi\sigma) = \int_{[0,\infty)} p^b(dt) \text{Ⓐ}_t \text{Ⓛ}_\sigma f$$

where Ⓐ$_t$ is the averaging operator (22.38). If p is absolutely continuous with density $p_0(\sigma)$, then

$$\text{(22.41)} \qquad p^b(dt) = \pi \sinh t\, p_0(e^{\frac{1}{2}tH})dt. \qquad ///$$

In fact the averaging operator Ⓐ$_t$ is just the Poisson integral of f evaluated at i and taken over the unique circle passing through the points $e^t i, e^{-t} i$ and symmetric about the y-axis. To see this consider

$$\text{(22.42)} \qquad Uz = (z-i)(z+i)^{-1}$$

which maps $\underline{\underline{X}}$ onto the open unit disc

$$\underline{\underline{D}} = \{\zeta : |\zeta| < 1\}$$

and takes i into 0, together with its inverse

$$U^{-1}\zeta = (\zeta+1)(i\zeta - i)^{-1}.$$

Also define $U^{\#}f$ on $\underline{\underline{D}}$ by

$$U^{\#}f(Uz) = f(z).$$

Then

$$Ⓐ_t f = (2\pi)^{-1}\int_{-\pi}^{+\pi} d\theta U^{\#}f(UT_{r_\theta} U^{-1}Ue^t i)$$

$$= (2\pi)^{-1}\int_{-\pi}^{+\pi} d\theta U^{\#}f(e^{2i\theta}\tanh\tfrac{1}{2}t)$$

is certainly the Poisson integral of $U^{\#}f$ evaluated at 0 for the circle of radius $\tanh\frac{1}{2}t$ centered at the origin. Our assertion follows since the conformal mapping U must preserve Poisson integrals. This has the immediate corollary

Theorem 22.7. Let P be represented by (21.33) with p a bi-invariant probability and let f be a bounded or nonnegative harmonic function on $\underline{\underline{X}}$. Then Pf = f. ///

The converse to Theorem 22.7 follows from the work of H. Furstenberg [19, Chapter III]. We will present a proof in Section 24 which is a specialization of the one given by Furstenberg.

Finally we look at the operation (22.35) in the special case when p, p' are bi-invariant. The basic result is

Theorem 22.8. Let p, p' be bounded bi-invariant measures on $\underline{\underline{G}}$ and let p^b, $p^{b'}$ be the measures on $[0, \infty)$ determined by (22.40). Then for $f \geq 0$ on $\underline{\underline{X}}$

(22.43) $$\int \{p*p'\}(d\sigma)f(\Pi\sigma) = \int p^{b}(ds)\int p^{b'}(dt)(2\pi)^{-1}\int_{-\pi}^{+\pi} d\theta \, Ⓐ_{r(s,t,\theta)} f$$

where $r(s,t,\theta)$ *is determined by the equation*

(22.44) $$\cosh r(s,t,\theta) = \cos^{2}\theta \cosh(s+t) + \sin^{2}\theta \cosh(s-t). \qquad ///$$

Proof. It's clear that the left side of (22.43)

$$= \int p^{b}(ds)\int p^{b'}(dt)(2\pi)^{-2}\int_{-\pi}^{+\pi} d\varphi \int_{-\pi}^{+\pi} d\theta f(\Pi[r_{\varphi} e^{\frac{1}{2}sH} r_{\theta} e^{\frac{1}{2}tH}).$$

The function $\frac{1}{2}\operatorname{trace}\sigma\sigma^{t}$ parametrizes the double cosets of $\underline{\underline{G}}$ with respect to $\underline{\underline{K}}$, and therefore we need only check that

$$\tfrac{1}{2}\operatorname{tr}(e^{\frac{1}{2}rH})(e^{\frac{1}{2}rH})^{t} = \cosh r$$

$$\tfrac{1}{2}\operatorname{tr}(e^{\frac{1}{2}sH} r_{\theta} e^{\frac{1}{2}tH})(e^{\frac{1}{2}sH} r_{\theta} e^{\frac{1}{2}tH})^{t} = \cos^{2}\theta\cosh(s+t) + \sin^{2}\theta\sinh(s-t). \qquad ///$$

This has the direct corollary

Corollary 22.9. *If* p, p' *are bounded bi-invariant measures on* $\underline{\underline{G}}$, *then their convolution* $p*p'$ *is absolutely continuous.*

Since the convolution of two square integrable functions on $\underline{\underline{G}}$ is continuous, and since the convolution of any nontrivial even function with itself is strictly positive at the identity, we have in addition

Corollary 22.10. *If* p *is a bounded bi-invariant measure, then its fourth power* p^{4} *is absolutely continuous with density bounded away from zero in a neighborhood of* 1. ///

The last corollary will be needed in Section 24.

23. Invariant Semigroups.

Let P_t, $t > 0$, be a submarkovian semigroup on $\underline{\underline{X}}$ which commutes with the action of $\underline{\underline{G}}$ on $\underline{\underline{X}}$. It follows from Lemma 22.4 and 22.5 that there is no loss of generality in viewing the P_t as bounded symmetric operators on $L^2(\underline{\underline{X}}, d_i z)$ which can be represented

$$P_t f(\Pi\sigma) = \{\Pi^* f\} * p_t(\sigma) \tag{23.1}$$

where each p_t is a bi-invariant subprobability on $\underline{\underline{G}}$. It is easy to see that $P_t 1 = e^{-tq} 1$ with $q \geq 0$, and so again there is no real loss of generality in assuming that each p_t is a probability and therefore each P_t is strictly Markov. We assume further that the semigroup P_t, $t > 0$, is strongly continuous on $L^2(\underline{\underline{X}}, d_i z)$. This is equivalent to the regularity condition imposed in [28, Sect. 10], that the p_t form a vaguely continuous semigroup on $\underline{\underline{G}}$. We establish here a complete classification of the possibilities. This was derived in a more general context first by G. A. Hunt [28, Sections 8, 9, 10] and then by R. Getoor [22] and R. Gangolli [20] where in addition some deeper results are established. Our approach is closer in spirit to [28] than to [22] or [20] since we do not use the Plancherel formula. The main difference with [28] is that we rely on Dirichlet space techniques and in particular we apply a slight extension of the theorem of Beurling and Deny (stated as Theorem 16.1 in this volume).

Let G_λ, $\lambda > 0$, be the resolvent operators, and let $(\underline{\underline{F}}, E)$ be the Dirichlet space associated with the semigroup P_t, $t > 0$. Our first result is

Lemma 23.1. The Dirichlet space $(\underline{\underline{F}}, E)$ is regular. ///

Proof. It is clear from (23.1) that the operators P_t preserve $C_0(\underline{\underline{X}})$, the Banach space of continuous functions vanishing at infinity. Therefore it suffices to observe that the set of functions $G_1 g$ with g running over $C_{com}(\underline{\underline{X}})$ is E_1 dense in $\underline{\underline{F}}$ and uniformly dense in $C_0(\underline{\underline{X}})$. ///

The right invariant differential operators

(23.2) $$Ⓓ'_X F(\sigma) = (d/dt)f(e^{tX}\sigma)|_{t=0}$$

can be viewed as acting on functions f on $\underline{\underline{X}}$ via

(23.2') $$Ⓓ'_X f(\Pi\sigma) = (d/dt)f(\Pi e^{tX}\sigma)|_{t=0}.$$

Imitating Hunt's technique in [28] we introduce C_1', the Banach space of functions f on $\underline{\underline{X}}$ such that

$$f, Ⓓ'_H f, Ⓓ'_{X^+} f$$

all exist and belong to $C_0(\underline{\underline{X}})$ with norm the sum of the uniform norms for these functions. Since the P_t commute with all of the operators (23.2') they form also a strongly continuous semigroup on C_1', and this is enough to guarantee that $C_1' \cap \underline{\underline{F}}$ is dense in C_1' relative to the C_1' norm. It follows in particular that in a neighborhood of every point in $\underline{\underline{X}}$ there exists a pair $u_1, u_2 \in \underline{\underline{F}}$ which form a local C^1 coordinate system, and this is enough for applying Theorem 16.1. Thus at least for $f \in \underline{\underline{F}} \cap C_1'$

(23.3) $$E(f,f) = D(f,f) + \tfrac{1}{2}J\langle f,f\rangle$$

where the diffusion form D can be represented

(23.4) $$D(f,f) = \Sigma_{i,j=1}^{2} \int \nu_{ij}(dz)\partial_i f(z)\partial_j f(z)$$

with $\partial_i f$ denoting the usual partial derivative. It is clear that (23.4) must be G-invariant, and we show now that this forces

$$D(f,f) = \tfrac{1}{2}\alpha\int_{-\infty}^{+\infty}dx\int_{0}^{\infty}dy\{\text{grad}\, f(x+iy)\}^2 \tag{23.5}$$

with $\alpha \geq 0$. Consider first the two functions

$$f(z) = x\ ;\ g(z) = \log y, \tag{23.6}$$

and observe that in the notation of SMP, Section 11,

$$\langle A_c f\rangle(dz) = \nu_{11}(dz) \tag{23.7}$$

$$\langle A_c g\rangle(dz) = y^{-2}\nu_{22}(dz)$$

$$\langle A_c f, g\rangle(dz) = 2y^{-1}\nu_{12}(dz).$$

Of course x, y are defined on $\underline{\underline{X}}$ by $z = x+iy$. Since

$$\textcircled{L}_{h_a}\, g(z) = \log y + \log a^2$$

$$\textcircled{L}_{n_b}\, g(z) = \log y,$$

it follows that $\langle A_c g\rangle(dz)$ is invariant under the action of the subgroup $\underline{\underline{P}}$ on $\underline{\underline{X}}$ and by uniqueness of left Haar measure on $\underline{\underline{P}}$

$$\nu_{22}(dz) = A_{22}dxdy \tag{23.8.1}$$

with A_{22} constant. Since

$$L_{h_a} f(z) = a^2 x$$

$$L_{n_b} f(z) = x+b,$$

the measure $y^{-2}\nu_{11}(dz)$ is invariant under $\underline{\underline{P}}$, and therefore

$$\nu_{11}(dz) = A_{11}dxdy. \tag{23.8.2}$$

Similarly $y^{-1}\nu_{12}(dz)$ is invariant, and so

(23.8.3) $$\nu_{12}(dz) = A_{12}dxdy.$$

Thus there exists a constant matrix A such that for any $f \in C^1_{com}(\underline{\underline{X}}) \cap \underline{\underline{F}}$

$$D(f,f) = \int_{-\infty}^{+\infty}dx\int_0^{\infty}dy\, \mathrm{grad}\, f(z)^t A\, \mathrm{grad}\, f(z),$$

and we need only show that A must be a multiple of the identity. But this is easily established by considering the action of $\underline{\underline{K}}$ on $\underline{\underline{X}}$. For example, $\mathrm{grad}\, f(i)^t A\, \mathrm{grad}\, f(i)$ is the limit of certain spherical averages of $<A_c f>(dz)$, and this must remain unchanged if we replace f by $Ⓛ_{r_\theta} f$ which gives in the limit $\mathrm{grad}\, f(i)^t T'_{r_\theta}(i)^t A T'_{r_\theta}(i) \mathrm{grad}\, f(i)$, and since $T'_{r_\theta}(i)$ can be a general rotation, it must be that A is a multiple of the identity. This establishes (23.5).

Remark. We cannot be sure at this stage that f, g belong to $\underline{\underline{F}}^{loc}$. But this apparent difficulty can be circumvented either by approximating f, g by $\underline{\underline{F}} \cap C_1'$ in the C_1' norm, or else by applying our results below for $J<f,f>$ which will guarantee that $\underline{\underline{F}}$ contains $C^1_{com}(\underline{\underline{X}})$. ///

We show next that

(23.9) $$J<f,f> = \int_{\underline{\underline{G}}} d\sigma \int_{\underline{\underline{G}}} J(d\tau)\{f(\Pi\sigma) - f(\Pi\sigma\tau)\}^2$$

where J is a unique bi-invariant Radon measure on $\underline{\underline{G}} \backslash \underline{\underline{K}}$ (set theoretic difference) satisfying

(23.10) $$\int_{\underline{\underline{G}}} J(d\sigma)u^2(\Pi\sigma, i)\{1+u^2(\Pi\sigma, i)\}^{-1} < +\infty$$

Here $u(z,z')$ is the $\underline{\underline{G}}$ invariant metric defined on $\underline{\underline{X}}$ by

(23.11) $$u(z,z') = |z-z'|^2/4yy'.$$

We know from general principles that the bilinear form J can be represented

$$J\langle f,f\rangle = \iint_{\underline{\underline{X}}\times\underline{\underline{X}}} J(dz,dz')\{f(z)-f(z')\}^2$$

where $J(dz,dz')$ is a Radon measure on the complement of the diagonal in $\underline{\underline{X}}\times\underline{\underline{X}}$ and is determined by

$$\iint_{\underline{\underline{X}}\times\underline{\underline{X}}} J(dz,dz')f(z)g(z') = -E(f,g)$$

for $f,g \in \underline{\underline{F}}\cap C_{com}(\underline{\underline{X}})$ having disjoint supports. For $\varepsilon > 0$ let

$$(23.12) \qquad J_\varepsilon(dz,dz') = J(dz,dz')I(u(z,z') > \varepsilon),$$

and consider the functional

$$\ell_\varepsilon(f) = \iint_{\underline{\underline{X}}\times\underline{\underline{X}}} J_\varepsilon(dz,dz')f(z)$$

which by regularity converges at least for $f \in C_{com}(\underline{\underline{X}})$. Clearly ℓ_ε commutes with the action of $\underline{\underline{G}}$, and this guarantees that ℓ_ε corresponds to a multiple of the invariant measure $d_i z$ on $\underline{\underline{X}}$, and therefore

$$J_\varepsilon(dz,dz') = d_i z j_\varepsilon(z,dz')$$

where the $j_\varepsilon(z,\cdot)$ are bounded measures on $\underline{\underline{X}}$, all having the same mass and satisfying the usual measurability conditions. Now it is easy to check that the corresponding operator

$$\text{Ⓙ}_\varepsilon g(z) = \int j_\varepsilon(z,dz')g(z')$$

commutes with the action of $\underline{\underline{G}}$, and so (23.9) follows after applying Lemma 22.4 and then passing to the limit $\varepsilon \downarrow 0$. Finally, the integrability condition (23.10) for $u(\pi\sigma,i) \geq 1$ follows directly from regularity and for $u(\pi\sigma,i) < 1$ follows since (23.9) must converge for a pair of functions forming a local C^1 coordinate system.

We have proved the direct part of the following theorem.

Theorem 23.2. Let P_t, $t > 0$, be a strongly continuous symmetric Markovian semigroup on $L^2(\underline{\underline{X}}, d_i z)$ which commutes with the action of $\underline{\underline{G}}$, and let $(\underline{\underline{F}}, E)$ be the associated Dirichlet space. Then $(\underline{\underline{F}}, E)$ is regular, $C^1_{com}(\underline{\underline{X}})$ is dense in $\underline{\underline{F}}$, and for $f \in C^1_{com}(\underline{\underline{X}})$

(23.13)
$$E(f, f) = \tfrac{1}{2}\alpha \iint_{\underline{\underline{X}}} dx dy \{\operatorname{grad} f(x, y)\}^2$$
$$+ \tfrac{1}{2} \int_{\underline{\underline{G}}} d\sigma \int_{\underline{\underline{G}}} J(d\tau)\{f(\pi\sigma) - f(\pi\sigma\tau)\}^2$$

where $\alpha \geq 0$ and where J is a uniquely determined bi-invariant Radon measure on $\underline{\underline{G}}\backslash\underline{\underline{K}}$ satisfying the integrability condition (23.10). ///

The proof of the converse to Theorem 23.2 is routine, and we omit it.

Getoor and Gangolli prove the deeper result that every infinitely divisible bi-invariant probability can be imbedded in a weakly continuous semigroup.

24. A Characterization of Harmonic Functions.

The results established here are a special case of those established by H. Furstenberg in Chapter III in [19]. Our argument follows Furstenberg's except for simplifications made possible by the fact that we are dealing with a special case.

The metric on $\underline{\underline{X}}$ is always understood to be the invariant metric $u(z, z')$ defined by (23.11). Our first result is

Lemma 24.1. *If Q is a neighborhood of 1 in $\underline{\underline{G}}$, then there exists an $\varepsilon > 0$ such that for every $z \in \underline{\underline{X}}$ the set $\{T_\sigma z : \sigma \in Q\}$ contains an ε-neighborhood of z.* ///

The proof which follows is identical with the one given on pp. 358-359 in [19] except that for this special case there is no need to consider the adjoint action of $\underline{\underline{G}}$ on its Lie algebra $sl(2, \underline{\underline{R}})$. Also we avoid the error in formula (2) on p. 359.

Proof. For $\delta > 0$ let Ω be the set of symmetric matrices in $sl(2, \underline{\underline{R}})$ with both eigenvalues less than δ in absolute value, and let $\Omega^{\cdot}$ be its boundary. It suffices to show that the images $\Pi \sigma^{-1} e^{\Omega} \sigma$ all cover a fixed neighborhood of i. For this it suffices to show first that for each $\sigma \in \underline{\underline{G}}$ the mapping $\psi_\sigma(\tau) = \Pi \sigma^{-1} \tau \sigma$ is nonsingular on the submanifold of symmetric matrices in $\underline{\underline{G}}$ and second that all the points in the image of the boundary $\psi_\sigma(e^{\Omega^{\cdot}})$ are a uniform distance away from i. For the purposes of this proof only we identify the homogeneous space $\underline{\underline{G}}/\underline{\underline{K}}$ not with $\underline{\underline{X}}$, but rather with the set $\underline{\underline{R}}$ of positive definite 2×2 matrices of determined 1 via the mapping $\Pi_1 \rho = \rho \rho^t$. Thus instead of ψ_σ we work with the mapping

$$\varphi_\sigma(\tau) = \sigma^{-1} \tau \sigma \sigma^t \tau^t \sigma^{-1t}.$$

Nonsingularity will follow if we can show that

$$(24.1)\qquad (d/dt)\varphi_\sigma(e^{X_0+tX})|_{t=0} \neq 0$$

whenever X_0, X are symmetric matrices in $sl(2,\underline{\underline{R}})$ with $X \neq 0$. Clearly

$$(24.2)\qquad Y = (d/dt)e^{X_0+tX}|_{t=0}$$

is symmetric and (24.1)

$$= \sigma^{-1}Y\sigma\sigma^t e^{X_0}\sigma^{-1t} + \sigma^{-1}e^{X_0}\sigma\sigma^t Y\sigma^{-1t}$$

$$= AB+BA$$

where

$$B = \sigma^{-1}e^{X_0}\sigma\sigma^t e^{X_0}\sigma^{-1t}$$

belongs to $\underline{\underline{R}}$ and where

$$A = \sigma^{-1}Ye^{-X_0}\sigma = \sigma^{-1}e^{\frac{1}{2}X_0}\{e^{-\frac{1}{2}X_0}Ye^{-\frac{1}{2}X_0}\}e^{-\frac{1}{2}X_0}\sigma$$

is equivalent to a symmetric matrix. Also

$$AB = B^{-\frac{1}{2}}\{B^{\frac{1}{2}}AB^{\frac{1}{2}}\}B^{\frac{1}{2}}$$

is equivalent to a symmetric matrix. If (24.1) fails, then AB is antisymmetric, which is impossible unless $A = 0$ and therefore $X = 0$. This establishes nonsingularity of φ_σ. If X belongs to the boundary Ω', then it must have eigenvalues $\pm\delta$, and therefore e^X has eigenvalues $e^{\pm\delta}$. If v is the eigenvector corresponding to e^δ, then

$$\{v^t\sigma\}\varphi_\sigma(e^X)\{\sigma^t v\} = e^{2\delta}v^t\sigma\sigma^t v,$$

and therefore $\varphi_\sigma(e^X)$ has one eigenvector at least as large as $e^{2\delta}$ which suffices to complete the proof. ///

It is clear that if $u(z, z') < 2\varepsilon$, then there exists at least one $z'' \in \underline{\underline{X}}$ such that $u(z, z'') < \varepsilon$ and $u(z', z') < \varepsilon$. Therefore an obvious induction argument establishes

Corollary 24.2. *Let Q be an arbitrary open set in $\underline{\underline{G}}$, and let $z, z' \in \underline{\underline{X}}$. Then there exists a positive integer n depending only on the distance $u(z, z'')$ such that $T_{Q^n} z \cap T_{Q^n} z' \neq \phi$.* ///

Before stating the last preliminary result we introduce some notations which are somewhat looser than what we have used up to this point. For $\sigma \in \underline{\underline{G}}$ and $z \in \underline{\underline{X}}$ we write σz in place of $T_\sigma z$. For μ a probability on $\underline{\underline{G}}$, and for $\sigma \in \underline{\underline{G}}$, $z \in \underline{\underline{X}}$ we let $\mu * \sigma$ be the convolution of μ with point mass concentrated at σ, and we let $\mu * z$ be the probability on $\underline{\underline{X}}$ determined by

$$\int_{\underline{\underline{X}}} \mu * z(dz') f(z') = \int_{\underline{\underline{G}}} \mu(d\sigma) f(\sigma z).$$

Lemma 24.3. *Let μ be a bi-invariant probability on $\underline{\underline{G}}$, and let $c > 0$. Then there exists $\varepsilon > 0$ and an integer n such that if $z_1, z_2 \in \underline{\underline{X}}$ and $u(z_1, z_2) < c$, then*

$$\int_{\underline{\underline{X}}} \{\mu^n * z_1\} \wedge \{\mu^n * z_2\}(dw) \geq \varepsilon. \tag{24.3}$$

Proof. By Corollary 22.9 we can assume that μ is absolutely continuous with density bounded away from zero on a neighborhood Q_1 of 1. If Q_2 is a second neighborhood of 1 with compact closure contained in Q_1, then for every n the convolution product μ^n has density bounded away from 0 on Q_2^n. Choose ℓ

depending only on c such that $Q_2^\ell z_1 \cap Q_2^\ell z_2 \neq \phi$, and take $n = \ell+1$. Let π' be the restriction of Haar measure on $\underline{\underline{G}}$ to Q_2, and let π be the restriction to Q_2^n. We will be done if we can show that πz_1 and πz_2 dominate the same translate of π'. But if $\sigma_1, \sigma_2 \in Q_2^\ell$ are such that $\sigma_1 z_1 = \sigma_2 z_2$, then $\pi' \sigma_1$ and $\pi' \sigma_2$ are both dominated by π, and therefore $\pi' \sigma_1 z_1 = \pi' \sigma_2 z_2$ is one such translate. ///

The next theorem is the main result in this section although afterwards we will reformulate it to get a statement which is suitable for application in Section 25.

Theorem 24.4. *Let μ be a bi-invariant probability on $\underline{\underline{G}}$, and let f be a bounded measurable function on $\underline{\underline{X}}$ such that for all $z \in \underline{\underline{X}}$*

(24.4) $$f(z) = \int_{\underline{\underline{G}}} \mu(d\sigma) f(\sigma z).$$

Then f is a constant function. ///

Proof. Let X_n, $n \geq 1$, be a sequence of mutually independent random variables taking values in the group $\underline{\underline{G}}$ and each having μ as its distribution, and let W_n, $n \geq 1$, denote the partial products $W_n = X_n X_{n-1} \ldots X_1$. The condition (24.4) guarantees that for each $z \in \underline{\underline{X}}$ the sequence $f(W_n z)$, $n \geq 1$, is a martingale. Thus for fixed z the limit

(24.5) $$\psi(z) = \operatorname{Lim}_{n \uparrow \infty} f(W_n z)$$

exists both almost everywhere and in the mean, and

$$f(z) = Ⓔ\psi(z).$$

To prove the theorem it suffices to show that for every $\delta > 0$ and for every pair z, z' in $\underline{\underline{X}}$ the set of sample paths for which

$$(24.6) \qquad |f(W_N z)-f(W_N z')| > 2\delta$$

for infinitely many N has probability zero. Choose n and $\varepsilon > 0$ such that (24.3) is valid whenever $u(z_1, z_2) \leq u(z, z')$. In particular (24.3) is valid for $z_1 = W_N z$ and $z_2 = W_N z'$ so that (24.6) implies at least one of the inequalities

$$(24.7) \qquad \mu^n * W_N z\{w : |f(w)-f(W_N z)| > \delta\} \geq \tfrac{1}{2}\varepsilon$$

$$\mu^n * W_N z'\{w : |f(w)-f(W_N z')| > \delta\} \geq \tfrac{1}{2}\varepsilon.$$

Now consider the function g defined on $\underline{\underline{X}}$ by

$$g(v) = E\{f(W_n v)-f(v)\}^2,$$

and let $\textcircled{F}_N$ denote the σ-algebra generated by the variables $X_1, \ldots, X_N$. The estimate

$$g(W_N z) = \textcircled{F}_N\{f(W_{N+n} z)-f(W_N z)\}^2$$

$$\leq \textcircled{F}_N\{\psi(z)-f(W_N z)\}^2$$

$$= \textcircled{F}_N \psi^2(z)-\{f(W_N z)\}^2$$

together with (24.5) guarantees that with probability one $g(W_N z) \to 0$. Similarly $g(W_N z') \to 0$. It follows from Tchelychev's inequality that with probability one, both inequalities (24.7) are violated for sufficiently large N, and therefore (24.6) can occur for at most finitely many N. ///

After reformulating Theorem 24.4 in terms of right invariant functions on $\underline{\underline{G}}$ and then replacing right cosets by left cosets, we get

Corollary 24.5. Let μ be a bi-invariant probability on $\underline{\underline{G}}$, and let F be a bounded left invariant measurable function on $\underline{\underline{G}}$ such that for all $\sigma \in \underline{\underline{G}}$

$$F(\sigma) = \int_{\underline{\underline{G}}} \mu(d\tau) F(\sigma\tau).$$

Then F is a constant function. ///

Now it is a relatively easy matter to prove

Theorem 24.6. Let μ be a bi-invariant probability on $\underline{\underline{G}}$, and let f be a bounded measurable function on $\underline{\underline{X}}$ such that for all $\sigma \in \underline{\underline{G}}$

$$f(\Pi\sigma) = \int_{\underline{\underline{G}}} \mu(d\tau) f(\Pi\sigma\tau). \tag{24.8}$$

Then f is harmonic on $\underline{\underline{X}}$. ///

Proof. Fix $\sigma \in \underline{\underline{G}}$, and define F on $\underline{\underline{G}}$ by

$$F(\tau) = (2\pi)^{-1} \int_{-\pi}^{+\pi} d\theta f(\Pi\sigma r_\theta \tau).$$

Then F satisfies the hypotheses of Corollary 24.5 and therefore must be constant. Thus

$$f(\pi\sigma) = (2\pi)^{-1} \int_{-\pi}^{+\pi} d\theta f(\Pi\sigma r_\theta \tau)$$

for any $\tau \in \underline{\underline{G}}$, and this guarantees that (24.8) is valid for every bi-invariant probability on $\underline{\underline{G}}$. The special case when μ is absolutely continuous with continuous density having compact support guarantees in turn that f is continuous. Now fix an open disc D with closure contained in $\underline{\underline{X}}$, and let $f^\sim$ be the unique harmonic function in D having the same boundary values as f. Then (24.8) is valid for $f - f^\sim$ whenever $\Pi\sigma \in D$ and μ has sufficiently small support (depending on $\Pi\sigma$). This is enough to guarantee that $f - f^\sim$ cannot have an extremum in D without being constant, and so the only possibility is $f = f^\sim$ on D. ///

25. Invariant Dirichlet Spaces.

Let $m(z)$ be an everywhere positive integrable function on $\underline{\underline{X}}$. In this section we construct a class of Dirichlet spaces $(\underline{\underline{F}}^{\sim}, E^{\sim})$ on $L^2(\underline{\underline{X}}, m \cdot d_i z)$ which are invariant in the following sense. If $f \in \underline{\underline{F}}_b^{\sim}$ and if $\sigma \in \underline{\underline{G}}$, then also the translated function $Ⓛ_\sigma f \in \underline{\underline{F}}^{\sim}$ and

$$(25.1) \qquad E^{\sim}(Ⓛ_\sigma f, Ⓛ_\sigma f) = E^{\sim}(f, f).$$

In general $(\underline{\underline{F}}^{\sim}, E^{\sim})$ is not regular on $\underline{\underline{X}}$, and therefore it cannot be transformed via random time change into a Diriclet space on $L^2(\underline{\underline{X}}, d_i z)$ which is associated with a $\underline{\underline{G}}$-invariant semigroup as in Section 23.

We fix once and for all one of the Dirichlet space $(\underline{\underline{F}}, E)$ described in Theorem 23.2. It follows directly from Theorem 24.6 that a bounded function is harmonic for $(\underline{\underline{F}}, E)$ if and only if it is harmonic in the classicial sense. Thus $(\underline{\underline{F}}, E)$ is transient, and its terminal boundary can be identified with the real line $\underline{\underline{R}}$, adjoined to $\underline{\underline{X}}$ in the usual way as a boundary. The transformations T_σ, $\sigma \in \underline{\underline{G}}$, are extended to $\underline{\underline{X}} \cup \underline{\underline{R}}$ in the usual way. The operator H is defined by the Poisson kernel for the half plane exactly as in Section 15. The excursion form defined on functions φ on $\underline{\underline{R}}$ by

$$(25.2) \qquad N(\varphi, \varphi) = E(H\varphi, H\varphi)$$

must be $\underline{\underline{G}}$-invariant. For the special case when $J = 0$ we can apply the results in Section 15 to conclude that

$$(25.3) \qquad N(\varphi, \varphi) = c \int_{-\infty}^{+\infty} ds \int_{-\infty}^{+\infty} dt (s-t)^{-2} \{\varphi(s) - \varphi(t)\}^2$$

where c is the constant

$$(25.4) \qquad c = (4\pi)^{-1} \alpha.$$

In fact every $\underline{\underline{G}}$-invariant Dirichlet form on $\underline{\underline{R}}$ must be of the form (25.3), and therefore this is the only possibility in general. To prove this directly and at the same time determine the constant c, we consider for $t > 0$ the expression

$$\int_{\underline{\underline{G}}} d\sigma (2\pi)^{-1} \int_{-\pi}^{+\pi} d\theta \{f(\Pi\sigma r_\theta e^{\frac{1}{2}tH}) - f(\Pi\sigma)\}^2 \tag{25.5}$$

where f is a bounded harmonic function on $\underline{\underline{X}}$. Since the interior integral is just the Poisson integral evaluated at $z = \Pi\sigma$ for the circle determined by the condition $u(x+iy, z) = u(ie^t, i)$, a special case of formula (11.15) in SMP gives

$$(2\pi)^{-1} \int_{-\pi}^{+\pi} d\theta \{f(\Pi\sigma r_\theta e^{\frac{1}{2}tH}) - f(z)\}^2 \tag{25.6}$$

$$= \tfrac{1}{2} \int\int dx dy I[u(x+iy, z) < u(e^t i, i)] N^{z,t}(z, x+iy) \{\mathrm{grad}\, f(x+iy)\}^2$$

where $N^{z,t}$ is the classical Green function for the (x, y) domain determined by the interior condition. Thus (25.5) can also be written as

$$\tfrac{1}{2} \int_{-\infty}^{+\infty} dx \int_0^\infty dy \{\mathrm{grad}\, f(x+iy)\}^2 \int_{\underline{\underline{X}}} d_i z I[u(x+iy, z) \leq u(e^t i, i)] N^{z,t}(z, x+iy). \tag{25.5'}$$

But using the invariant properties of $d_i z$ and also of the Green function, it is easy to see that the interior integral is independent of $x+iy$, and indeed it is given by

$$\beta_t = (2\pi) \int_0^{\tanh\frac{1}{2}t} dr\, r\, 2(1-r^2)^{-2} \pi^{-1} \log\{\tanh\tfrac{1}{2}t / r\} \tag{25.7}$$

$$= 4 \int_0^{\tanh\frac{1}{2}t} dr\, r (1-r^2)^{-2} \log\{\tanh\tfrac{1}{2}t / r\}.$$

To verify (25.7) it suffices to consider the special case i and to observe that (22.42) maps the circle in question into the circle of radius $\tanh\frac{1}{2}t$, that $d_i z$ maps into $2(1-r^2) r dr d\theta$, and that $\pi^{-1} \log\{\tanh\frac{1}{2}t, r\}$ is the correct Green function. It follows that in the general case (25.3) is valid with the constant c given by

$$c = (4\pi)^{-1} \{\alpha + \int_0^\infty J_0(dt) \beta_t\} \tag{25.4'}$$

where β_t is given by (25.7) and where J_0 is determined by the formula

(25.8) $$\int_{\underline{\underline{G}}} J(d\sigma)f(\Pi\sigma) = \int_0^\infty J_0(dt)(2\pi)^{-1}\int_{-\pi}^{+\pi} d\theta f(\Pi\sigma r_\theta e^{\frac{1}{2}tH}).$$

The integrability condition (23.10) which is required by Theorem 23.2 is equivalent to

(23.10') $$\int_0^1 J_0(dt)t^2 + \int_1^\infty J_0(dt) < +\infty.$$

The constant β_t is asymptotically equivalent to t^2 as $t\downarrow 0$, but it becomes unbounded as $t\uparrow\infty$. Thus (23.10') does not imply convergence in (25.4'). If we put $u = \tanh\frac{1}{2}t$, then as $t\uparrow\infty$ the constant β_t is asymptotically equivalent to

$$\int_0^u dr(1-r)^{-2}\log(u/r) \sim \int_0^u dr(1-r)^{-2}(u-r)$$

$$= (u-1)\int_0^u dr(1-r)^{-2} + \int_0^u dr(1-r)^{-1}$$

$$\sim -\log(1-u),$$

and it follows that if (23.10') is satisfied, then convergence in (25.4') is equivalent to

(25.9) $$-\int_1^\infty J_0(dt)\log\{1-\tanh\tfrac{1}{2}t\} < +\infty.$$

We summarize in

Theorem 25.1. *If the integrability condition (25.9) is satisfied, then the excursion form is given by (25.3) with c determined by (25.4'). A function φ belongs to the extended excursion space $\underline{\underline{N}}_{(e)}$ if and only if (25.3) converges. If the integrability condition (25.9) fails, then the excursion space $\underline{\underline{N}}$ is the one-dimensional subspace of constants on $\underline{\underline{R}}$ and the excursion form vanishes.* ///

Of course (25.9) is very important for the classification of extensions of

$(\underline{\underline{F}}, E)$, whether or not they are $\underline{\underline{G}}$ invariant (in the restricted sense described at the beginning of this section). It is always true that if f belongs to the reflected space $\underline{\underline{F}}^{ref}$, then the boundary function γf exists in an appropriate L^1 sense (see Theorem 15.1) and belongs to the excursion space $\underline{\underline{N}}$ of Section 15. If (25.9) fails, then γf must be a constant $\gamma f(\infty)$. The only possibility then is that $\underline{\underline{F}}$ has co-dimension 1 in $\underline{\underline{F}}^{\sim}$, and if $f \in \underline{\underline{F}}^{\sim}$, then $f - \gamma f(\infty) 1 \in \underline{\underline{F}}$, and

$$E^{\sim}(f, f) = E(f - \gamma f(\infty)1, f - \gamma f(\infty)1) + a\,\gamma f(\infty)^2$$

$$= E(f, f) + a\,\gamma f(\infty)^2$$

with $a \geq 0$. Of course $(\underline{\underline{F}}^{\sim}, E^{\sim})$ is $\underline{\underline{G}}$ invariant. This possibility also exists if (25.9) is satisfied, but in addition there is another class of $\underline{\underline{G}}$ invariant spaces. The function space $\underline{\underline{F}}^{\sim}$ includes all bounded harmonic functions in the reflected space of Section 15, and

$$E^{\sim}(f, f) = E(f, f) + aN(\gamma f, \gamma f)$$

$$= E(f - H\gamma f, f - H\gamma f) + (a+c)N(\gamma f, \gamma f)$$

with $a \geq 0$. This case is always conservative whereas the previous is conservative only for $a = 0$.

Bibliography

1. N. I. Akhiezer, On some inversion formulas for singular integrals (in Russian), Bull. Acad. Sci. USSR Ser. Math., Vol. 9(1945).

2. A. Beurling and J. Deny, Dirichlet Spaces, Proc. Nat. Acad. Sci. U.S.A. 45(1959), 208-215.

3. J. Bretagnolle, Résultats de Kesten sur les processus à acroissements independants, Seminaire de Probabilites V, Lecture Notes in Math. 191,

4. H. Cartan, Sur les fondements de la théorie du potential, Bull. Soc. Math. France 691(1941), 71-96.

5. ________, Théorie générale du balayage en potential newtonien, Ann. Univ. Grenoble, Sect. Sci. Math. Phys. (N.S.) 22(1946), 221-280.

6. G. Choquet and J. Deny, Sur l'equation de convolution $\mu = \mu * \sigma$, C. R. Acad. Sci. Paris 250(1960), 799-801.

7. K. L. Chung, A Course in Probability Theory, 2nd ed., Academic Press, 1974.

8. ________, Probabilistic approach to the equillibrium problem in potential theory, to appear.

9. ________, Maxima in Brownian excursions, to appear in the Bull. Amer. Math. Soc.

10. E. B. Dynkin, Wanderings of a Markov process, Theory of Prob. and App. 16(1971), 401-428.

11. J. Elliott, Boundary value problems and semi-groups associated with certain integro-differential equations, TAMS 76(1954), 300-331.

12. ________, Dirichlet spaces associated with integro-differential operators, parts 1 and 2, Ill. J. of Math., 9(1965), 87-98; 10(1966), 66-89.

13. ________, Dirichlet spaces and boundary conditions for submarkovian resolvents, J. Math. Anal. App., 36(1971), 251-282.

14. W. Feller, Generalized second order differential operators and their lateral conditions, Ill. J. of Math. 4(1957), 459-503.

15. ________, On the intrinsic form for second order differential operators, Ill. J. Math. 2(1958), 1-18.

16. M. Fukushima, On boundary conditions for multidimensional Brownian motion with symmetric resolvents, J. Math. Soc. Japan 21(1969), 485-526.

17. ________, On the generation of Markov processes by symmetric forms. Proceedings of the Second Japan-USSR Symposium on Probability Theory, Lecture Notes in Mathematics 330(1973), 46-79.

18. ________, Dirichlet spaces and strong Markov processes, T.A.M.S. 162(1971), 183-224.

19. H. Furstenberg, A Poisson formula for semi-simple Lie groups, Ann. of Math. 77(1963), 335-386.

20. R. Gangolli, Isotropic infinitely divisible measures on symmetric spaces, Acta Math. 111(1964), 213-246.

21. I. M. Gelfand, Spherical functions on symmetric spaces, Doklady Akad. Nauk. S.S.S.R. 70(1950), 5-8.

22. R. K. Getoor, Infinitely divisible probabilities on the hyperbolic plane, Pac. J. Math. 11(1961), 1287-1308.

23. ________ and M. J. Sharpe, Last exit decompositions and distributions, Indiana Math. J. 23(1973), 377-404.

24. ________, First passage times for symmetric stable processes in space, TAMS 101(1961), 75-90.

25. B. Haller, Verteilungsfunktioner und ihre Auszeichnung durch Funktionalgleichungen, Mitteilungen der Vereinigung schweizerischer Versicherungsmathematiker, vol. 45(1945), 97-163. Translated by R. E. Kalaba for the Rand Corp. as T-27, 1953.

26. G. Hamel, Integralgleichungen, Springer 1949.

27. S. Helgason, Differential Geometry and Symmetric Spaces, Academic Press, New York 1962.

28. G. A. Hunt, Semi-groups of measures on Lie groups, TAMS 81(1956), 264-293.

29. ________, Martingales et Processus de Markov, Dunod, Paris 1966.

30. N. Ikeda and S. Watanabe, The local structure of a class of diffusions and related problems, Proc. of the Second Japan-USSR Symposium on Probability, Lect. Notes in Math. 330(1973), 124-169.

31. K. Ito, Poisson point processes attached to Markov processes, Proc. 6th Berk. Symp., III(1971), 225-240.

32. ______ and H. P. McKean, Diffusion Processes and their Sample Paths, Academic Press, 1969.

33. H. Kesten, Hitting probabilities of single points for processes with stationary independent increments, Mem. Amer. Math. Soc. 93(1969).

34. H. Kunita, General boundary conditions for multi-dimensional diffusion processes, J. Math. Kyoto Univ. 10(1970), 273-335.

35. N. S. Landkof, Foundations of Modern Potential Theory, Springer Verlag (1972). (English trans.)

36. S. Lang, $SL_2(R)$, Addison Wesley, 1975.

37. B. Maisonneuve and P. A. Meyer, Ensembles aleatories markoviens homogènes, Seminaire de Probabilités VIII, Lect. Notes in Math. 381(1974),

38. P. A. Meyer, Processus de Poisson ponctuels, apres K. Ito, Seminaire de Probabilites V, Lect. Notes in Math. (1971).

39. P. W. Millar, Exit properties of stochastic processes with stationary independent increments, TAMS 178(1973), 459-479.

40. ____________, On the first passage distributions of processes with independent increments, Ann. of Prob.

41. M. Motoo, Application of additive functionals to the boundary theory of Markov processes, 5th Berkeley symp. (1967), II, part 2.

42. K. R. Parthasarathy, Probability Measures on Metric Spaces, Academic Press 1967.

43. G. Polya and G. Szegö, Über den transfiniten Durchmesser (Kapazitäts-konstante) von ebenen und raumlichen Punktmengen, J. Reine Angen. Math. 165(1931), 4-49.

44. A. O. Pittenger and C. T. Shih, Coterminal families and the strong Markov property, Bull. Am. Math. Soc. 78(1972), 439-443.

45. F. Riesz and B. S. Nagy, Functional Analysis, Ungar, New York 1955.

46. M. Silverstein, Classification of stable symmetric Markov chains, Ind. Univ. Math. J. 24(1974), 29-77.

47. ____________, Symmetric Markov Processes, Lect. Notes in Math. 426, Springer-Verlag 1974.

48. H. Sohngen, Die Lösunger der Integralgleichung $f(x) = (2\pi)^{-1}\int_{-a}^{+a} f(\xi)dz/(\xi-x)$ und deren Anwendung in der Traflügeltheorie, Math. Zeit. 45(1939), 245-264.

49. E. M. Stein, Singular Integrals and Differentiability Properties of Functions, Princeton Univ. Press, 1970.

50. ____________ and G. Weiss, Introduction to Fourier analysis on Euclidean spaces, Princeton Univ. Press, 1971.

51. F. G. Tricomi, Integral Equations, New York 1957.

52. S. Watanabe, On stable processes with boundary conditions, J. Math. Soc. Japan, 14(1962), 170-198.

53. ___________, Construction of diffusion processes with Wentzell's boundary conditions by means of Poisson point processes, recent preprint.

54. A. Zygmund, Trigonometric Series, Cambridge University Press (1959).

55. N. Ikeda and S. Watanabe, The local structure of diffusion processes, Seminar on Probability Vol. 35(1971), (Japanese).

56. M. Fukushima, Local property of Dirichlet forms and continuity of sample paths, preprint.

Index

Corrections for SMP

1. p. 20.8: Condition (i) in Theorem 20.2 is incorrect. See the remark following Corollary 4.4 in this volume for a discussion.

2. p. 24.1: Sentence at bottom is misleading. See Section 13 in this volume.

3. p. 23.8, line 5: The derivative does exist. See Section 12 in this volume.

4. p. 23.7, line 8 from b: same as 3.

5. p. 24.3, line 3: This is false.

6. p. 11.7, line 3 from b: $G<A_t f>(x) - G<A_t f'>(x)$.

7. p. 20.7, line 4: $\textcircled{E} e^{-u(\zeta-\zeta^*)}$ etc.

8. p. 20.7, lines 3, 4 from b: replace $e^{-u(\zeta-\zeta^*)}$ by $1-e^{-u(\zeta-\zeta^*)}$.